The Welfare of Sheep

Animal Welfare

VOLUME 6

Series Editor

Clive Phillips, *Professor of Animal Welfare, Centre for Animal Welfare and Ethics, School of Veterinary Science, University of Queensland, Australia*

Titles published in this series:

Volume 1: The Welfare of Horses
Natalie Waran
ISBN 1-4020-0766-3

Volume 2: The Welfare of Laboratory Animals
Eila Kaliste
ISBN 1-4020-2270-0

Volume 3: The Welfare of Cats
Irene Rochlitz
ISBN 978-1-4020-3226-4

Volume 4: The Welfare of Dogs
Kevin Stafford
ISBN 978-1-4020-4361-1

Volume 5: The Welfare of Cattle
Jeffrey Rushen, Anne Marie de Passillé,
Marina A.G. von Keyserlingk and
Daniel M. Weary
ISBN 978-1-4020-6557-6

Cathy M. Dwyer

The Welfare of Sheep

Springer

Dr. Cathy M. Dwyer
Scottish Agricultural College
Sustainable Livestock System Group
Animal Behaviour and Welfare
King's Buildings
Edinburgh
United Kingdom EH9 3JG
cathy.dwyer@sac.ac.uk

ISBN 978-1-4020-8552-9 ISBN 978-1-4020-8553-6 (eBook)

Library of Congress Control Number: 2008927061

Printed on acid-free paper

9 8 7 6 5 4 3 2 1

springer.com

Animal Welfare Series Preface

Animal welfare is attracting increasing interest worldwide, but particularly from those in developed countries, who now have the knowledge and resources to be able to improve the welfare of farm animals. The increased attention given to farm animal welfare in the West derives largely from the fact that the relentless pursuit of financial reward and efficiency has led to the development of intensive animal production systems that disturb the conscience of many consumers. In developing countries, human survival is still a daily uncertainty, so that provision for animal welfare has to be balanced against human welfare. Welfare is usually provided for only if it supports the output of the animal, be it food, work, clothing, sport or companionship. In reality there are resources for all if they are properly husbanded in both developing and developed countries. The inequitable division of the world's riches creates physical and psychological poverty for humans and animals alike in many sectors of the world. Livestock are the world's biggest land user (FAO, 2002) and the population is increasing rapidly to meet the need of an expanding human population. Populations of farm animals managed by humans are therefore increasing worldwide, and in some regions there is a tendency to allocate fewer resources, such as labour, to each animal with potentially adverse consequences on the animals' welfare. Land is one of the most important resources for sheep production, as it mostly utilises marginal areas and competes not with other forms of agriculture but with forestry and land for recreation.

Increased attention to welfare issues is also evident for companion, laboratory, wild and zoo animals. The key issues of provision of adequate food, water, a suitable environment, companionship and health remain as important as they are for farm animals. Of increasing importance is the ethical management of breeding programmes, now that genetic manipulation is easier but there is less tolerance of deliberate breeding of animals that are not suited to their environment. However, the quest for producing novel genotypes has fascinated breeders and scientists for centuries, and where dog and cat breeders produced a variety of extreme forms with adverse effects on their welfare in earlier times, nowadays the quest is pursued in the laboratory, where the laboratory mouse is genetically manipulated with even more dramatic effects.

The intimate connection between animal, owner or manager that was a feature of the animal management in the past is rare nowadays in the animal industries, having been superseded by technologically efficient production systems, in which animals on farms and in laboratories are tended by fewer and fewer humans in the drive to increase labour efficiency. In today's busy lifestyle, pets too may suffer from reduced contact with humans, although their value in providing companionship for the sick and the elderly is increasingly recognised. Consumers also rarely have any contact with the animals that produce their food. In this estranged, efficient world man struggles to find the moral imperatives to determine the level of welfare that he should afford to animals within his charge. Some aim for what they believe to be the highest levels of welfare provision, such as certain owners of companion animals, others deliberately or through ignorance keep animals in impoverished conditions because it is most profitable to do so. Religious beliefs and directives encouraging us to care for animals have been cast aside in an act of supreme human self-confidence, stemming largely from the accelerating pace of scientific development. Instead, today's moral codes are derived as much from media reports of animal abuse and the assurances that we receive in supermarkets that animals used for the products that we purchase were not exploited in any way. The young have always been exhorted to be kind to animals, through exposure to fables whose moral message was the benevolent treatment of animals. Such messages are today enlivened by the powerful images of modern technology, but essentially still alert children to the wrongs associated with cruelty to animals.

This Animal Welfare series has been designed to provide academic texts discussing the provision for the welfare of the major animal species that are managed and cared for by humans. They are not detailed blue-prints for the management of each species, rather they describe and consider the major welfare concerns, often in relation to similar species or the wild progenitors of the managed animals. Welfare is also considered in relation to the animal's needs, concentrating on nutrition, behaviour, reproduction and the physical and social environment. Economic effects of animal welfare provision are addressed where relevant, and key areas identified that require further research.

In this volume, Dr Cathy M. Dwyer has drawn on her extensive experience of research in sheep management systems to gather a team of experts who describe aspects of sheep welfare from a variety of different perspectives. Dr Dwyer herself has contributed to several of these chapters, which is invaluable for this topic, since she is one of the world's leading researchers into the welfare of extensively-kept sheep. In contrast to earlier volumes of this series, which concentrated on intensively managed animals, this volume explores in detail the welfare concerns in situations where labour and other management inputs are at low levels, usually for economic reasons. Although not often considered to be a cause for serious concern in the past, primarily because of the apparent naturalness of the production systems, it becomes clear in this book that extensive sheep production can also suffer from major welfare problems. In fact, it is increasingly recognised that adequate nutrition, health and environmental comfort are particularly difficult to assure in systems occupying harsh terrains and extreme climatic regions. Despite these real concerns, the areas of concern in intensive systems, such as space availability, abnormal behaviours, social structure and fear of humans are often less of an issue in extensive sheep production systems. At a time when livestock management systems are increasingly questioned for their impact on the environment, it is an opportune time for this volume to explore the issues surrounding the welfare of sheep in detail.

With the growing pace of knowledge in this relatively new field of research, it is hoped that this volume in the series will provide a timely and much-needed text for researchers, lecturers, leading sheep farmers and veterinarians, advisors and students. My thanks are particularly due to the publishers for their support, and to the authors and editors of the series for their hard work in producing the texts.

Clive Phillips
Series Editor
Professor of Animal Welfare and Director,
Centre for Animal Welfare and Ethics,
School of Veterinary Science,
University of Queensland,
Australia

Reference

Food and Agriculture Organisation (2002). http://www.fao.org/ag/aga/index_en.htm.

Preface

Concern for the welfare of farmed livestock, and the scientific research that this has sparked, has been increasing since the 1960s when public attention was drawn to confined conditions under which some animals were kept (Harrison, 1964). For much of this period attention has focused on those animals typically kept in confined and restrictive housing (initially pigs and poultry and, latterly, dairy cows). The livestock species traditionally managed extensively have received relatively little attention. Much of the concern for animal welfare that arose in the 1960s was related to the behavioural restriction and unnatural environments that the animals were living in, thus the apparent naturalness and freedom of behavioural expression afforded to extensively managed animals suggested that there were few welfare concerns for these animals. Freedom to express natural behaviour is, however, only one of the universally-accepted welfare definition, the Five Freedoms (Brambell, 1965), and an extensive environment may not serve the animal well in meeting the other four aspects of welfare. In his book, A Cool Eye Towards Eden, John Webster paints a vivid picture of a flock of aged ewes outwintered on poorly drained pasture where animals are chronically underfed, many are chronically lame, they suffer frequent cold stress and often frightened and injured by domestic dogs, yet do have the freedom to engage in natural behaviour, such as panic and flight (Webster, 1994). He argues that in this, admittedly extreme, example the intensity of animal suffering may be as great as or greater than that of a battery chicken. The aim of this book, therefore, is to consider the welfare of this important livestock species, and to assess the needs and requirements of sheep for good welfare, not just for behavioural expression, but also for other aspects of welfare.

In this book, my co-authors and I have considered the welfare of the sheep from the perspective of evolution and ecological environmental requirements, the behavioural patterns and cognitive abilities of the sheep, health, management, breeding and economics. Perhaps uniquely amongst livestock species, the sheep is kept for a variety of uses (ranging from meat and milk to fibre and portage) and in a diversity of management systems, often traditional and specific to a region or environment. The ability of these different systems to provide good welfare for the sheep is addressed, and suggests that different aspects of welfare are emphasised in different situations. Thus much could be learnt about providing good welfare by looking to other systems that may provide facets of management that might be more generally incorporated. The book concludes by placing welfare in general, and that of the sheep in particular, into the wider context of society and global trade, and considers the pressures facing farming and offers potential solutions to improve welfare. In reading this book, I hope the reader will gain or enhance their understanding of the often complex lives of sheep, their fundamental place in maintaining many communities, and develop, as I have, a respect and concern for the welfare of this often overlooked species. Those who have worked with sheep often come to understand and appreciate the rich behavioural and emotional repertoire of the sheep. I hope that, in reading this book, those readers who have not had that opportunity will also come to see something of the 'point of view' of the sheep.

In editing this book I am indebted to all the contributing authors for their hard work and great patience, who have produced diverse chapters that have explored the welfare of the sheep from many perspectives and provided a fascinating insight into the life and times of the sheep. Their patience, understanding and support during the long genesis of this book have greatly aided the final production of this volume.

Edinburgh, UK Cathy M. Dwyer

References

Brambell, F. W. R. (1965) *Report of the Technical committee to enquire into the welfare of animals kept under intensive livestock husbandry systems.* Her Majesty's Stationery Office, London, United Kingdom.

Harrison, R. (1964) *Animal Machines.* Robinson and Watkins, United Kingdom.

Webster, J. (1994) *A Cool Eye Towards Eden.* Blackwell Science, Oxford, United Kingdom.

Contents

Contributors

S. Agenäs
Swedish University of Agricultural Science, Uppsala, Sweden, sigrid.agenas@huv.slu.se

D. Blache
School of Animal Biology of Natural and Agricultural Sciences, University of Western Australia, Australia, dbla@cyllene.uwa.edu.au

J. Conington
Sustainable Livestock Systems Group, SAC, Edinburgh, EH26 0PH, UK, jo.conington@sac.ac.uk

C.M. Dwyer
Aimal Behaviour and Welfare, Sustainable Livestock Systems Group, Scottish Agricultural College, Edinburgh, EH9 3JG, UK, cathy.dwyer@sac.ac.uk

P. Goddard
Macaulay Institute, Craigiebuckler, Aberdeen, AB15 8QH, UK, p.goddard@macaulay.ac.uk

J.P. Hogan
Centre for Animal Welfare and Ethics, University of Queensland, Australia, jim.hogan81@hotmail.com

I.D. Ivanov
Research Institute of Agricultural Science-NIGO, 6000 Stara Zagora, Bulgaria, iv.dimitrov@dir.bg

K.M. Kendrick
Cognitive and Behavioural Neuroscience, The Babraham Institute, Babraham, Cambridge, CB22 3AT, UK, keith.kendrick@bbsrc.ac.uk

R.J. Kilgour
NSW Department of Primary Industries, Agricultural Research Centre, Trangie, NSW 2823, Australia, bob.kilgour@dpi.nsw.gov.au

A.B. Lawrence
Sustainable Livestock Systems Group, SAC, Edinburgh, EH26 0PH, UK, alistair.lawrence@sac.ac.uk

C.E. Milne
Land Economy Group, SAC, Aberdeen, UK, cath.milne@sac.ac.uk

R. Nowak
Equipe Comportement, Neurobiologie, Adaptation, Unité de Physiologie de la Reproduction et des Comportements, INRA, Nouzilly, France, nowak@tours.inra.fr

C.J.C. Phillips
Centre for Animal Welfare and Ethics, University of Queensland, Australia, c.phillips@uq.edu.au

R.H. Porter
Equipe Comportement, Neurobiologie, Adaptation, Unité de Physiologie de la Reproduction et des Comportements, INRA, Nouzilly, France, porter.rh@gmail.com

P.A. Roger
Veterinary Consultancy Services, Victoria Cottage, Reeth, Richmond, North Yorkshire, DL11 6SZ, UK, paul.roger@btinternet.com

J.M. Santarossa
Land Economy Group, SAC, Aberdeen, UK, john.santarossa@sac.ac.uk

A.W. Stott
Land Economy Group, SAC, Aberdeen, UK, alistair.stott@sac.ac.uk

T. Waterhouse
Sustainable Livestock Systems Group, Scottish Agricultural College, Edinburgh, EH9 3JG, UK, tony.waterhouse@sac.ac.uk

Chapter 1
Introduction to Animal Welfare and the Sheep

C.M. Dwyer and A.B. Lawrence

Abstract Concerns for the lives of animals have been voiced for centuries, with concerns about the welfare of agricultural animals increasing since the 1960s. Animal welfare concerns arise for many reasons: care about the quality of lives of animals, concerns about human health, product quality, the environment, and trade and marketing issues. Some of these concerns, therefore, include animal welfare as part of a package of issues involving 'green' or ethical living, whereas others may arise through direct impacts on animal welfare as a consequence of, for example, trade issues. A consensus on the definition of welfare has not been reached, however definitions have been proposed based on (i) the ability of the animal to perform natural behaviour, (ii) the animals' subjective experiences, or (iii) the biological functioning of the animal. Integrated hypotheses suggest that all are important but that different concerns may arise depending on the interaction of the animal with the environment. For example, use of ethological knowledge gained from the existing species of wild sheep can help to determine how far genetic selection of domestic sheep has altered their behaviour from that of the wild progenitors. Investigation of how different the modern farming environment is from that in which sheep first evolved will help determine where mismatches exist and where suffering might be expected to occur. Animal welfare concerns have tended to focus on those animals that are kept in confinement agriculture (e.g. pigs and poultry). Extensively managed species have received less attention, often as these animals are perceived to be free to engage in natural behaviour, because farming is considered more traditional or because the ruminant is considered to be 'tough'. However, welfare concerns do occur in sheep systems, for example, arising from the lack of inspection in extensive systems, surgical procedures, or management practices.

Keywords Sheep · Welfare · Extensive · Natural behaviours · Feelings · Biological function

---

C.M. Dwyer
Animal Behaviour and Welfare, Sustainable Livestock Systems Group, SAC, Edinburgh, UK
e-mail: cathy.dwyer@sac.ac.uk

C.M. Dwyer (ed.), *The Welfare of Sheep*,

1.1 Introduction

1.1.1 A Brief History of Animal Welfare

Concerns for the lives of animals have been present for as long as humans and animals have co-existed. Enshrined in Eastern religions and the animal mythology of many cultures are the concepts of respect for animal life, living in harmony with nature and enjoying the co-operation of animals for human survival. In the West, however, arguments have been made, particularly by Descartes and Kant (1600 and 1700s), for the uniqueness of humans and this was held to justify the use of animals by man for any purpose. By the early 19th century evidence was amassing to challenge these opinions with the evolutionary theories of Charles Darwin being pivotal to changing attitudes. In Western philosophy, also, writers such as Herman Daggett in 1791 and Henry Salt in 1892 were advocating rights for animals, and these arguments continue today in the writings of, for example, Tom Regan and Peter Singer. These concerns became more crystallised and expressed in legal terms, in the UK, with the first animal welfare law: 'The Ill-treatments of Horses Act' of 1822, and the founding of the Royal Society for the Protection of Animals in 1824. Darwin, with other leading biologists, argued for more humane treatment of animals (e.g. with the publication of his book 'The Expression of the Emotions in Man and Animals' in 1872), and was instrumental in the setting up of a Royal Commission in 1875, which led to the Cruelty to Animals Act (1876). Other acts of parliament followed, specifically related to the prevention of cruelty to animals, culminating in the Protection of Animals Act in 1911, which is still in force to this day. This law, the 'grandfather' of all other animal welfare legislation in the UK, essentially sets out to protect all animals from unnecessary suffering whether through omission or commission. Although this law has a broad brief, encompassing all animals whether captive or not, it's initial concerns were to regulate the use of animals in medical experiments. The good treatment and husbandry of farm animals was considered to be an integral part of the success of livestock farming, thus it was in the interests of both the farmer, and his livestock, for the animals to be treated well.

Alongside the changes in legislation and sensibilities regarding animal welfare in the west, the late 19th century also saw a shift in agricultural practices resulting in the New Agriculture. This period saw an increase in agricultural production, promoted by an increased use of selective animal breeding (pioneered by the sheep farmer Robert Bakewell in the 18th century)[1] and crop growing strategies (such as

[1] Robert Bakewel (1725–1795) is generally considered the father of modern animal breeding, and the first to use selective breeding for meat production (previously cattle and sheep had been used largely for labour and wool respectively) and to improve carcase quality. He is largely credited with the first production of distinct sheep breeds by separating males from females for the first time, and using in-breeding to exaggerate characteristics he considered desirable. Starting from the old Lincolnshire sheep he created the New Leicester – a large longwool breed with fatty forequarters to meet the then popular taste for fatty mutton. In addition to his animal breeding (which seem to have been carried out in some secrecy to avoid public controversy arising from prejudice against

those advocated by Lord 'Turnip' Townshend)[2]. These developments were associated with an increased interest in rigorous scientific evaluation, a more universal access to education and the increased requirement for efficient food production from the rapidly urbanising population following the Industrial Revolution. In the first half of the 20th century the drive for increased local food production during the two World Wars, typified by poster slogans such as 'Dig for Victory', galvanised agricultural production across the Western world. This provided the additional motivation to increase production, alongside the growth of the use of science in agriculture (principally genetics, nutrition, and hygiene). Many of the scientific societies for animal production (in the UK, Europe, Australia and New Zealand) were founded in the 1940s and early 1950s, reflecting the increased application of science to food animal production. Thus science, education, the motivation to produce more food and increased mechanisation (occurring in all sectors of society) were the drivers for the move from 'animal husbandry' towards intensified animal production.

In 1964, the publication of 'Animal Machines' by Ruth Harrison was hugely influential in the UK and Europe in raising awareness and concern for the welfare of farmed animals. Her book was an exposé of what she termed 'factory farming' and drew attention to the use of animals purely as 'products' and the close confinement of many agricultural animals but also emphasised the risks to human health of feeding antibiotics, growth stimulants and hormones to farm animals. The book promoted such an intense public reaction that the British Government commissioned Professor Roger Brambell to investigate intensive farming practices in Europe. The Brambell Committee Report (published in 1965) defined animal welfare both in terms of mental well-being as well as the animal's physical state, and is perhaps best known for providing a list of principles for rearing farm animals, which have since become known as the Five Freedoms (see below). These two events ushered in a new era of farm animal welfare with the Agriculture Act of 1968, its accompanying Codes of Recommendation for the Welfare of Livestock and the setting up of the Farm Animal Welfare Council (FAWC) in 1979. Elsewhere in Europe similar investigations into the welfare of intensively farmed animals were also taking place (e.g. the Husbandary and Animal Welfare Committee in The Netherlands, 1975), and in 1976 the Council of Europe drew up the European Convention on the Protection of Animals kept for Farming Purposes. A landmark decision took place in 1997 (the Treaty of Amsterdam) that animals should be defined as 'sentient creatures' in European law and no longer just as agricultural products. Elsewhere, such as the Animal Welfare Act in New Zealand (brought into law at the beginning of the 21st

---

'close' breeding), he also pioneered changes in animal husbandry, designing raised platforms for his cattle winter stalls to prevent them lying in their own manure, and doing away with the need for straw bedding.

[2] Lord Charles Townshend (1674–1738) retired from politics in 1730 to concentrate on the development of agriculture and was known colloquially as 'Turnip' for his introduction of the turnip into the Norfolk crop rotation system. Norfolk had become the focus for agricultural improvements, largely through his efforts, and through the uptake of ideas from France and Belgium.

century), moves are taking place to extend animal welfare legislation beyond the absence of cruelty by placing emphasis on care, animal husbandry and prevention of suffering by reference to the Five Freedoms.

In Europe, although the pressures were brought to bear by public concerns, the main route to improving farm animal welfare has been through legislation. In North America, however, a recent concern for animal welfare has been brought about by very different means. The infamous McLibel trial, brought by McDonalds against two members of London Greenpeace in 1994 and concluded in 1997, brought animal welfare issues, particularly slaughter handling and conditions, to the forefront of the public conscience (McSpotlight 1998). In efforts to redress the balance the fast food industry has been instrumental in beginning an improvement in animal welfare in the USA by setting up scientifically-based Animal Welfare councils and codes of practice for its suppliers. This may have had a knock-on effect on legislation. The United States has had legislation covering humane methods of slaughter since 1958, in 2002 President George Bush signed the Farm Security and Rural Investment Act including a resolution that act be fully enforced[3]. This broad ranging bill, encompassing subsidy payments, conservation and trade, supports sustainable agriculture and introduces animal welfare provisions at a Federal level.

At a global level, the Office International des Epizooties (OIE, also known as the World Organisation for Animal Health) identified animal welfare as an important priority area in 2001 and established a permanent Working Group on Animal Welfare in 2002. Following a conference in Paris in 2004, the OIE adopted four animal welfare standards in 2005 covering the transport of live animals by land and by sea, and the slaughter of animals for meat or disease control purposes. Welfare standards for the housing and management of animals kept for food production are set to follow. Thus, animal welfare is now seen as a global concern, requiring standards for appropriate welfare to be applied in all countries.

1.1.2 Why be Concerned About Animal Welfare?

The foregoing short discussion of the major events in the development of concern for farm animal welfare has touched on several of the reasons why concern for animal welfare has continues to be an issue. These concerns appear to be consumer-driven, it is the action of the general public and their perception of welfare that drives legislative and other animal welfare changes. So why are we, as consumers, concerned about animal welfare?

---

[3] This history has concentrated mainly on the development of concern for farm animal welfare in Europe. For a more detailed discussion of farm animal welfare in the USA see *Farm Animal Welfare: The focus of animal protection in the USA in the 21st century* by Rowan, O'Brien, Thayer & Patronek (1999) available on line at http://www.tufts.edu/vet/cfa/faw.pdf. Discussion of developments in animal welfare in New Zealand can be found in Stafford et al. (2002).

Broadly speaking farm animal welfare concerns fall into four main camps:

- Ethical or moral concerns about the lives of animals;
- Concerns about human health and product quality, and beliefs that improving the lives of animals on farm will have associated benefits for these other areas of concern;
- Environmental and biodiversity concerns, where animal welfare is seen as part of a package of concerns about modern farming practices and how we treat the planet;
- Trading and marketing concerns, either where there is concern for animal welfare arising due to the impact of these issues, or where animal welfare can be used as a marketing tool to leverage higher prices.

Discussion of the differing philosophical positions underlying why we should, or should not, be concerned about the lives of animals are beyond the scope of this chapter and the reader is referred to other texts. Broadly, there are three main philosophical positions in our dealings with animals (see Appleby 1999):

(1) *Consequentialism* (such as utilitarianism), which argues that it is the consequences of our actions, rather than the actions themselves, which are of moral concern. This argument is widely used to argue that we should act to produce the greatest good and cause the least harm. The philosophy of animal liberation (Singer 1975) uses these arguments to stress that, generally, the benefits are to humans and the costs to animals, and advocates equal rights to all sentient beings (see Fraser 1999 for a more detailed assessment of these arguments).
(2) *Deontology*, which focuses on the actions and whether it is morally right to use animals for certain things, regardless of the consequences. These arguments lead to discussions of our duties and the rights of animals (Regan 1983).
(3) *Agent-centred ethics*, which argues that is it neither the action nor the consequences that is important but the agent involved.

In animal welfare science, writers have proposed hybrid views, e.g. Sandøe et al. (1997), combining elements of utilitarianism and deontology (such that mostly it is the consequences that guide actions, but that there are things that may not be done, regardless of the beneficial consequences). Other philosophers have emphasised the care aspects of animal husbandry and welfare (see Fraser 1999). Bernard Rollin (1990) argues for an extension of animal welfare beyond merely the prevention of cruelty or harm. He argues that, in democratic societies, our 'consensus social ethic' (the excepted moral norms of rights and behaviours) should also be extended to animals. Thus the needs, desires and predilections of animals matter as much to the animal as our own do to us, and therefore the fundamental nature and interests of the animal (it's *telos* – of which more later) should be encoded and protected. Thus what we feel it is acceptable to do with, or to, animals depends on our ethical position.

In 'Animal Machines' Ruth Harrison also expressed concern for the impact on human health of the growth promoters and antibiotics fed to food animals. More recently human health concerns are also being expressed about manipulations of the animal's genome (although these may also be through moral concern for the animal's quality of life). For example, the use of recombinant (derived from *Escherichia coli*) bovine somatotrophin (rbST) to increase milk yields in dairy cattle is causing concern in Europe for it's potential cancer risk through elevations of insulin-like growth factor-I (IGF-I) in milk and because milk composition may affect the triggering of allergic reactions[4]. Incidentally, rbST is also associated with health risks in the cows too, as treated dairy cows suffer an increased incidence of clinical mastitis, foot and leg problems and reproductive effects, probably secondary to increased milk yield[5]. Under some circumstances poor welfare can also lead to poor meat quality (Gregory 1993). Thus concerns about the way farm animals are kept can be expanded to encompass concerns about the potential consequences to the consumer, in the absence of any particular concern for the quality of life of the animal.

Concern for animal welfare is also part of a wider raft of concerns encompassing sustainability, climate change, protection of the environment and rural communities, biodiversity, maintenance of family-run farms (an 'agrarian ideal', Fraser 2001) and the production of 'real foods'. This diverse spectrum of interests range from concerns about how to handle the waste products produced by modern intensive agriculture to a view of animal welfare within an agricultural concept of working in harmony with the land. Improved animal welfare, seen as access to fields, the ability to express natural behaviours, using natural feedstuffs, management without antibiotics and routine drug administration (The Soil Association requirements), form part of the organic ideal of healthy land, food and people.

At a local level in some countries, animal products produced to high welfare standards command a premium price in relation to conventionally reared food animals. Thus some actions to improve animal welfare, for example requirements demanded by retailers, may be related to maintaining market share and meeting consumer requirements, rather than ethical interests in animal welfare *per se*. Of course, these routes to improved animal welfare are driven by the consumers ethical or other interests in animal welfare so can not be completely divorced from other categories of concern for animal welfare. Market forces may also impact on animal welfare at the level of meat processing as bruised or blemished carcases, as may occur with rough handling, are scored as lower quality and hence of lower value. On a more global level, animal welfare has been charged with being simply a mechanism for trade protection and a barrier to free trade. The conflicting views of the European Union, which sees animals as sentient and not to be treated as commodities (see above), and the World Trade Organisation, which classes animals simply as commodities or resources, make these issues difficult to resolve. However, as described in the previ-

---

[4] The outcome of discussions on the health aspects of rbST have been summarised in an online report (March 1999) produced by the European Commission for Food Safety (From the farm to the fork) and can be found at http://europa.eu.int/comm/food/fs/sc/scv/out19_en.html#_Toc446393145.

[5] Discussions about the animal welfare aspects of rbST use, produced by the European Commission Scientific Committee on Animal Health and Animal Welfare (March 1999), can also be found online at http://europa.eu.int/comm/food/fs/sc/scah/out21_en.pdf.

ous section, Europe has a long history of concern for animal welfare, and actions to improve animal welfare have largely been consumer-driven and science-based. This suggests that concern for animal welfare exists primarily through ethical and moral issues surrounding the quality of animal lives, or through concerns about human health and protection of the environment.

1.2 Welfare Definitions

1.2.1 Can we define Animal Welfare?

Most people would probably agree that concern for animal welfare[6] is a good thing, however whether they would agree on what actually *is* animal welfare is quite another matter. This lack of consensus over a universal definition of animal welfare has been a thorn in the side of animal welfare science for over a decade. Once we add to that the differing philosophical positions underpinning our moral concerns for animal welfare (as alluded to above), discussions about how (or even if) animal welfare can be measured, and arguments about the objectivity and value-free nature of scientific assessments of animal welfare (or not; e.g. Tannenbaum 1991; Rollin 1996) then we seem to be deep in a quagmire through which little progress can be made. In an attempt to separate these issues some writers have proposed that animal welfare acts as a 'bridging concept linking science to ethics' (Fraser et al. 1997). The conception of animal welfare thus needs to be both accessible to scientific method and to reflect the ethical concerns of society. Fraser (1999) then suggests that animal welfare values can be divided into 'descriptive statements', which describe some property of a housing system, the environment, the animal etc., 'evaluative statements', which gives value to that statement (that it is better, worse, more important etc. for the animal's quality of life), and 'prescriptive statements', which reflect ethical concerns and what should or should not be done to that animal. In this scheme animal welfare is seen as an evaluative concept, where we attempt to scale the animal's perception of its quality of life. Although this links animal welfare to ethics, and potentially separates what is and is not accessible to scientific enquiry we still need some methods of measurement that defines the animal's perception of what is 'better' or 'worse' for it's welfare.

There are some general concepts about animal welfare that most people accept: (i) that animal welfare is a property of the animal (rather than of the environment, or something given to the animal); (ii) that animal welfare concerns are 'quality of life' concerns; and (iii) that welfare exists on a continuum from very poor to very good.

---

[6] Throughout this chapter the term animal welfare is used to encompass both welfare and well-being. Some authors (e.g. Tannenbaum 1991; Gonyou 1993) have suggested different usages for these terms, however we suggest that, for the sake of simplicity, having one poorly defined term is better than two.

One of the first definitions of farm animal welfare, that proposed by the Brambell Report (paragraph 25), defined animal welfare thus:

> Welfare is a wide term that embraces both the physical and mental well-being of the animal. Any attempt to evaluate welfare therefore must take into account the scientific evidence available concerning the feelings of animals that can be derived from their structure and functions and also from their behaviour.

In this definition animal welfare was explicitly defined as being composed of both physical and mental aspects of quality of life, and extending beyond the absence of disease. This definition was supplemented by a proscribed list of freedoms that should be extended to all farm animals, the well-known Five Freedoms (as used by the codes of recommendations for the welfare of livestock of many countries):

- Freedom from thirst, hunger and malnutrition – by ready access to fresh water and a diet to maintain full health and vigour.
- Freedom from thermal or physical distress – by providing an appropriate environment including shelter and a comfortable resting area.
- Freedom from pain, injury and disease – by prevention or by rapid diagnosis and treatment.
- Freedom to display most normal patterns of behaviour – by providing sufficient space, proper facilities and company of the animals' own kind.
- Freedom from fear and distress – by ensuring conditions and treatment to avoid mental suffering.

These concepts contain elements of the animal's health status, emotional state, and physical and behavioural functioning, and are, sometimes in a modified form, incorporated into the welfare codes of farm animals in many countries. In attempts to derive measurable components to describe an animal's welfare state three main schools of thought have arisen:

1.2.1.1 Natural-Living Based Definitions of Animal Welfare

This welfare definition suggests that good welfare depends on the animal being able to live a 'natural' life and be allowed to express its evolved behaviour patterns. This picks up on the views expressed in the Brambell Report (paragraph 37):

> . . . we disapprove of a degree of confinement of an animal which necessarily frustrates most of the major activities which make up its natural behaviour.

However, some early proponents of this definition extended this from 'most of the major activities' to hold that to prevent suffering an animal needs 'to perform all the behaviours of its repertoire' (Kiley-Worthington 1989). However, as many behaviours have evolved as an adaptation to deal with an adverse situation (distress calls in isolation, fleeing from a predator and so on), it seems that performance of the whole behavioural repertoire is not necessary, only those parts of it that the animal perceives to be important (Dawkins 1998). The natural-living definition has been reworked by Rollin (1990; 1993) who proposes that welfare, in addition to the control of pain and suffering, should also include nurturing and fulfilling the animals nature or, as he terms it, its *telos* (also see above). He suggests that it is the 'wants and desires' of animals (including humans) that separates them from plants or even cars (which can have needs) and thus why we have moral concerns about the welfare of animals. An ethical parallel of the 'natural-living' definition is the concept of animal integrity, which has been defined as

> the wholeness and completeness of the species specific balance of the creature, as well as the animal's capacity to maintain itself independently in an environment suitable to the species (Rutger & Heeger 1999).

These ideas suggest we should not infringe the animal's physical wholeness (such as castration or tail-docking), and also create conditions where the animal has a life that accords with their species-specific capacities and adaptation patterns: conditions where the animal can be fulfilled and flourish. These ideas extend the concept of welfare beyond the absence of suffering to include concepts of pleasure, contentment or positive experiences (e.g. see Mench 1998; Fraser & Duncan 1998 for discussion).

1.2.1.2 Feelings-Based Definitions of Animal Welfare

This welfare definition, which we were edging towards at the end of the previous passage, argues that animal welfare concerns are, in fact, concerns about the subjective experience, the 'feelings', of the animal involved (see for example: Dawkins 1980; 1990; 1998; Duncan & Petherick 1991; Duncan 1993; 1996). What distinguishes an animal from a plant is its sentience, and its capacity to experience pain, fear, distress, pleasure etc., and thus it should be the experience of those emotional states that plays a central role in the determination of its welfare. The role of feelings in welfare was stated in the Brambell Report (see quote from paragraph 25 above) which also concluded (paragraph 28):

> We accept that although pain, suffering and stress are certainly not identical in animals and men, there are sound reasons for believing they are substantial in domestic animals and that there is no justification for disregarding them ... We accept that animals can experience emotions such as rage, fear, apprehension, frustration and pleasure ...

Thus feelings, particularly the experience of pain and suffering, have always been part of the definition of welfare. This definition is probably closest to the public perception of animal welfare, and the reason that farm animal welfare first received public attention. Within this definition there are variations of views, from the relatively narrow view that welfare is *only* about feelings (Duncan 1993; 1996), such that welfare measurement, rather than being concerned with biological functioning (see below) should be concerned with the animal's affective experience of that biological functioning. This, he argues, is the evolved and cognitive experience (the wants and desires mentioned earlier) of having biological needs. Other writers have also argued for an evolutionary basis to feelings (Baxter 1983; Dawkins 1998; Fraser & Duncan 1998) as they play an important role in motivating the animal to respond to situations which will increase fitness, either immediately (e.g. escaping a predator) or in the long term (e.g. play or exploratory behaviours). However, many animal welfare scientists find it hard to accept a view of animal welfare, based solely on feelings, such that an animal with a disease condition that has not yet begun to cause it feelings of pain or discomfort would be regarded as being in a state of good welfare. Others have argued that feelings are transitory or incidental and not necessarily of relevance to welfare, what matters are the long-term impacts on functional design (Barnard & Hurst 1996). Other anomalies arise (as Duncan (1996) acknowledges) as drug taking could conceivably by described as improving welfare, by promoting pleasurable feelings, when it's continued use is likely to lead to suffering and reduced welfare.

The measurement of feelings and subjective states (requirements if we are going to use this definition to assess welfare) are, of course, far from straightforward and deal with the thorny issues of consciousness, cognition, and accessing the subjective experience of others. Similar arguments can be levelled at the 'natural-living' definition of welfare, that it is hard to determine precisely what is natural in both the animal's behaviour and what constitutes a natural environment for a particular species. Some authors have used the difficulty of assessing animal emotion as arguments against these definitions of welfare altogether (Moberg 1985; McGlone 1993), preferring to confine assessment of animal welfare to biological functioning and physiological states that are readily scientifically accessible, such as stress.

1.2.1.3 Biological-Functioning Based Definitions of Welfare

The biological-functioning based welfare definition looks primarily at the animal's physiological responses, particularly the functioning of the hypothalamic-pituitary-adrenal (HPA) axis, the sympathetic-adrenal medullary system (SAM), immune function, health, and agricultural productivity measures. The HPA axis is a neuroendocrine system that registers changes in homeostasis and triggers a cascade reaction to deal with the change. The SAM is an autonomic system that brings about changes in heart rate, metabolic rate, respiratory rate etc., in response to stressors. Within this definition there are, as before, variations in interpretation of what constitutes welfare. At one extreme are views that suggest the only measures of unacceptable welfare should be where the survival and reproduction of the animal are compromised (e.g. McGlone 1993). However, the productivity of an animal has been widely criticised as a sensitive welfare measure, and was rejected by the Brambell Committee (1965; paragraph 30):

> ... a satisfactory growth or egg production rate is a reliable guide to the welfare of the animal in certain respects – for example that it is being well-fed – but it is inadequate in other respects. Growth, on occasion, can be a pathological symptom, although it is more often a mark of health. Growth rate and condition ... are not inconsistent with periods of acute, but transitory, physical or mental suffering.

Alternative views, dealing with biological functioning, have been expressed by Broom (1986) who defines the welfare of an individual as 'its state as regards its attempts to cope with its environment'; by Wiepkema and Koolhaas (1993) who define welfare in terms of the ability of the animal to control and predict environmental events; and Moberg (1996) who defines poor welfare as when the animal suffers from sufficient stress to elicit a prepathological stress state. Broadly speaking, these welfare definitions all deal with how the animal perceives and deals with the stressors it encounters in its daily life. Animals have evolved to be constantly monitoring the environment (not always consciously, responses to thermal disturbances, for example, may elicit physiological alterations that the animal is unaware of) and reacting to minor deviations from set points using species specific behavioural and physiological mechanisms. Thus when the animal is able to predict and control events, and adjust to disturbances using species typical responses, the welfare of the animal is not threatened. For these authors, welfare declines when the animal's responses are no longer sufficient and a consequence of the altered biological function, in attempts to deal with the stressor, is a depression of the immune system such that the animal becomes more susceptible to disease.

This view of animal welfare is not necessarily incompatible with either of the other interpretations outlined above. An animal living in a natural environment may be able to express more of its evolved species typical responses to a stressor than an animal in confinement, however its biological functioning may still be overwhelmed in the presence of some stressors. Thus the biological functioning argument may concur with the natural-living definitions view of animal welfare in some states, but not all. Likewise, Broom (1998) has expanded upon his initial welfare definition to suggest that feelings are important parts of the animal's coping system. However, (as discussed by Fraser et al. 1997) there may be evolved adaptations that have no function in an agricultural environment (foraging, for example) thus the animal may be highly motivated to perform a behaviour with no opportunity to do so. The animal may then experience, for example, frustration or fear in the absence of effects on its agricultural productivity because those now non-functional behaviours were important aspects of biological fitness in the environment in which it evolved.

1.2.1.4 Towards a Consensus View?

At their most extreme there seems to be little consensus on what constitutes animal welfare, and use of differing definitions could lead to completely different conclusions being drawn about the welfare of an individual. In reality, many animal welfare scientists use operational definitions that might comprise parts of some or all of the three definitions. Many of the concepts can be brought together, particularly when we view both feelings and biological functioning as evolved and adaptive responses of the animal to its environment. Thus feelings can be related to both the natural living and biological functioning arguments if we conceive of feelings as evolutionary mechanisms designed to enhance fitness. In the natural environment the animal may be able to deal with minor stressors through evolved mechanisms (even though this may involve short-term negative emotions, such as fear at the presence of a predator) and to be able to express positive emotional states through play, exploration and social encounters. Animal integrity, or *telos*, can be related to the biological functioning definition through allowing the animal to display its species typical adaptations in an environment in which it has evolved those adaptations.

An integrated hypothesis for animal welfare has been proposed (Fraser et al. 1997), which seeks to draw out the important elements of each of the animal welfare conceptions and integrate them into a single definition. The essential features of this hypothesis are the animal (made up of all the adaptations that it has evolved) and the environment (comprising a series of challenges that the animal experiences). In the natural environment in which the animal has evolved (with the proviso that it may be difficult to ascertain exactly what that was after thousands of years of domestication) we can imagine that there is an almost perfect overlap between these two, that is the animal possesses the adaptations that allow it to meet the challenges that occur in that environment (Fig. 1.1a). This does not mean, however, that the animal is in a constant state of good welfare (here, then, we begin to deviate from the natural-living definition of welfare). There may be occasions when the animal's adaptations are insufficient to cope with the challenges it experiences. For example, in times of drought there may be an acute shortage of food where, despite expressing

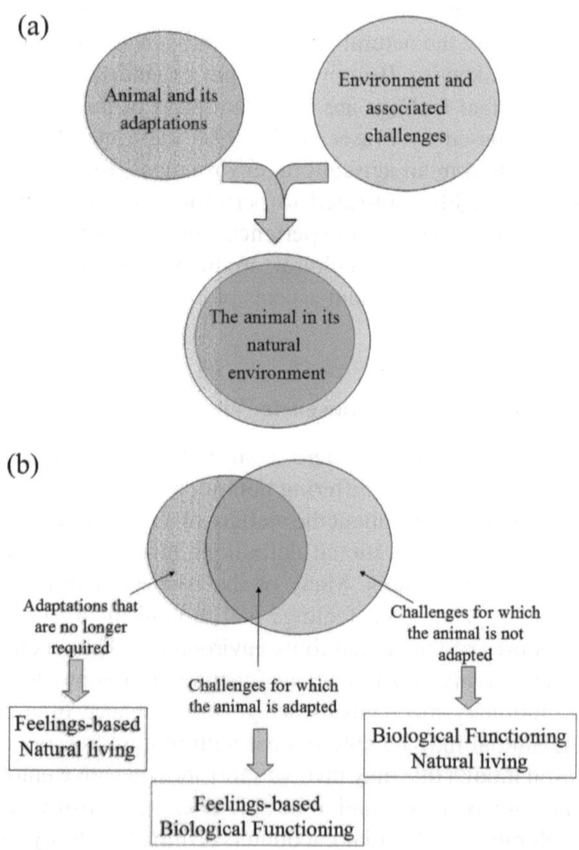

Fig. 1.1 Model of an integrated hypothesis of welfare concerns (after Fraser et al. 1997). (**a**) The animal living within the environment with a full set of adaptations to meet challenges presented by the environment. (**b**) The potential mismatch that can occur when the animal is domesticated and the environment may be less than optimal. Different classes of welfare concern then arise depending on the region of mismatch

evolved foraging behaviours, the animal may experience hunger, and show changes in biological functioning associated with malnutrition (tissue mobilisation, slowed growth, altered reproductive function etc.). Under these circumstances, although the welfare state of the animal is not good, we would expect that the animal's biological functioning would be reflected in its subjective feelings.

With domestication and intensive agriculture we can envisage that there may be increasing mismatch between the animal and the environment (Fig. 1.1b). This mismatch leads to different types of quality of life concerns dependent on where the mismatch occurs. The animal may have adaptations that are no longer required in the environment in which it now lives, but are associated with strong reinforcing affective experiences, both positive and negative. In the sheep, flocking and the presence of social companions are important anti-predator defences (see Chapter 2). Thus a socially isolated sheep may experience feelings of fear and panic, appropriate for an isolated sheep in the wild, without any actual threat to its biological functioning or fitness. Under these types of mismatch between animal and environment the animal may experience negative affective states (or fail to experience positive states) without there necessarily being any impact on its biological function. The other area of mismatch concerns where the environmental challenges differ from those in which the animal has evolved, thus the animal has no adaptations to deal with the challenge. Animals may typically fail to show avoidance on exposure to environmental toxins or overeat when given access to highly concentrated feed if they do not have adaptive mechanisms to deal with these challenges. Thus biological functioning may be impaired under these circumstances in the absence of, at least to begin with, the animals experiencing negative emotional states.

Fraser et al. (1997) thus suggest that the different welfare definitions could be conceived as reflecting the impact on the animal's quality of life in different parts of the model. Feelings-based concerns or concerns about the animal's subjective experiences, would occur primarily where the animal has adaptations that are no longer required, although these animals may show normal biological functioning, and in the overlap where subjective experiences may be associated with impaired function. Biological functioning concerns also occur in the overlap but are additionally seen where the environment is providing new challenges for which the animal has not evolved adaptations. Welfare concerns perceived as natural-living definitions occur in either condition when the animal and environment are not matched.

This model provides a conceptual framework which seems to address most welfare concerns: in general we imagine that the welfare of an animal is poor if it is in a state of ill-health (or at least heading that way) and if it is experiencing negative feelings. We may also feel that to deny the animal the opportunity for positive feelings might be an infringement of good welfare. The ways in which we might attempt to ascertain whether the animal is in an environment where it is strongly motivated to perform adaptive behaviours but cannot, or where the environmental challenges are greater than the animals adaptation will be discussed later in this Chapter.

1.3 Public Perception of the Welfare of Sheep

Much of the public concern for animal welfare has been directed towards animals other than the sheep, with pigs and poultry probably being the focus of greatest concern and research effort. Exceptions to this arise only when there is a highly visible challenge to sheep welfare, such as the suffering of the sheep trapped on the Como Express in 2003, or high levels of lamb mortality in years of extreme spring snowfall, and a regenerally short-lived. The stimulus for increased concern for farm animal welfare, and pressure to change the way farmed animals are kept, has been the increase in intensive agriculture and confinement. Although there are many different systems of sheep farming (see Chapter 6), they can nearly all be loosely classed under a definition of extensive where the animal spends at least part of the year outdoors, and gets some of its food from the environment. Being outdoors, in particular, has many positive associations with good animal welfare, health, naturalness and traditional agriculture. Contrasting these images with those of animals reared indoors in crates and cages and we might rapidly conclude that extensive or outdoor agriculture is a more animal friendly rearing environment. In this we may well be correct, although, for example, the continuing drought conditions in Australia might begin to make outdoor animal raising less appealing from a welfare perspective. However, we should not extrapolate from this comparison of indoor and outdoor agriculture to conclude that there are few welfare concerns in sheep production. Some of the differing perceptions surrounding the welfare of the sheep will be considered in this section.

1.3.1 Importance of Performing Natural Behaviours

Public perception of animal welfare places great weight on the ability of the animal to perform natural behaviours. In comparison to the hen in a battery cage or the pig in a gestation crate the sheep can move about, forage, engage in social behaviour and rarely, if ever under these conditions, show behavioural abnormalities, such as stereotypy. However, as argued by Webster (1994) and discussed above, most definitions of animal welfare extend beyond simply being given the freedom to behave naturally. This is only one of the Five Freedoms and extensive animals may still experience other threats to good welfare: hunger, thirst, thermal and physical discomfort, pain, injury, disease, fear or distress. For example, Webster proposes a hypothetical example where sheep are wintered on a poorly drained pasture with little shelter where they are chronically underfed, are forced to stand and lie in rain and mud, suffer from untreated chronic foot rot, and are regularly frightened by uncontrolled dogs. These sheep do, however, have the freedom to express their natural behaviour, even if this is predominantly panic and flight, but this can hardly be seen as an example of animals in good welfare.

The assumption that the extensive animal can show its full behavioural repertoire can also be challenged. Even if extensive, animals are generally not kept in habitats that resemble those in which their wild ancestors may have evolved. With the possible exception of hefted hill sheep in parts of the UK and range-managed sheep (e.g. in USA and Australia), most sheep may still be confined, sometimes at relatively high stocking density, may be exposed to limited numbers or types of plant species and may be kept in relatively featureless paddocks. In certain situations sheep may be highly motivated to perform various behaviours, such as seeking isolation (e.g. at lambing) or cover (e.g. if threatened) which will not be possible even in an outdoor environment. Some of these factors will be elaborated on in later Chapters, however, as an example, the frequency of alarm behaviours shown by Merinos has been reported to increase with a decrease in physical complexity of the environment (Stolba et al. 1990). Thus being able to move around and being outdoors do not automatically equate to an animal being able to perform all the behaviours that it perceives to be important, perhaps only those behaviours that *we* perceive to be important.

1.3.2 Responsibility Issues

Many of the threats to the Five Freedoms that can face an extensive animal come from the environment: rain, snow, wind, thermal extremes, lack of feed, predation (see Chapter 2). There is a tendency to perceive these as 'natural' or 'fate' and that these are outside our responsibility. The RSPCA in the UK, and the Animal Act of 1911, consider acts of cruelty (or causing unnecessary suffering) to occur both by abuse or commission and by neglect or omission. Thus, failing to provide feed or shelter to an animal kept on a hillside with little grazing and no natural shelter could be seen as causing unnecessary suffering by omission, assuming that the shepherd was able to provide that feed or shelter but chose not to. In addition, many of the decisions affecting the lives of the animals will have been made by man (e.g. the land and plants the sheep will have access to, the sheep genotypes that will use the land, the flock structure, etc.) and will have a direct effect on the ability of the animal to cope with the natural environmental situations. Thus it is not sufficient to conclude that, for example, lamb mortality is a 'natural' death and therefore outside our concern for good welfare.

1.3.3 Traditional Farming Practices

There is a strong perception that sheep farming and extensive farming systems retain the most traditional elements of agriculture. Since it is the more intensified modern agricultural systems that are considered to be worst for the animal's welfare, the converse might argue that traditional forms of agriculture are best for welfare. Part of this assumption lies in the belief that traditional agricultural practices depend on good husbandry and stockpeople, and hence better animal care, to be productive. Whilst this relationship may sometimes be the case, in places like the European Union many extensive sheep farms are not economically viable without subsidy. If subsidy is paid on a per head basis then this emancipates the care of the animal from economic productivity, the only financial benefit to the farmer is for the animal to be alive. The lack of individual monetary worth of a sheep may mean that the costs of shepherding, supplementary feeding, and veterinary care may exceed the financial return on the animal. Linking subsidy to production in a way that encourages good husbandry is essential to ensure that subsidised farming is associated with good welfare.

Adopting very traditional farming practices may, in some instances, negate some of the welfare benefits that research and improvements in, for example, nutrition and health have brought, albeit with the aim of improving productivity. As an example, scientific advances in reproductive management of pregnant ewes through the use of ultrasound scanning to determine litter size and the provision of better nutrition at critical times in pregnancy have halved hill sheep ewe and lamb mortality rates over the last four decades (Waterhouse 1996). Although these practices have undoubtedly improved productivity, they have also improved the lives of the animals as well. Belief in traditional methods, particularly as they are often accompanied by a fatalistic acceptance of misfortune (e.g. high levels of neonate mortality, lameness), do not inevitably lead to improved animal welfare.

Within traditional farming practices management interventions occur that can cause the animal to experience pain, fear and distress, at least temporarily and even if their ultimate aim is to improve animal welfare. For example, castration and tail-docking without anaesthetic cause pain and distress, working sheep with dogs and shearing cause high levels of fear and occasionally result in injuries and death. However these practices have been carried out for centuries and are accepted and unquestioned by the general public. As pointed out by Kilgour (1985) the absence of a tail in sheep is so much part of the public perception that, in books, sheep are rarely illustrated with a full tail. Likewise the idea that sheep should be worked with dogs is so much part of our perception of traditional sheep farming that we even have competitions to demonstrate how well the dogs can move sheep. Whilst some of these practices may improve animal welfare in the long term (reduction of fly strike in tailed sheep, for example) it would be unlikely that a plan to use a predator to manage free range hens would be considered an ethically acceptable practice, whereas this is accepted in traditional sheep farming. In the same way high levels of lameness in sheep (in the UK around 10% of sheep annually are lame with footrot, Royal Veterinary College survey, 1999) are perceived as being an integral part of sheep farming and the pain and chronic suffering associated with lameness under-emphasised.

1.3.4 Characteristics of Sheep

Ruminants, and sheep in particular, are frequently described as 'stoical' or 'physically tough' (Webster 1994). Unlike pigs and poultry, they are considerably more resistant to thermal extremes and the presence of the rumen means they can survive for longer periods without access to food and water. Physically, they are capable of surviving under conditions, for example during transport, which would result in high levels of mortality in other animals. Clearly there are strong financial and moral pressures to effect change to a system which causes high levels of mortality, that are not so easily brought to bear where the animals can survive the insult. However, because these things do not kill them, can we conclude that they do not suffer?

Ecologically, the sheep has evolved as a predated animal and as such has developed subtle behaviour patterns to avoid communicating disease, injury or physical impairment to a watching predator. To the casual observer it may be that the first indication that anything is wrong with a sheep is its death. Sheep are sometimes described as behaviourally cryptic, where their behaviour may not be readily interpreted. In particular the sheep is not particularly vocal in response to stressors (with the exception of lambs separated from their mothers where there is a clear functional purpose to vocalisation), and vocalisation is inhibited in the presence of predators. A vocal commentary is an integral part of our assessment of the internal state of other humans and animals, thus it is all too easy to conclude that the animal that does not complain does not suffer.

We have already discussed the relatively low monetary worth of sheep. In addition, they may also be perceived as having relatively low intrinsic worth. Sheep are generally perceived as being rather stupid in comparison to pigs for example, even by members of the public with little or no direct experience with either animal. In a survey of staff and students at a university in the USA (Davis & Cheeke 1998) sheep were consistently rated as being of lower intelligence than dogs, cats, horses, pigs and cows, and were ranked only slightly above chickens and turkeys. Does this matter for animal welfare? After all, as pointed out by the utilitarian philosopher, Jeremy Bentham, in 1789 'the question is not can they reason, nor can they talk, but can they suffer?', a sentiment reiterated by Dawkins (2001) as 'you don't need to be very clever to feel pain or hunger or fear'. It is the effect on public pressure for animal welfare change that perceived relative intelligence may have the greatest impact. For example, a farm animal that is perceived to be of lesser intelligence may be considered to have lower intrinsic worth, and therefore be less likely to have its welfare protected. In the survey by Davis and Cheeke many of the respondents felt that animal intelligence should influence how they were kept. About half the respondents considered that more intelligent animals needed better care to prevent boredom. Rather worryingly, some of the respondents to the survey, who clearly had a very low opinion of animal abilities, considered that animals of perceived low intelligence required extra husbandry attention to prevent them from killing or injuring themselves! A caveat: there is in fact no scientific evidence to support the public perception of the stupidity of sheep, relative to other domestic animals. As should become clear in later Chapters sheep show well-developed abilities to learn about the environment, have a highly organised and complex social structure and, having evolved as a predated animal and therefore needing to constantly outwit predators to survive, might be expected to be the most likely to have evolved consciousness (Griffen 2001).

1.4 Specific Welfare Issues Pertaining to the Sheep

Unlike many other farmed animals, sheep are maintained under a variety of conditions, even within the same country (this is expanded on in Chapter 6). Sheep farming may be extremely extensive, such as the range management systems of Norway and USA or Scottish hill sheep where, although highly managed, the sheep may be treated virtually as wild animals for weeks or months at a time. Other extensive systems include the nomadic pastoralism practised in some countries of Europe, Asia and North Africa where the sheep are free to wander but are accompanied by a herder. At the other end of the scale dairy sheep, or finishing lambs, may be kept indoors for all or much of the day, usually group housed at relative high stocking density in straw-bedded pens, fed on concentrates and will have high levels of human contact. The sort of welfare issues that cause concern in any of these systems will differ from intensive confinement agriculture but may also differ between the differing management systems. Generally, concerns about sheep welfare can be seen as falling into three major areas, the relative prevalence of any area of welfare concern may change with different systems of sheep farming.

1.4.1 Problems Connected to Extensive Systems

As we have argued above, although the extensive environment allows the animal much greater freedom to express its behavioural repertoire, this does come with some costs. Animals are exposed to much greater environmental challenges than animals maintained in temperature and humidity controlled housing. This environmental variability is not, of itself, likely to cause poor welfare, and may even be an important and neglected aspect of good welfare (Appleby 1996). However, prolonged exposure to extreme environmental conditions, particularly if they are accompanied by other challenges (undernutrition, poor body condition, lack of shelter, for example), may be a source of chronic stress. In addition extensively managed animals in particular may suffer similar predation risks to wild animals. These issues are dealt with in more detail in following chapters. Extensively managed animals also differ from intensively managed animals in the frequency of interactions with stockpersons, and those interactions that do occur are often aversive.

An important part of assessing welfare is clearly to inspect the animals on a regular basis and, generally, failure to do this can lead to prosecutions for neglect and animal cruelty. However the nature of extensive systems means that the degree of inspection is likely to be less than in other systems. For example, in a modelling exercise Waterhouse (1996) demonstrated that it is almost impossible for a single farmer to observe all ewes in a 800 strong flock at lambing time when the area available to the sheep exceeded 800 hectares (at this level it required the shepherd to cover 40 km per day and spend over 10 h just to observe the sheep once without considering the time needed to provide care to mother and offspring if required). Does a lower level of inspection carry a welfare cost to the sheep when they are able to access feed, water and shelter without human intervention? Arguably a sheep living a semi-wild existence on an extensive farm, and unaccustomed to a regular human (and sheep-dog) presence, may find a daily inspection more stressful than being left alone. Certainly the UK law appears to recognise this since the daily inspection of livestock is required only in systems where the health and welfare of the livestock depend on frequent human inspection (Welfare of Livestock Regulations 1994). Shepherds on extensive farms are required only to inspect flocks 'as often as is necessary'. Lack of regular inspection, however, leaves extensive sheep open to the possibility of chronic and untreated distress, disease or injury. The most common problems are obstetric difficulties and related problems around lambing time (e.g. vaginal prolapse, poor ewe-lamb bonding, mastitis), fly strike in warm and humid summer months, lameness, and parasitic infestation. Current trends towards the genetic selection of sheep to be 'robust' or better able to take care of themselves may improve sheep welfare if the frequency of inspection is low, however a further reduction in inspection frequency can leave those animals that do experience welfare problems particularly vulnerable.

Opinion is divided on the benefits of supervision at lambing time on ewe and lamb survival in extensive systems. Generally it is agreed that the ideal is for the ewe to give birth unaided and for ewe and lamb bonding to occur without human intervention. However, intervention decisions must be made whilst the ewe is in labour, based on assumptions and previous experience, to determine whether a ewe or lamb will survive if left alone and thus it can be difficult to know whether these interventions are genuinely helpful. Fear, stress and disturbance are known to cause involuntary suppression of uterine contractions in mammals during labour, presumably to be able to deal with the presence of a predator or to escape from a stressor before giving birth. For ewes unaccustomed to human presence, close supervision may act as a source of stress and unnecessarily delay or prolong parturition. A prolonged labour affects the expression of ewe maternal behaviour (Dwyer & Lawrence 1998), impairs lamb behavioural development (Dwyer 2003) and reduces lamb survival (Haughey 1993). Thus a low stress environment for lambing ewes is likely to be associated with better welfare for the ewe, and improved lamb survival. A recent review of shepherding during lambing of extensive flocks in New Zealand concluded that there was little evidence to suggest that shepherding inputs ensured either easy births or promoted ewe-lamb bonding (Fisher & Mellor 2002). However, in the UK, Pattinson et al. (1994) concluded that shepherding did increase lamb survival to weaning. In the USA, lamb survival was improved by shed lambing compared to unsupervised range lambing (Burfening & van Horn 1993), although a UK survey suggests that intensive rearing increases peri- and postnatal mortality (primarily due to infection) although stillbirths decreased (Binns et al. 2002). These differences may be attributable to the breeds of ewe, particularly the use of 'easy-care' sheep in New Zealand which have been selected for several generations (both naturally and artificially) for their ability to survive and to raise lambs unaided, and different management strategies (these are discussed further in later Chapters).

Although the role of inspection in improving the welfare of lambing ewes and their lambs is not clear, it should be remembered that there are many other situations, some occurring around lambing, where timely intervention may vastly reduce suffering. In particular a number of relatively common diseases and parasitic infections (notably foot rot, fly strike and sheep scab) are associated with pain and suffering and, if left untreated, are a major source of welfare compromise in the sheep. The effects of disease on welfare are expanded on in Chapter 5.

1.4.2 Surgical Procedures (or Mutilations)

Sheep are routinely subjected to a number of surgical procedures that may not be directly for therapeutic benefit (hence the term mutilation), usually without anaesthesia. Many of these procedures (especially castration and tail docking) have been carried out for centuries, however there has been considerable recent research that suggests both procedures are associated with considerable acute pain and may also result in chronic pain responses. Some of these procedures are carried out to allow animals to be managed in a particular environment, either because farmers are unwilling or unable to manage the animals in a way that does not require the intervention. Often these procedures, such as tail docking or mulesing, are justified on the basis that they prevent other welfare problems. An ethical dilemma remains, nonetheless, that if the mutilation must be carried out for sheep to be managed in a particular environment then is it acceptable that the animals are managed in that environment at all? The most frequently practised procedures are briefly described below, along with a welfare assessment based on the schema provided above (Fig. 1.1).

1.4.2.1 Procedures to Identify Animals

The ability to permanently identify individual animals within the group can have a number of important benefits for animal welfare as well as for the traceability of animals for human and animal health. The ability to identify individuals is essential if animals are to be selected for disease resistance or survivability traits such as attentive maternal behaviour or active neonatal lamb behaviours. The Codes of Welfare for Sheep of the UK, New Zealand and Australia suggest that ear notches or punches, ear tags, ear tattoos, and horn brands are suitable methods for identifying sheep, with horn brands being preferred if possible. However, the most common forms of identification are usually via removal of parts of the ear or ear tagging. Electronic transponders are now also available and subcutaneous injection or intra-oral deposition of a rumen bolus may prove to be the least painful method of identifying animals. Applying a cost/benefit analysis to animal identification might conclude that there are health and welfare benefits to being able to individually recognise sheep that offset the acute pain caused by placing ear tags. However, improved methods of identification that reduce or eliminate pain would clearly be preferable.

In the EU, in response to outbreaks of Foot and Mouth disease and BSE, identification of all sheep is required, with electronic identification planned from 2008.

This is primarily to allow traceability of animals, and to identify the animals' place of origin. However, this may bring welfare benefits if it limits the number, duration and complexity of journeys that sheep make between farms and markets.

1.4.2.2 Tail-Docking

As described above, tail-docking is such a frequent procedure in sheep that it is largely accepted without question, although some organic producers are starting to emphasise that their lambs are never docked. Lambs' tails are shortened to reduce the amount of faecal soiling of wool and so to reduce the incidence of fly strike. Fly strike clearly represents a major welfare challenge in infected sheep and, again, a cost/benefit analysis might conclude that tail docking is justified on the basis that this procedure reduces the possibility that the sheep will experience a worse welfare compromise later in life. However, a welfare cost:benefit analysis would also require an assessment of the likelihood of achieving the benefit against the known cost (the pain of tail-docking). In fact, in UK law, the farmer or stockperson are required to do just that by considering whether tail docking within a particular flock is necessary, and tail docking may be carried out *only* if failure to do so would lead to subsequent welfare problems. Hill ewes, for example, may never encounter the warm, humid conditions of lowland farms where fly strike may become a problem. In addition, the welfare 'cost' can be reduced by using methods to reduce the pain of tail removal. As sheep do not appear to use their tails for communication or expression, or to remove flies, and there is little evidence that sheep experience phantom limb pain from the stump, the main welfare concerns (except for the animal integrity arguments posed above) are with the method rather than the absence of the tail *per se*. Use of analgesia or anaesthesia at tail-docking would make this practice more acceptable on welfare grounds.

1.4.2.3 Castration

Male lambs traditionally have been castrated for management purposes to prevent indiscriminate breeding and to improve meat quality. However, with selection for faster growth in many breeds, lambs may often reach slaughter weight before sexual maturity making castration unnecessary. In addition, there may be some production benefits in leaving male lambs entire as ram lambs grow faster and produce leaner carcasses than castrated males (Ritar et al. 1988; Gregory 1998). As with tail-docking, both UK law and the New Zealand Code of Recommendations for sheep welfare require that farmers should consider whether castration is necessary and it should not be carried out unless it has significant management advantages. A recent survey of New Zealand farmers suggests that nearly 40% of male lambs in that country are now left entire (Tarbotton et al. 2002). The welfare cost/benefit equation is less supportive of castration when we consider that it is the lamb that experiences all the costs and, probably, the farm management system that achieves all the benefits. There are some suggestions that indiscriminate breeding can carry some welfare costs, for example ewe lambs that conceive grow more slowly and tend to have much higher lamb mortality than ewes which become pregnant at an older age. In addition, in a feral sheep population, castrated males had a much greater survivability than either entire rams or ewes (Jewell 1997). However, although reproduction does carry longevity costs it is hard to see that an animal would choose to refrain from breeding in order to improve survival, since this would carry other costs in terms of biological fitness. Furthermore, as farm animals are never retained for their maximum lifespan, these do not appear to be defensible arguments in favour of castration, and certainly not castration without anaesthesia.

1.4.2.4 Mulesing

This surgical procedure is usually confined to Merino sheep and their crosses (under the New Zealand Codes of Recommendations this procedure can only be carried out on Merino or Merino-dominant animals), and involves removing wool-bearing skin from the breech and tail regions without anaesthesia. Merinos have excessively wrinkled folds of skin in this area which are particularly prone to becoming soiled and susceptible to fly strike. As the wound heals bare, less wrinkly and wool-free scar tissue grows which reduces the chances of a sheep succumbing to fly strike. Similar welfare arguments to those made for tail docking are also relevant here – fly strike is considered such a challenge to good welfare that permanent protection from its consequences is likely to improve welfare. Mulesing certainly reduces the incidence of fly strike and appears to reduce lamb deaths (Dunlop and Johnston 1985). Recent evidence, however, suggests that alternative practices, such as those used on organic farms, might also be as effective in reducing the incidence of fly strike (Morris 2000). Lee and Fisher (2007) argue that flystrike rates could be kept at present rates by increased use of chemical preventatives and increased inspection (and likely increasing welfare), but this would require producers to invest time and resources in alternative methods of flystrike control.

Mulesing is known to cause pain both during the procedure and whilst healing. Protection from one form of welfare challenge by imposition of another is a difficult compromise, which provision of analgesia and/or anaesthesia might make more acceptable. As with tail docking the welfare issues are predominantly with the pain and distress experienced by the animal during a surgical procedure without anaesthesia rather than the procedure itself. However, since the use of anaesthesia and analgesics seems to suppress only the acute response to the procedure (Paull et al. 2007) rather than the long term discomfort or chronic pain responses, it is arguable that even with anaesthesia the procedure will still result in pain. Of consideration, also, is how much of the susceptibility of Merinos to the problem of fly strike may have been created by breeding Merinos for excessive wool development, and keeping sheep in regions in which they are not particularly well adapted. Merinos originate from Spanish flocks so, although well-adapted for hot, dry countries, may fare less well in warm, moist climates with an abundance of flies. Alternative solutions to mulesing might also come from breeding for welfare goals, such as sheep with less wrinkles or wool in the breech area (e.g. as proposed by Scobie et al. 1999). This area is expanded on in Chapter 10.

1.4.2.5 Other Procedures

There are a number of other surgical procedures that have been perpetrated against sheep over the years. UK law specifically prohibits tooth-grinding[7] or breaking, freeze dagging, short tail docking (where insufficient tail is left to cover the vulva), electro-immobilisation, or any penile operations unless these are carried out as part of treatment for disease or injury. New Zealand and Australian Codes of Recommendations also suggest that there is little or no benefit to be obtained, either directly or indirectly, for many of these procedures, and they are likely to be associated considerable pain and distress. For example, sheep are known to find electro-immobilisation[8] to be extremely aversive, and considerably more stressful than being manually restrained (Rushen 1986). Both New Zealand and Australia do, however, permit a procedure known as 'pizzle-dropping' to be carried out on Merino or Merino-dominant wethers (castrated males). The aim of the procedure is to prevent fly strike of the prepuce following bacterial infection, and involves cutting the skin that holds the prepuce against the belly so that it hangs free. As with mulesing there is an increased risk of fly strike of the wound during healing, and other combinations of disease prevention measures may be equally effective.

The UK, Australian and New Zealand Codes of Recommendations are broadly in agreement that there is no valid reason for dehorning or disbudding sheep, except when horns are ingrowing and likely to cause pain or distress if left untreated. Since hot branding of horns is considered the most welfare friendly method to mark horned sheep (see above), there are several welfare reasons for not removing horns. Under UK law only the insensitive tip of a horn may be removed by a lay person, complete dishorning requires a general anaesthetic and can only be carried out by a veterinary surgeon.

1.4.3 Management Practices

The effect on sheep welfare of many management practices are discussed in more detail in Chapter 8 so specific practices are not elaborated on here. However, some general principles are pertinent. Firstly, with the exception of poultry and some goat breeds, sheep are smaller and more defenceless than other farmed species. They

[7] Tooth-grinding, or trimming of the incisor teeth, was advocated in the 1980s, in some countries, as it was believed to be a solution to teeth loss or periodontal disease. Teeth were ground or cut level with the dental pad, which was thought to improve bite efficiency and the longevity of the ewe. Subsequent studies suggested that these claims were not substantiated as feed intakes and weight gains were not improved. Instead exposure of the pulp cavity was likely to cause considerable pain, and one study found 90% of sheep had exposed pulp cavities in at least one tooth (Denholm & Vizard 1986).

[8] Electro-immobilisation involves the passing of a low voltage pulsed current through the body between two electrodes placed in different locations on the body. This causes tetanic contraction of the skeletal muscles between the electrodes and results in immobilization although not loss of consciousness or sensation in most animals.

are, therefore, more vulnerable to physical abuse (to be dropped, thrown, dragged, physically assaulted etc.) than larger animals where size makes these impossible, or where a greater risk of injury to themselves may make handlers treat the animal with more circumspection. Additionally, most individual sheep are of low financial value, thus farmers may be reluctant to spend money on individual veterinary treatment (where the cost of a farm visit by the veterinary surgeon may exceed the value of the sheep to be treated), or to handle animals with particular care.

In comparison with other farm species, sheep are more frequently mustered or gathered, yarded, transported and marketed. Each of these handling practices is associated with compound stressors: mixing with unfamiliar animals, exposure to unfamiliar environments, crowding or high stocking density, dehydration, heat stress, food deprivation, movement achieved through fear stimuli, often with the use of dogs, and risks of physical injury and smothering. Transportation may also include noise, vibration, unfamiliar motions, trans-shipping (movement from one vehicle to another), sea journeys, and ventilation stressors. Journey structures can be complex and diverse, such that the animal will be exposed to a multitude of these stressors, with complexity increasing as duration of travel increases (Murray et al. 2000). The complexity of the stressor exposure make it difficult for experimental studies to accurately reproduce the experiences of sheep to assess the impact on their welfare. Surveys of mortality rates suggest that transport of slaughter lambs in the UK is associated with very low mortality (0.02%) although transport from Australia to the Middle East (journeys involving considerable sea travel) is associated with several fold higher mortality (2.2%; Kent 1997). However, as we have discussed above, the ability of the sheep to survive adversity does not necessarily imply lack of suffering. Each of the individual factors that may form part of the experiences of movement, transportation and marketing are known to act as stressors. These include mixing, unfamiliarity and crowding, (sheep appear to find the presence of unfamiliar animals stressful, preferring animals of the same breed as themselves: Bouissou et al. 1996; Kendrick et al. 1996), food and water deprivation, fear, handling and the potential for injury.

To return to our original diagram of welfare compromise in sheep, it is clear that sheep cannot have evolved mechanisms for dealing with many of these procedures. In some cases the evolved mechanisms (e.g. flight) are used to bring about the management intervention (e.g. gathering). Thus the challenges the sheep experiences are likely to exceed their ability to cope, resulting in changes in biological functioning associated with negative emotional states. Frequently the assessment of the animals response to these procedures is monitored by measuring components of the biological response to, for example, the handling associated with shearing (this will be discussed in more detail below). An important component of the animal's response to many of these procedures is the negative experience of fear, a potent psychological stressor, of novelty and of humans. This is particularly relevant for extensive animals, where their experience of human contact is very low, and confined largely to unpleasant and aversive experiences. Taming or gentling, where sheep are stroked and hand-fed over a number of days, produces animals that approach a human more readily, have shorter flight distances and lower heart rates than unhandled sheep (Hargreaves & Hutson 1990; Mateo et al. 1991). Gentling, therefore, apparently reduces the fear component of handling. However, gentled sheep have similar struggling responses to restraint as sheep that have not been gentled (Mateo et al. 1991) or a similar aversion to sham-shearing (Hargreaves & Hutson 1990), although early handling of lambs may reduce the fear responses to subsequent isolation, and restraint (Uetake et al. 2000). Tamed sheep also seem to show less responsiveness to transport than untamed sheep (Hall et al. 1998). Positive human contact, leading to a reduction in fearfulness, may reduce the stressfulness of handling procedures that are not intrinsically aversive and where the main stressor is handling. For extensively-managed sheep, however, their lack of experience of human contact means that handling procedures are likely to be more stressful, and hence a greater welfare cost for each interaction, than for intensively-reared animals. For intensively-managed sheep the frequency of negative interactions may lead to an increase in the cumulative welfare cost, even if each interaction is less stressful than for extensively-managed sheep.

1.5 Recognition of Welfare Problems in the Sheep

To identify welfare problems we need to have ways of assessing whether an animal is showing signs of altered biological functioning, whether it can express natural behaviour and/or whether it is experiencing negative emotional states. The latter of these is probably the most problematic, but measures or indicators of all these welfare definitions are not without their own difficulties of interpretation. Opinions are divided about the usefulness and accuracy of each type of measure, although generally it is agreed that there is no one simple measure that can describe an animal's welfare state, and the best methods of assessment are through composite and integrated measures. Experimentally, measures can be divided into physiological, production (both measures of biological functioning, albeit at different stages in the animals response) and behavioural. From these measures some inferences about the animals' ability to cope with the challenges, and sometimes their emotional state, are drawn, and our 'evaluative' statements (as described above) produced. In this section we will look briefly at the different experimental measures that are used, the reader is referred to more specialised texts for more detailed descriptions of the methodologies (see for example, Moberg & Mench 2000).

1.5.1 Neuroendocrine Measures Relating to Welfare

As mentioned above (Section 1.2.1.3), the physiological responses measured in welfare assessment are generally those relating to stress and the functioning of the HPA system, the SAM system or some assessment of immune function. The HPA axis is a neuroendocrine cascade, triggered from the hypothalamus in response to the detection of some perturbation, either external (e.g. approaching predator) or internal (e.g. fear) that prepares the body to mount a fight or flight response. Whether the threat actually exists or not is immaterial to the response, it is the perception of a threat that triggers the response. A hormone called corticotrophin-releasing hormone (CRH) is released from the hypothalamus and travels to the anterior pituitary where it brings about the release of adrenocorticotropic hormone (ACTH). ACTH acts on the adrenal cortex to produce the steroid glucocorticoids (cortisol in sheep). The glucocorticoids are known as the 'anti-stress' hormones and act to increase carbohydrate metabolism (providing energy to mount the fight/flight response) as well as feeding back to suppress their own synthesis and release. The responses of this cascade can be assessed indirectly by monitoring the amounts of (usually) cortisol in bodily fluids (blood, saliva, milk, urine, etc.). The autonomic SAM system is likewise activated on detection of some threat or change in the system and causes release of catecholamines (adrenaline and noradrenaline[9]) from the adrenal medulla in response to sympathetic neural signals. The catecholamines influence circulatory and metabolic systems affecting heart rate and cardiac output, blood pressure, body temperature, and glycolysis. This system responds more quickly than the HPA axis (as it is predominantly neural rather than endocrine) and can be monitored directly by measurement of, for example, changes in heart rate, respiration rate, temperature or blood pressure. So far this is all relatively straightforward, the difficulties arise in deciding what a change, and what level of change, might mean for the animals' biological functioning (we are also leaving aside difficulties of measurement without altering the very physiological parameters of interest).

Moberg (2000) argues that the responses outlined above are part of the normal biological functioning of the animal, no animal remains in a constant unvaried state and minor deviations and corrections are part of the normal response to environmental variability. For example, in the wild, a sheep may detect the approach of a predator, which activates the different systems: heart rate increases and blood flow to the skin may decrease so that blood can preferentially deliver oxygen to the limb muscles, increased glycolysis in the liver initially provides energy for flight, followed shortly after by an increase in carbohydrate metabolism triggered by cortisol release. If the sheep successfully evades the predator, the threat stimulus is effectively no longer present and heart rate and blood flow return to normal, cortisol feeds back onto the higher parts of the cascade mechanism to inhibit its own production and the animal resumes normal feeding and other behavioural patterns. The animal may experience fear initially at their first perception of the threat but we might conclude that the animal has successfully used species-typical means to deal with the potential threat, and therefore there has been little negative consequences on its welfare. Thus the initial rise in heart rate, and subsequent increase in cortisol, would tell us that the animal has responded to some alteration in the environment but doesn't necessarily tell us anything about its welfare. Moberg (2000) suggests that an animal's welfare is compromised only when these changes cause alterations in

[9] These are also known by their synonyms, epinephrine and norepinephrine, particularly in North America.

biological functioning leading to a prepathological state. Using our example above, if the sheep was persistently chased, or had no adequate areas to escape into, we might see a sustained catabolic response, the animal is unable to engage in normal feeding or reproductive behaviour and its biological functioning, and welfare, is threatened.

So, transient rises in heart rate or cortisol indicate the sheep is showing normal homeostatic responses but how do we determine what response is likely to lead to prepathological states? Cortisol responses are known to show considerable individual variation, being affected by, for example, breed, age, experience, physiological status and sex. Thus setting 'cut-off' cortisol values above which a prepathological state must be present is impossible. An important feature of any homeostatic system is feedback of hormones to regulate their own production, thus the normal cortisol response consists of an increase in cortisol production, followed by a decline. This decrease in cortisol release has sometimes been interpreted in experiments as habituation, by a reduction in the perception of a threat of the stimulus, and thus that the animal is less stressed and in better welfare. However, the animal may still be perceiving the threat in the same way at the level of higher brain areas, and it is just the physiological responsiveness that has diminished (Smith & Dobson 2002). Prolonged exposure to a stressor, or repeated exposure to stressors (chronic stress), causes a specific form of adaptation where a biochemical down-regulation of areas of the HPA response occurs altering the control systems for stress responses (Jensen et al. 1996; Terlouw et al. 1997). This is believed to help the animal maintain sensitivity to further acute stressors or disease responses under conditions of chronic stress, and because exposure to excessive amounts of cortisol can be detrimental. Thus lower levels than normal of cortisol may also be indicative of chronic stress and poor welfare. Finally, cortisol is also known to be elevated in times of excitement, when the animal may be experiencing extremely positive emotions, and is not exclusively confined to negative emotional states. Thus elevation in cortisol cannot always be interpreted as a symptom of an animal experiencing a decline in welfare without corroborating behavioural or other physiological evidence to support this conclusion.

One of the well-described responses of the animal to repeated or intense stress exposure is the suppression of immune function (Wiepkema and Koolhaas 1993; Moberg 1996). In sheep psychological stressors (for example, restraint and isolation) cause an alteration in the blood profile of cells involved in mounting an immune response. The main features of this response are an increase in the number of neutrophils (the white blood cells or leukocytes involved in phagocytosis) and a decrease in the number of lymphocytes (cells involved in specific immunity – they either secrete antibodies or participate in cell-mediated immune responses) in the blood (Minton & Blecha 1990; Coppinger et al. 1991). Stressed sheep also show a reduced lymphocyte blastogenic response when challenged with specific mitogens (Minton et al. 1992; 1995). The exact mechanism underlying the immunosuppressive effects of stress are not yet clear, however stressed sheep do not mount as efficient a response to pathogen challenges as unstressed animals. Thus stress-induced changes in immune function indicate an animal entering a prepathological state where the animal is more susceptible to environmental pathogens and hence at a greater risk of disease.

1.5.2 Behavioural Responses Relating to Welfare

Some of the difficulties in determining what a cortisol rise, or other physiological response, means can be alleviated by adding behavioural measures to the assessment. Thus it may be possible to determine whether elevated cortisol is associated with positive or negative emotional states by observing the behaviour of the animal (approaching or avoiding the stressor, for example), or whether a decline in cortisol is habituation or biochemical down-regulation. However, behavioural measures do not always correlate well with physiological measures and caution is also often necessary in interpreting behaviour. As argued by Rushen (2000), the neuroendocrine and motivational control of behaviour is complex, and frequently poorly understood. For example, dam-reared and isolation-reared lambs show very different behaviours within the same test (Moberg & Wood 1982), although their physiological responses are identical. Whilst behaviours that stem from anti-predator responses may be simpler to explain in terms of motivation (since sheep will be primarily motivated to flee), understanding the motivational basis of behaviours of an animal confronted with shearing or transport, which has no evolutionary basis, is more difficult. Behaviours can also show the same inter-animal variability as described for physiological measures and, as they will have evolved functionally to cope with a particular stressor, will not be common to all welfare challenges. Some of the ways that behaviour has been used to assess welfare are described below:

1.5.2.1 Anti-Predator Behaviours

The motivation of anti-predator responses, largely fear responses, are the most straightforward behaviours to explain functionally, and may be helpful in determining stress responses. The behavioural responses of the wild ancestors of domestic sheep to the threat of predation are characterised predominantly by vigilance, flocking, flight to cover and behavioural inhibition once refuge has been reached. Thus both behavioural activity and immobility can form part of the behavioural response of the sheep to fear (as seen experimentally, e.g. Romeyer & Bouissou 1992). Vigilance and flight distance are affected by the animal's assessment of risk and are influenced by the environment, social group size, age, sex and reproductive conditions, as well as previous experience of a potential predator (e.g. man). More specifically, vigilance behaviours are influenced by the closeness to escape terrain, social group size, predator density, whether it was night or day and, for maternal animals, whether their offspring were active or not (Woolf et al. 1970; Frid 1997; White & Berger 2001; Laundré et al. 2001). Flight behaviours were likewise influenced by escape terrain, familiarity with the type of predator approaching, age and sex (Berger 1991; Bleich et al. 1997; Bleich 1999; Martinetto & Cugnasse 2001). As these are graded responses, relying on the animals' assessment of the risks they face, they start to tell us something about the underlying emotional state of the sheep, assuming that negative feelings have evolved as a component of the anti-predator behavioural strategy (see above). However, difficulties in interpretation arise when these behaviours are used to assess stress responses under artificial conditions, for example exposure to a novel object or environment (Rushen 2000). Does an increase in locomotion indicate a flight response and fear, or exploration of a new environment in a relaxed and confident animal?

1.5.2.2 Tests of Aversion and Preference

In addition to behavioural assessments of sheep exposed to various potential stressors as described above, behaviours can be used to learn more about what a sheep might find stressful by aversion learning techniques (as mentioned above) and preference testing. These procedures set out to ask the animal what it wants or how it feels about various procedures rather than relying on interpretation of behaviours expressed. What distinguishes a pleasurable from an unpleasant experience is whether the animal seeks to avoid being exposed to the experience again. Aversion learning paradigms are based on this assumption, and that the animal can learn predictive relationships between events (Rushen 1996), where these have been used in sheep the relationship is generally between an unpleasant experience and a particular place. In these techniques the animal is repeatedly walked down a race to a handling pen where various procedures (e.g. restraint, isolation, inversion, transport, sham-shearing) are applied. The increased aversiveness of the handling pen by the procedures carried out there can be measured by the willingness of the sheep to return to the pen when compared to control animals that were just returned to the home pen after completing the race. Measures such as the time taken to enter the pen, the amount of time spent in the race, or the amount of pushing required to move the animal along the race allow a ranking to be made of how aversive the sheep finds the various procedures. These techniques are often easier to interpret in terms of animal welfare than taking behavioural or physiological measures, particularly for management procedures when the motivational basis of a behaviour may be obscure. Aversion learning has, for example, demonstrated that sheep find electro-immobilisation to be more stressful than manual restraint (Rushen 1986).

Preference testing, on the other hand, asks an animal what it likes and makes inferences about welfare based on an animal's choice when offered two or more alternatives. This technique has mostly been used with confined animals to ask questions about the types of housing they might prefer and only rarely applied to sheep. However, we might conceivably want to use these techniques to ask sheep questions about shelter, amount of human contact or about types of foods, for example. A variation on this type of test, an 'operant' test, is to get the animal to do some sort of task, such as pushing a panel or lever, which is rewarded by getting access to something we think the animal might want. This has been used with sheep where they were asked to push through a weighted door to get access to feed after various periods of deprivation (Jackson et al. 1999). Interpretation of preference tests are complicated by a number of issues, particularly previous experiences and learning, the number of choices the animal is offered (are they choosing only the best of several poor choices?), and behavioural wisdom, i.e. does the animal really choose what is best for it, especially if these are offered under highly artificial circumstances? An extension of these methodologies, Consumer Demand studies (Dawkins 1990), applies economic arguments in an attempt to separate the important behaviours from those that are less important or luxuries. These studies might use a similar operant task as in preference tests but the amount of work an animal has to do increases (e.g. by increasing the weights on the door that the sheep has to push through). A demand curve is then constructed for different commodities as an animal is likely to continue to push through heavier doors for an essential commodity, e.g. food, whereas they may rapidly stop pushing on a heavy door for something inessential, e.g. the opportunity for play. These have rarely been used to ask welfare questions of sheep, but in other species have revealed that, for example, the opportunity to root for a pig or to nest for a hen are behaviours that the animal finds important.

1.5.2.3 Behavioural Indicators of Good Welfare

Preference and aversion learning tests then start to tell us the minimum requirements of an animal for reasonable welfare, or how to avoid poor welfare. However, as we have argued above, the ability to express good welfare and behavioural 'wants' are also important in animal welfare. Thus part of the use of behaviours may be as indicators of good welfare by performance of non-essential parts of an animals repertoire. Play behaviour in young lambs is a good example of a non-essential behaviour that is sensitive to environmental perturbations. In young animals play behaviour can be affected by a reduction in energy intake or poor diet quality (Müller-Schwarze et al. 1982; Reale et al. 1999), by a risky environment (Berger 1979), bad weather (Rasa 1984) and by pain (Berger 1980; Kent et al. 2000). Thus the frequency of play behaviour can be an indicator of good welfare in the young lamb, or its absence as a potential indicator of poor welfare.

1.5.2.4 Behavioural Abnormalities

For intensively housed animals there are a number of behavioural abnormalities that are frequently taken to be indicative of poor welfare, or suboptimal housing. These consist of vacuum behaviours (where the behaviour may be performed in the absence of the normal releasers or substrates that would elicit that behaviour under natural conditions), injurious behaviours or self-directed behaviours (where the behaviour is directed inappropriately) and stereotypic, or repetitive, functionless behaviours. We would consider these behaviours to fall into the region in the welfare diagram (Fig. 1.1) where the animal possesses behavioural adaptations that are no longer required, but which it may, nevertheless, be highly motivated to perform. Furthermore, we might expect the animal to experience negative emotional states if it is unable to perform these behaviours. Sheep appear to be less likely to perform stereotypies than other species (Houpt 1987; Lawrence & Rushen 1993), which may be due to the lower frequency with which sheep are kept in intensive housing. An alternative hypothesis is that rumination acts to alleviate some of the experience of the stress conditions in a similar way to stereotypies (Fraser & Broom 1990). Individually housed sheep have, however, been shown to demonstrate stereotypical oral behaviours and locomotor behaviours (Done-Currie et al. 1984; Marsden & Wood-Gush 1986a,b; Yurtman et al. 2002).

Sheep also show other forms of abnormal behaviours, including wool-pulling and redirected sucking. Wool-pulling occurs exclusively in indoor-housed sheep within restrictive enclosures, and disappears when the sheep are turned out. It is most frequent at high stocking density and eliminated by increasing space per animal (Fraser & Broom 1990). Provision of roughage also reduces the incidence of wool-biting (Vasseur et al. 2006), suggesting that this behaviour occurs as a result of behavioural restriction, where sheep lack oral or other forms of stimulation. Redirected sucking occurs in artificially reared lambs where lambs suck the navels and scrotums of other lambs (Stephens & Baldwin 1971). Lambs separated from their dams for 48 hours in the first few days of birth, before being raised by their dams, also show a propensity to re-directed sucking even at 2 months of age (Markowitz et al. 1998). Isolate-reared lambs, when stressed, show a 'flank-touching' behaviour, characterised by the lamb reaching back and touching its own flank with its muzzle, which is not seen in either dam- or peer-reared lambs under the same conditions (Moberg & Wood 1982). Ewes frequently nuzzle the lamb's rump, particularly when the lamb is sucking, and artificially reared lambs in groups also turn their bodies to touch rumps whilst sucking (Stephens & Baldwin 1971) suggesting that this may function as a comfort behaviour in lambs.

1.5.2.5 Qualitative Assessment

An alternative and novel use of behaviour as an indicator of welfare is qualitative assessment of an animal's style of behaving. Whilst previous behavioural and physiological methodologies described above tries to explain the animal's state by inference (e.g. an animal that runs is 'fearful'), there is no certainty that a particular behavioural of physiological measure actually reflects this experience (Rushen 1990). Qualitative assessment attempts to access these emotional experiences more directly by measuring the 'whole animal' state, integrating information from all the behaviours of the animal (Wemelsfelder et al. 2001). Thus what is recorded is not what the animal does but how the behaviours are carried out. To return to our example above of the sheep moving about in a novel environment, we might be able to distinguish between the nervous, anxious, fearful animal and the calm, relaxed, exploratory sheep by considering the quality of the behaviours (are movements smooth or jerky, are the steps long and relaxed or short and abrupt, etc.), rather than the behaviours themselves. This form of welfare assessment is only just starting to be applied to the sheep (Wemelsfelder & Farish 2004) but could be a useful tool to understand sheep behavioural responses.

1.5.3 Physiological and Production Responses Relating to Welfare

Both behavioural and neuroendocrinological measurements have difficulties of interpretation in welfare terms, even when carried out experimentally, which can become even more difficult to use when taken out onto farms or markets. For this reason production type responses become more attractive as indicators of welfare, and have a certain appeal in that they are the easiest to measure and carry an economic weight also. However, as discussed above, they are insensitive to short term, but possibly intense, suffering and could be seen as reflecting a prolonged period of poor welfare, thereby triggering action too late in the chain from altered biological function to the presence of a pathology (Moberg 2000). Nevertheless, poor welfare may have an impact on these responses, and prevention of these economic detriments may spur welfare improvements at an earlier stage. Furthermore, much of the antipathy to the use of production measures as indicators of welfare come from their use in intensive confinement agriculture. For more extensively-managed animals, such as the sheep, where the animal is far less supported by management interventions during periods of stress, production measures can potentially serve as welfare indicators. Although good production should not be taken as an indicator that welfare is good, poor production (against a background of what should be achievable with particular genotypes in a specific environment) may suggest that welfare problems are present.

1.5.3.1 Effects on Reproduction

As the hypothalamus and pituitary are intimately involved in both stress responses and reproductive function it is perhaps not surprising that reproduction can be influenced by stress. In both males and females reproduction is controlled via the release of gonadotrophin-releasing hormone (GnRH) produced from the hypothalamus that acts on the anterior pituitary to produce luteinising hormone (LH), this then acts on either the ovaries or testes to produce oestrogen or testosterone respectively. Oestrogen secretion is important for follicular development and the release of the ovum at oestrus in the ewe. The female reproductive system appears to be particularly affected by stress with actions at every stage in the cycle. Oestrus, and oestrus behaviour, can be blocked or delayed by stress (Ehnert & Moberg 1991). Stress affects the release of both GnRH and LH in rams and ewes, thus the effects are seen at the level of the hypothalamus or higher brain areas and at the level of the pituitary (Matteri et al. 1984; Dobson & Smith 2000). Restraint, confinement or transport suppress follicular growth and development by blocking or delay of the preovulatory surge of LH (Rasmussen & Malven 1983; Dobson & Smith 1995; Dobson et al. 1999a, b). This then leads to a reduction in oestradiol production by the slower growing follicles. Sub-fertility thus occurs when the sheep experiences increasing amounts of stress. Stress reduces the expression of oestrus behaviour in ewes and libido in rams, increases the number of barren ewes, reduces milk yield and composition, and impairs maternal behaviour (Bush & Lind 1973; Kiley-Worthington 1977; Knight et al. 1988; Sabrh et al. 1992; Sevi et al. 1999).

1.5.3.2 Effects on Growth, Wool and Meat Quality

At its simplest stress is likely to have an impact on growth as one of the functions of cortisol is to change nutrient partitioning, increasing carbohydrate metabolism as a fuel to mount the stress response, and therefore decreasing the amount of substrate available to the tissues for growth. This is, however, an overly simplistic view as stress, particularly if prolonged or severe, also interacts with gut motility and absorption, appetite and activity level (Elsasser et al. 2000). In common with reproductive function and stress responses, secretion of growth hormone from the anterior pituitary is controlled by factors released from the hypothalamus. Sheep under various stressful conditions (high stocking density, isolation, artificial rearing) are known to have a reduced growth rate (Gonyou et al. 1985; Abdel-Rahman et al. 2000; Napolitano et al. 2002). These effects may be secondary to alterations in feed intake, and will also be less sensitive in adults rather than young, growing animals. Relationships between plasma cortisol and growth rate are weak in sheep (Purchas et al. 1980) and infusing cortisol does not affect basal growth hormone (GH) release, although the release of GH from the pituitary in response to exogenous growth-hormone releasing factor is attentuated (Thompson et al. 1995). Impaired wool growth is a symptom of both chronic lameness and parasitism, although whether these effects are seen in psychologically stressed animals is not known. However, sheep given exogenous cortisol have reduced wool growth, reduced staple-strength and increased fibre-shedding (Ansari-Renani & Hynd 2001; Schlink et al. 2002). Restraint in isolation, or rough transport, causes increases in muscle pH and reduced glucose and lactate concentrations, thereby increasing the propensity for dark-cutting meat (Apple et al. 1995; Ruiz-de-la-Torre et al. 2001). There is, therefore, some evidence that chronic stress impairs growth rate, wool growth, feed conversion efficiency and meat quality, however this is inconclusive, particularly in animals highly selected for growth rate or wool production. For example, Marsden and Wood-Gush (1986b) showed high levels of stereotypy or abnormal behaviours in sheep without effects on growth rate.

1.5.3.3 Effects on Parasitism and Disease Resistance

As described above, one of the prepathological stress responses is an alteration in immune function. The interest in the stress effects on immunity were largely triggered by observations that animals under environmental and psychological stress were more likely to succumb to disease (reviewed by Blecha 2000). Stress factors are believed to have an effect on the resistance of sheep to worm burdens (MacKay 1974), which may be due to the influence of stress on immunity. For example, stress associated with transport and indoor housing caused a sustained elevation of faecal egg counts in comparison to pastured sheep when both were infected with *Dicrocoelium dendriticum* (Sotiraki et al. 1999). Weaned lambs also had significantly greater faecal egg counts than control lambs, which remained with their dams, after both were experimentally infected with *Haemonchus contortus* and *Trichostrongylus colubriformis* (Watson 1991). Control lambs also had earlier and stronger serum antibody responses than the weaned lambs. Sheep under nutritional stress, such as low feed intake during the high protein demands of lactation, also show higher faecal egg counts than well fed ewes (Houdijk et al. 2001).

Thus, prolonged or severe stress in sheep may be accompanied by other indicators of poor welfare, including a reduced disease resistance, reduced reproductive capacity, reductions in growth and growth efficiency and an inability to deal effectively with worm burdens.

1.6 Conclusions

Historically animal welfare concerns have increased over the last 50 years, in line with the increase in confinement agriculture, and have primarily been directed towards pigs, poultry and dairy calves. The welfare of sheep, and other extensively farmed species, has been less of a focus for welfare attention. This appears to be due to particular concerns for 'naturalness' and the ability of extensive sheep to display many of their natural behaviours. The issue of living an 'outdoor' life also seems to be an important factor in equating extensive farming with good animal welfare, with traditional farming practices seen as being bound up with good animal husbandry. However, the naturalness or freedom to express natural behaviour is only one part of a welfare definition and other aspects of good health and welfare may be overlooked. If we consider the Five Freedoms, extensive agriculture does not necessarily protect the sheep from violations of any of the other four freedoms (hunger and thirst, thermal and physical discomfort, pain, injury and disease, fear and distress). In fact, the likelihood of an animal experiencing prolonged, or severe, exposure to any of these other threats to good welfare may be greater in extensive rather than intensive farming.

Sheep are well adapted for some extreme environments, both behaviourally and physiologically, and are be able to cope well in some difficult situations. However, under artificial conditions such as transport, these adaptations make it hard to assess whether the sheep is suffering, particularly as sheep are remarkably tough and can survive under conditions that other mammals cannot. The welfare framework outlined here depends on appreciating the behavioural characteristics of the sheep, and the environments in which they evolved. The first Chapters in this book deal with the adaptations that sheep possess, both behavioural and neuroendocrine, and how they have evolved, as a means to understanding where welfare challenges are likely to originate. The main areas of concern for animal welfare are aspects of the systems in which they are kept, management practices and health issues, particularly in animals where the amount of inspection they receive may allow disease or injury to go untreated for prolonged periods. These specific aspects of welfare of the sheep are expanded upon in the following Chapters.

Finally, if sheep are to remain in extensive environments, we need to find novel ways of ensuring that their welfare is good, whilst allowing the farmers who keep them there to make a living from their production. In some cases, management practices that cause stress and pain (e.g. shearing, dipping, tail-docking) are designed with the long term welfare of the sheep in mind. The use of novel, scientifically-based methods, be they management solutions or genetically-based breeding solutions to some of these welfare issues are explored in the final Chapters.

References

Abdel-Rahman, M. A., Ahmed, M. M. & Sotohy, S. A. (2000) Acclimatization of sheep to housing stress of metabolism crates with reference to their behavioural and endocrinological aspects. *Assiut Veterinary Medical Journal* 43: 37–50.

Ansari-Renani, H. R. & Hynd, P. I. (2001) Cortisol-induced follicle shutdown is related to staple strength in Merino sheep *Livestock Production Science* 69: 279–289.

Apple, J. K., Dikeman, M. E., Minton, J. E., McMurphy, R. M., Fedde, M. R., Leith, D. E. & Unruh, J. A. (1995) Effects of restraint and isolation stress and epidural blockade on endocrine and blood metabolite status, muscle glycogen metabolism, and incidence of dark-cutting longissimus muscle of sheep. *Journal of Animal Science* 73: 2295–2307.

Appleby, M. C. (1996) Can we extrapolate from intensive to extensive conditions. *Applied Animal Behaviour Science* 49: 23–27.

Appleby, M C (1999) Tower of Babel: Variation in ethical approaches, concepts of welfare and attitudes to genetic manipulation. *Animal Welfare* 8: 381–390.

Barnard, C. J. & Hurst, J. L. (1996) Welfare by design: the natural selection of welfare criteria. *Animal Welfare* 5: 405–433.

Baxter, M. R. (1983) Ethology in environmental design for animal production. *Applied Animal Ethology* 9: 207–220.

Berger, J. (1979) Social ontogeny and behavioural diversity: consequences for Bighorn sheep *Ovis canadensis* inhabiting desert and mountain environments. *Journal of Zoology, London* 188: 251–266.

Berger, J. (1980) The ecology, structure and functions of social play in bighorn sheep (*Ovis canadensis*). *Journal of Zoology* 192: 283–289.

Berger, J. (1991) Pregnancy incentives, predation constraints and habitat shifts: experimental and field evidence for wild bighorn sheep. *Animal Behaviour* 41: 61–77.

Bleich, V. C. (1999) Mountain sheep and coyotes: Patterns of predator evasion in a mountain ungulate. *Journal of Mammalogy* 80: 283–289.

Bleich, V. C., Bowyer, R. T. & Wehausen, J. D. (1997) Sexual segregation in mountain sheep: Resources or predation? *Wildlife Monographs* 134: 3–50.

Binns, S. H., Cox, I. J., Rizvi, S. & Green, L. E. (2002) Risk factors for lamb mortality on UK sheep farms. *Preventative Veterinary Medicine* 52: 287–303.

Blecha, F. (2000) Immune system response to stress. In '*The Biology of Animal Stress: Basic Principles and Implications for Animal Welfare*' (Eds. G. Moberg & J.A. Mench), CAB International Publishing, pp. 111–121.

Bouissou, M.-F., Porter, R. H., Boyle, L. & Ferreira, G (1996) Influence of a conspecific image of own vs. different breed on fear reactions of ewes. *Behavioural Processes* 38: 37–44.

Brambell, F. W. R. (1965) Report of the Technical committee to enquire into the welfare of animals kept under intensive livestock husbandry systems. Her Majesty's Stationery Office, London, UK.

Broom, D. M. (1986) Indicators of poor welfare. *British Veterinary Journal* 142: 524–526.

Broom, D. M. (1998) Welfare, stress and the evolution of feelings. *Advances in the Study of Behavior* 27: 371–403.

Burfening, P. G. & van Horn, J. L. (1993) Comparison of range versus shed lambing in the Northern Great Plains. *Sheep Research Journal* 9: 86–90.

Bush, L. F. & Lind, R. A. (1973) Performance of ewes and lambs in confinement. *Journal of Animal Science* 36: 407–410.

Coppinger, T. R., Minton, J. E., Reddy, P. G. & Blecha, F. (1991) Repeated restraint and isolation stress in lambs increases pituitary-adrenal secretion and reduces cell-mediated immunity. *Journal of Animal Science* 69: 2808–2814.

Davis, S. L. & Cheeke, P. R. (1998) Do domestic animals have minds and the ability to think? A provisional survey of opinions on the question. *Journal of Animal Science* 76: 2072–2079.

Dawkins, M. S. (1980) *Animal Suffering: The Science of Animal Welfare,* Chapman and Hall, London, UK.

Dawkins, M. S. (1990) From an animal's point of view: motivation, fitness and animal welfare. *Behavioural and Brain Sciences* 13: 1–61.

Dawkins, M. S. (1998) Evolution and animal welfare. *The Quarterly Review of Biology* 73: 305–328.

Dawkins, M. S. (2001) Who needs consciousness? *Animal Welfare* 10: S19–29.

Denholm, L. J. & Vizard, A. L. (1986) Trimming the incisor teeth of sheep: another view. *Veterinary Record* 119: 182–184.

Dobson, H. & Smith, R. F. (1995) Stress and reproduction in farm animals. *Journal of Reproduction and Fertility, Supplement* 49: 451–461.

Dobson, H., Tebble, J. E., Phogat, J. B. & Smith, R. F. (1999a) Effect of transport on pulsatile and surge secretion of LH in ewes in the breeding season. *Journal of Reproduction and Fertility* 116: 1–8.

Dobson, H., Tebble, J. E., Ozturk, M. & Smith, R. F. (1999b) Effect of transport on pulsatile LH release in ovariectomized ewes with or without prior steroid exposure at different times of year. *Journal of Reproduction and Fertility* 117: 213–222.

Dobson, H. & Smith, R. F. (2000) What is stress, and how does it affect reproduction? Animal Reproduction Science 60–61: 743–752.

Done-Currie, J. R., Hecker, J. F. & Wodzicka-Tomaszewska, M. (1984) Behaviour of sheep transferred from pasture to an animal house. *Applied Animal behaviour Science* 12: 121–130.

Duncan, I. J. H. (1993) Welfare is to do with what animals feel. *Journal of Agricultural and Environmental Ethics* 6(Supplement 2): 8–14.

Duncan, I. J. H. (1996) Animal welfare defined in terms of feelings. *Acta Agriculturae Scandinavia section A, Animal Science Supplement* 27: 29–35.

Duncan, I. J. H. & Petherick, J. C. (1991) The implications of cognitive processes for animal welfare. *Journal of Animal Science* 69: 5017–5022.

Dunlop, L. B.& Johnston, D. C. (1985) The effects of three mulesing treatments applied to lambs at marking and to ewes at first shearing on the prevention of blowfly strike. In '*Grazing animal welfare*'. Proceedings of a symposium held at the Bardon Professional Development Centre, Brisbane, 26–27th April 1985, pp. 39–47

Dwyer, C. M. (2003) Behavioural development in the neonatal lamb: Effect of maternal and birth-related factors. *Theriogenology* 59: 1027–1050.

Dwyer, C. M. & Lawrence, A. B. (1998) Variability in the expression of maternal behaviour in primiparous sheep: Effects of genotype and litter size. *Applied Animal Behaviour Science* 58: 311–330.

Ehnert, K. & Moberg, G. P. (1991) Disruption of oestrous behaviour in ewes by dexamethasone or management-related stress. *Journal of Animal Science* 69: 2988–2994.

Elsasser, T. H., Klasing, K. C., Filipov, N. & Thompson, F. (2000) The metabolic consequences of stress: Targets for stress and priorities for nutrient use. In '*The Biology of Animal Stress: Basic Principles and Implications for Animal Welfare*' (Eds. G. Moberg & J.A. Mench), CAB International Publishing, pp. 77–110.

Fisher, M. W. & Mellor, D. J. (2002) The welfare implications of shepherding during lambing in extensive New Zealand farming systems. *Animal Welfare* 11: 157–170.

Fraser, A. F. & Broom, D. M. (1990) *Farm Animal Behaviour and Welfare.* Bailliere Tindall, London, UK.

Fraser, D. (1999) Animal ethics and animal welfare science: bridging the two cultures. *Applied animal Behaviour Science* 65: 171–189.

Fraser, D. (2001) The culture and agriculture of animal production. *ANZCAART News* 14: 1–4.
Fraser, D. & Duncan, I. J. H. (1998) 'Pleasures', 'pains' and animal welfare: Toward a natural history of affect. *Animal Welfare* 7: 383–396.
Fraser, D., Weary, D. M., Pajor, E. A. & Milligan, B. N. (1997) A scientific conception of animal welfare that reflects ethical concerns. *Animal Welfare* 6: 187–205.
Frid, A. (1977) Vigilance by female Dall's sheep: interactions between predation risk factors. *Animal Behaviour* 53: 799–808.
Gonyou, H. W. (1993) Animal welfare: definitions and assessment. *Journal of Agricultural and Environmental Ethics* 6(Supplement 2): 37–43.
Gonyou, H. W., Stookey, J. M. & McNeal, L. G. (1985) Effects of double decking and space allowance on the performance and behavior of feeder lambs. *Journal of Animal Science* 60: 1110–1116.
Gregory, N. G. (1993) Welfare and product quality: the need to be humane. In *'Safety and Quality of Food from Animals'*, British Society for Animal Production Occasional Publication, 17: 51–56.
Gregory, N. G. (1998) *Animal Welfare and Meat Science.* CABI Publishing, Oxon, UK.
Griffin, D. R. (2001) *Animal Minds: Beyond Cognition to Consciousness.* University of Chicago Press, Chicago, UK.
Hall, S. J. G., Kirkpatrick, S. M. & Broom, D. M. (1998) Behavioural and physiological responses of sheep of different breeds to supplementary feeding, social mixing and taming, in the context of transport. *Animal Science* 67: 475–483.
Hargreaves, A. L. & Hutson, G. D. (1990) The effect of gentling on heart rate, flight distance and aversion of sheep to a handling procedure. *Applied Animal Behaviour Science* 26: 243–252.
Harrison, R. (1964) *Animal Machines.* Robinson and Watkins, UK.
Haughey, K. G. (1993) Perinatal lamb mortality – Its investigation, causes and control. *Irish Veterinary Journal* 46: 9–28.
Houdijk, J. G. M., Kyriazakis, I., Jackson, F. & Coop, R. L. (2001) The relationship between protein nutrition, reproductive effort and breakdown in immunity to *Teladorsagia circumcincta* in periparturient ewes. *Animal Science* 72: 595–606.
Houpt, K. A. (1987) Abnormal behaviour. In *'The Veterinary Clinics of North America 3, 2 Farm Animal Behaviour'* (Ed. E. O. Price), Saunders, Philadelphia, pp. 357–367.
Jackson, R. E., Waran, N. K. & Cockram, M. S. (1999) Methods for measuring feeding motivation in sheep. *Animal Welfare* 8: 53–63.
Jensen, K. H., Hansen, S. W. & Pedersen, L. J. (1996) The effect of long term stress on the hypothalamic-pituitary-adrenocortical axis and the role of the stressor. *Acta Agriculturae Scandinavia, Section A, Animal Science Supplement* 27: 40–45.
Jewell, P. A. (1997) Survival and behaviour of castrated Soay sheep (*Ovis aries*) in a feral island population on Hirta, St. Kilda, Scotland. *Journal of Zoology* 243: 623–636.
Kendrick, K. M., Atkins, K., Hinton, M. R., Heavens, P. & Keverne, E. B. (1996) Are faces special for sheep? Evidence from facial and object discrimination learning tests showing effects of inversion and social familiarity. *Behavioural Processes* 38: 19–35.
Kent, J. E. (1997) Stress in transported sheep. *Comparative Haematology International* 7: 163–166.
Kent, J. E., Jackson R. E., Molony V. & Hosie B. D. (2000) Effects of acute pain reduction methods on the chronic inflammatory lesions and behaviour of lambs castrated and tail docked with rubber rings at less than two days of age. *The Veterinary Journal* 160: 33–41.
Kiley-Worthington, M. (1977) *Behaviour Problems of Farm Animals.* Oriel Press Ltd, Northumberland, UK.
Kiley-Worthington, M. (1989) Ecological, ethological and ethically sound environments for animals: toward symbiosis. *Journal of Agricultural Ethics* 2: 323–347.
Kilgour, R. (1985) Animal welfare considerations – pastoral animals. *New Zealand Veterinary Journal* 33: 54–57.
Knight, T. W., Hall, D. R. H., Lynch, P. R. & Hockey, H. U. P. (1988) Effects of pre-joining shearing, stress, pasture allowance, and haemoglobin type on reproductive performance of Romney and Marshall Romney ewes. *New Zealand Journal of Agricultural Research* 31: 249–258.
Laundré, J. W., Hernandez, L. & Altendorf, K. B. (2001) Wolves, elk, and bison: reestablishing the "landscape of fear" in Yellowstone National Park, USA. *Canadian Journal of Zoology* 79: 1401–1409.

Lawrence, A. B. & Rushen, J. (1993) Introduction. In '*Stereotypic Animal Behaviour*' (Eds. A. B. Lawrence and J. Rushen), pp. 1–6. CAB International, Oxon, UK.
Lee, C. & Fisher, A. D. (2007) Welfare consequences of mulesing in sheep. *Australian Veterinary Journal* 85: 89–93.
Mackay, R. R. (1974) Observations on the control of nematodes in housed young female sheep. *Veterinary Record* 94: 73–77.
Markowitz, T. M., Dally, M. R., Gursky, K. & Price, E. O. (1998) Early handling increases lamb affinity for humans. *Animal Behaviour* 55: 573–587.
Marsden, D. & Wood-Gush, D. G. M. (1986a) A note on the behaviour of individually-penned sheep regarding their use for research purposes. *Animal Production* 42: 157–159.
Marsden, D. & Wood-Gush, D. G. M. (1986b) Abnormal behaviour versus production performance criteria in the assessment of welfare in housed lambs. *Applied Animal Behaviour Science* 15: 90–91.
Martinetto, K. & Cugnasse, J. M. (2001) Reaction distance in Mediterranean Mouflon (*Ovis gmelini musimon* x *Ovis* sp.) in the presence of hikers with a dog on the Caroux plateau (Herault, France). *Revue de Ecologie – La Terre et la Vie* 56: 231–242.
Mateo, J. M., Estep, D. Q. & McCann, J. S. (1991) Effects of differential handling on the behaviour of domestic ewes (*Ovis aries*). *Applied Animal Behaviour Science* 32: 45–54.
Matteri, R. L., Watson, J. G. & Moberg, G. P. (1984) Stress or acute adrenocorticotrophin treatment suppresses LHRH-induced LH release in the ram. *Journal of Reproduction and Fertility* 72: 385–393.
McGlone, J. J. (1993) What is animal welfare? *Journal of Agricultural and Environmental Ethics* 6(Supplement 2): 26–36.
McSpotlight (1998) The McLibel trial story. Available on line at: http://www.mcspotlight.org/case/trial/story.html.
Mench, J. A. (1998) Thirty years after Brambell: Whither animal welfare science? *Journal of Applied Animal Welfare Science* 1: 91–102.
Minton, J. E. & Blecha, F. (1990) Effect of acute stressors on endocrinological and immunological functions in lambs. *Journal of Animal Science* 68: 3145–3151.
Minton, J. E., Coppinger, T. R., Reddy, P. G., Davis, W. C. & Blecha, F. (1992) Repeated restraint and isolation stress alters adrenal and lymphocyte functions and some leukocyte differentiation antigens in lambs. *Journal of Animal Science* 70: 1126–1132
Minton, J. E., Apple, J. K., Parsons, K. M. & Blecha, F. (1995) Stress-associated concentrations of plasma cortisol cannot account for reduced lymphocyte function and changes in serum enzymes in lambs exposed to restraint and isolation stress. *Journal of Animal Science* 73: 812–817.
Moberg, G. P. (1985) Biological responses to stress: key to assessment of animal well-being? In '*Animal Stress*' (Ed. G. P. Moberg,) American Physiological Society, Bethesda, USA.
Moberg, G. P. (1996) Suffering from stress: An approach for evaluatin the welfare of an animal. *Acta Agriculturae Scandinavia section A, Animal Science Supplement* 27: 46–49.
Moberg, G. P. (2000) Biological response to stress: implications for animal welfare. In '*The Biology of Animal Stress: Basic Principles and Implications for Animal Welfare*' (Eds. G. Moberg & J.A. Mench), CAB International Publishing, pp. 1–21.
Moberg, G. A. & Mench, J. A. (Editors) (2000) The Biology of Animal Stress: Basic Principles and Implications for Animal Welfare CAB International Publishing.
Moberg, G. P. & Wood, V. A. (1982) Effect of differential rearing on the behavioural and adrenocortical response of lambs to a novel environment. *Applied Animal Ethology* 8: 269–279.
Morris, M. C. (2000) Ethical issues associated with sheep fly strike research, prevention, and control. *Journal of Agricultural & Environmental Ethics* 13: 205–217.
Müller-Schwarze, D., Stagge, B. & Müller-Schwarze, C. (1982) Play behaviour: persistence, decrease and energetic compensation during food shortage in deer fawns. *Science* 215: 85–87.
Murray, K. C., Davies, D. H., Cullinane, S. L., Eddison, J. C. & Kirk, J. A. (2000) Taking lambs to the slaughter: marketing channels, journey structures and possible consequences for welfare. *Animal Welfare* 9: 111–122.

Napolitano, F., Bragheir, A., Cifuni, G. F., Pacelli, C. & Girolami, A. (2002) Behaviour and meat production of organically farmed unweaned lambs. *Small Ruminant research* 43: 179–184.

Pattinson, S. E., Waterhouse, A., Ashworth, S. W. & Wyllie, J. T. B. (1994) Cost/benefit analysis of nutritional and shepherding inputs at lambing in an extensive hill sheep system. *Proccedings of the 45th Annual Meeting of the European Association for Animal Production,* Edinburgh, pp. 242.

Paull, D. R., Lee, C., Colditz, I. G., Atkinson, S. J. & Fisher, A. D. (2007) The effect of topical anaesthetic formulation, systemic flunixin and carprofen, singly or in combination, on cortisol and behavioural responses of merino lambs to mulesing. *Australian Veterainry Journal* 85: 98–106.

Purchas, R. W., Barton, R. A. & Kirton, A. H. (1980) Relationships of circulating cortisol levels with growth rate and meat tenderness of cattle and sheep. *Australian Journal of Agricultural Research* 31: 221–232.

Rasa, O. A. E. (1984) A motivational analysis of object play in juvenile dwarf mongooses (*Helogale undulata rufula*). *Animal Behaviour* 32: 579–589.

Rasmussen, D. D. & Malven, P. V. (1983) Effects of confinement stress on episodic secretion of LH in ovariectomized sheep. *Neuroendocrinology* 36: 392–396.

Reale, D., Bousses, P., & Chapuis, J. L. (1999) Nursing behaviour and mother-lamb relationships in mouflon under fluctuating population densities. *Behavioural Processes* 47: 81–94.

Regan, T. (1983) *The Case for Animal Rights.* University of California Press, Berkeley, USA.

Ritar, A. J., O'May, P. J., Williams, P. M., Gilbert, K. D. & Bond, E. M. (1988) Effects of castration and age on carcass composition of male lambs. *Proceedings of the Australian Society of Animal Production* 17: 457.

Rollin, B. E. (1990) *The Unheeded Cry.* Oxford University Press, Oxford, UK.

Rollin, B. E. (1993) Animal welfare, science and value. *Journal of Agricultural and Environmental Ethics* 6(supplement 2): 44–50.

Rollin, B. E. (1996) Animal welfare, animal rights and agriculture. *Journal of Animal Science* 68: 3456–3461.

Romeyer, A. & Bouissou, M-F (1992) Assessment of fear reactions in domestic sheep, and influence of breed and rearing conditions. *Applied Animal Behaviour Science* 34: 93–119.

Rowan, A., O'Brien, H., Thayer, L., & Patroneck, G. J. (1999) Farm Animal Welfare: The focus on animal protection in the USA in the 21st Century. Tufts Center for Animals and Public Policy, School of Veterinary Medicine, Tufts University, USA.

Ruiz-de-la-Torre, J. L., Velarde, A., Diestre, A., Gispert, M., Hall, S. J. G., Broom, D. M. & Manteca, X. (2001) Effects of vehicle movements during transport on the stress responses and meat quality of sheep. *Veterinary Record* 148: 227–229.

Rushen, J. (1986) Aversion of sheep to electro-immobilisation and physical restraint. *Applied Animal Behaviour Science* 15: 315–324.

Rushen, J. (1990) Use of aversion learning techniques to measure distress in sheep. *Applied Animal Behaviour Science* 28: 3–14.

Rushen, J. (1996). Using aversion learning techniques to assess the mental state, suffering and welfare of farm animals. *Journal of Animal Science* 74: 1990–1995.

Rushen, J. (2000) Some issues in the interpretation of behavioural responses to stress. In '*The Biology of Animal Stress: Basic Principles and Implications for Animal Welfare*' (Eds. G. Moberg & J.A. Mench), CAB International Publishing, pp. 23–42.

Rutgers, B. & Heeger R. (1999) Inherent worth and respect for animal integrity. In '*Recognising the Intrinsic Value of Nature*' (Eds. M. Dol, M. Fentener van Vlissingen, S. Kasanmoentalib, T. Visser, and H Zwart) Van Gorcum, pp. 41–53.

Sabrh, H. A., Shalash, M. R., Ashoub, M. R., Al-Wakeel, A. M. & Hassan, S. G. (1992) Effect of stocking density, age and body weight on some reproductive performance of Barki ewes. *Egyptian Journal of Veterinary Science* 29: 47–54.

Sandøe, P., Holtug, N. & Simonsen, H. B. (1997) Ethics. In '*Animal Welfare*' (Eds. M. Appleby & B. O. Hughes,), CAB International, Wallingford, UK. pp. 114–122.

Schlink, A. C., Wynn, P. C., Lea, J. M., Briegel, J. R. & Adams, N. R. (2002) Effect of cortisol acetate on wool quality in sheep selected for divergent staple strength. *Australian Journal of Agricultural Research* 53: 183–189.
Scobie, D. R., Bray, A. R. & O'Connell, D. (1999) A breeding goal to improve the welfare of sheep. *Animal Welfare* 8: 391–406.
Sevi, A., Massa, S., Annicchiarico, G., Dell'Aquila, S. & Muscio, A. (1999) Effect of stocking density on ewes' milk yield, udder health and microenvironment. *Journal of Dairy Research* 66: 489–499.
Singer, P. (1975) *Animal Liberation.* Avon Books, New York (2nd Edition published 1990).
Smith, R. F. & Dobson, H. (2002) Hormonal interactions within the hypothalamus and pituitary with respect to stress and reproduction in sheep. *Domestic Animal Endocrinology* 23: 75–85.
Sotiraki, S. T., Leontides, L. S. & Himonas, C. A. (1999) The effect of transportation and confinement stress on egg production by *Dicrocoelium dendriticum* in sheep. *Journal of Helminthology* 73: 337–339.
Stafford, K. J., Mellor, D. J. & Gregory, N. G. (2002) Advances in animal welfare in New Zealand. *New Zealand Veterinary Journal* 50 *Supplement*, pp. 17–21
Stephens, D. B. & Baldwin, B. A. (1971) Observations of the behaviour of groups of artificially reared lambs. *Research in Veterinary Science* 12: 219–224.
Stolba A, Hinch G N, Lynch J J, Adams D B, Munro R K & Davies H I 1990 Social organisation of Merino sheep of different ages, sex and family structure. *Applied Animal Behaviour Science* 27: 337–349.
Tarbotton, I. S., Bray, A. R. & Wilson, J. A. (2002). Incidence and perception of cryptorchid lambs in 2000. *Proceedings of the New Zealand Society of Animal Production* 62: 334–336.
Tannenbaum, J. (1991) Ethics and animal welfare: the inextricable connection. *Journal of the American Veterinary Medicine Association* 198: 1360–1376.
Terlouw, E. M. C., Schouten, W. G. P. & Ladewig, J. (1997) Physiology. In '*Animal Welfare*' (Eds. Appleby, M. C. & Hughes, B. O.) CAB International, Wallingford, UK. pp. 143–158.
Thompson, K., Coleman, E. S., Hudmon, A., Kemppainen, R. J., Soyoola, E. O. & Sartin, J. L. (1995) Effects of short-term cortisol infusion on growth hormone-releasing hormone stimulation of growth hormone release in sheep. *American Journal of Veterinary Research* 56: 1228–1231.
Uetake, K., Yamaguchi, S. & Tanaka, T. (2000) Psychological effects of early gentling on the subsequent ease of handling in lambs. *Animal Science Journal* 71: 515–519.
Vasseur, S. Paull, D. R., Atkinson, S. J., Colditz, I. G. & Fisher, A. D. (2006) Effects of dietary fibre and feeding frequency on wool biting and aggressive behaviour in housed Merino sheep. *Australian Journal of Experimental Agriculture* 46: 777–782.
Waterhouse, A. (1996) Animal welfare and sustainability of production under extensive conditions – A European perspective. *Applied Animal Behaviour Science* 49: 29–40.
Watson, D. L. (1991) Effect of weaning on antibody responses and nematode parasitism in Merino lambs. *Research in Veterinary Science* 51: 128–132.
Webster, J. (1994) *Animal Welfare: A Cool Eye towards Eden.* Blackwell Science, Oxford, UK.
Wemelsfelder, F. & Farish, M. (2004) Qualitative categories for the interpretation of sheep welfare: a review. *Animal Welfare* 13: 261–268.
Wemelsfelder, F., Hunter, T. E. A., Mendl, M. T. & Lawrence, A. B. (2001) Assessing the 'whole animal': a free choice profiling approach. *Animal Behaviour* 62: 209–220.
White, K. S. & Berger, J. (2001) Antipredator strategies of Alaskan moose: are maternal trade-offs influenced by offspring activity? *Canadian Journal of Zoology* 79: 2055–2062.
Wiepkema, P. R. & Koolhaas, J. M. (1993) Stress and animal welfare. *Animal Welfare* 2: 195–218.
Woolf, A., O'Shea, T. & Gilbert, D. L. (1970) Movements and behavior of Bighorn sheep on summer ranges in Yellowstone National Park. *Journal of Wildlife Management* 34: 446–450.
Yurtman, I. Y., Savas, T., Karaagac, F. & Coskuntuna, L. (2002) Effects of daily protein intake levels on the oral stereotypic behaviours in energy restricted lambs. *Applied Animal Behaviour Science* 77: 77–88.

Chapter 2
Environment and the Sheep

Breed Adaptations and Welfare Implications

C.M. Dwyer

Abstract The wild ancestors of domestic sheep have evolved specialisations to exploit a diverse range of habitats and can survive in extreme environments, from the desert to the Arctic and sub-Antarctic. They can cope with poor quality diets, foraging on a wide range of plant types including cactus, fruit, lichens and seaweed. A consistent feature of wild sheep habitats, however, in addition to food and water sources, is escape terrain, as their main defence against predators is flight to cover. Seasonal and diurnal migrations about their home range occur in response to forage availability and safety. In domestic sheep, when given the opportunity to express these behaviours, similar habitat preferences and movements about the home range occur. Selection for breed traits and adaptation in domestic sheep has led to breed differences in environmental adaptation, seasonality responses and ability to cope with low food availability. Behavioural differences are also seen between breeds in social behaviours, antipredator responses, fearfulness, shelter-seeking and grazing behaviours. In general, the more specialised and selected breeds show the greatest tolerance for crowding and are the least responsive to predators or other fear-eliciting stimuli. The environments in which domestic sheep are kept do not accurately represent the wild situation, thus the ability of sheep to cope with thermal extremes, poor food availability and predation by behavioural means may be impaired.

Keywords Wild sheep · Adaptation · Domestication · Predation · Habitat preferences · Welfare

2.1 Introduction

The sheep (genus *Ovis*) is distributed widely throughout the world. The sheep is an ungulate (or hoofed mammal), belonging to the highly successful order Artiodactyla, the family Bovidae (including true bovines, buffalo, goats and sheep), and the Tribe Caprini (comprising sheep and goats). The sheep is the most successful

C.M. Dwyer
Animal Behaviour and Welfare, Sustainable Livestock Systems Group, SAC, Edinburgh, UK
e-mail: cathy.dwyer@sac.ac.uk

C.M. Dwyer (ed.), *The Welfare of Sheep*,

Pleistocene mammal, with a distribution extending from Europe to Siberia and Alaska to South America. Although the wild ancestors of domestic sheep are generally found in hill and rugged country, they are highly adaptive and have successfully colonised a variety of terrain, including desert and island habitats. In Asia and Europe the sheep has competed for habitat with wild goats, resulting in sheep occupying lower mountain slopes and hills (Clutton-Brock 1987). In North America the absence of competition from goats has influenced the distribution of sheep, such as the mountain Bighorn, which range over the highest peaks.

The sheep was one of the first species to be domesticated, with evidence for the presence of domestic sheep in Iraq as early as 9000 BC (Ryder 1984; Hemmer 1990). Sheep were probably brought to Europe around 7000 BC by Neolithic farmers and were common in Western Asia by 3000 BC (Clutton-Brock 1987). The early domestication of the sheep may have proceeded through a series of initially unconscious connections between man and sheep. The first stage of domestication is represented by the formation of loose ties, for example by the sharing of watering places, but there is evidence of confinement and breeding control by Neolithic farmers, and the presence of 'breeds' by the Bronze Age (Ryder 1984). Early husbandry of sheep probably amounted to no more than the herding of animals by nomadic pastoralists where the natural behaviour and habitat use of the sheep was restrained only in that the leader was now man. Early agricultural settlements and the cultivation of crops meant that sheep could be kept in enclosures at night and some protection from predators could be provided. Ryder (1983) describes writings of the ancient Greeks, which note the herding of sheep to fresh mountain pastures in the spring, and penning of sheep in the winter where they were offered a range of feed stuffs (from barley, clover and alfalfa to oak leaves, figs and pressed grape residues from wine-making).

Domestication has resulted in sheep being managed and kept in ways that suit their human keepers. However, there is a great diversity in the management, habitat and feed stuffs that domestic sheep experience. Sheep may be free to range over vast areas of temperate hill and scrub in Europe and North America, kept on arid plains and desert conditions in Australia and North Africa, survive on seaweed in island habitats and are managed in relatively small fenced paddocks in Europe. This diversity of environment reflects their great adaptability to different environments but may also be related to the range of products that are produced from sheep, as husbandry systems vary for the different outputs. This chapter will consider the behavioural and physiological adaptations that the wild ancestors of sheep possess to successfully exploit various environments. The consequences of domestication, and the variety of sheep breeds so produced, for these adaptations will be explored. Finally, the environmental challenges faced by sheep and the consequences of these for the welfare of sheep will be considered.

2.2 Environmental Preferences and Adaptations of Wild Sheep

The ancestral origins of domestic sheep (*Ovis aries*) are unclear and it is possible that several wild *Ovis* species may have been domesticated or have contributed to modern domestic sheep breeds. There remain vestigial populations of seven main species of wild sheep (Ryder 1984; Clutton-Brock 1987; Hemmer 1990; Lynch et al. 1992). In Europe the mouflon (*O. musimon*) is found in Mediterranean regions extending into the Middle East and Iran with island populations in Corsica and Sardinia. This species is now often considered as a subspecies of *O. orientalis*, or the Asiatic mouflon, as described by some authors (Clutton-Brock 1987). Further east are found the larger urial (*O. vignei*) and argali (*O. ammon*) sheep species extending into Afghanistan and China, and the snow sheep (*O. nivicola*) in Siberia. In North America there are two further species of wild sheep, Dall's (*O. dalli*) and Bighorn sheep (*O. canadensis*). Other wild species, such as the blue sheep (*Pseudois nayaur*) in the Himalyas and Barbary sheep (*Ammotragus lervia)* in North Africa, appear to be intermediate between sheep and goats and have been described as goats with sheep-like characteristics. The *Ovis* species vary in diploid chromosome number from 52 pairs (snow sheep) to 56 (argali) and 58 pairs (urial) but can successfully interbreed in captivity (Ryder 1984). It is thought that, as the sheep was first domesticated in western Asia, the urial was the most likely species initially domesticated. Alternatively, it has been suggested the most likely progenitor of modern domestic sheep is the mouflon as it shares a similar number of chromosomes (54) with all domestic sheep (Hemmer 1990). However, hybridisation of different species of *Ovis* results in F2 generations with 54 chromosomes suggesting that this interpretation may be flawed. More recent analyses suggest that both the mouflon and the argali have had different influences on domestic sheep depending on the breed under study (Melinkova et al. 1995; Jugo & Vicario 2000). Thus the mouflon is most likely to have contributed to the development of European domestic sheep, and the urial and argali to Asiatic breeds, whereas the Bighorn has never been domesticated (Ryder 1991). Whether the mouflon is a true wild sheep, or a feral relic of earlier domestication has also been debated, however it does appear to be genetically related to all modern domestic European sheep. The wild ancestors of the domestic sheep provide a rich source of information to aid understanding of the behaviour and environmental adaptations of domestic sheep. The most studied populations of wild sheep are the Bighorns, although there has also been considerable recent interest in the European mouflon. In addition, the primitive feral Soay sheep of St Kilda, although once domesticated, have remained almost unchanged since the Bronze age, and continue to be the subject of extensive study.

2.2.1 Predation Risks

Predation risks and food availability are the major evolutionary forces shaping habitat use, social and foraging behaviour in ungulates. The main predators of sheep are wolves, coyotes, mountain lions, snow leopards, lynx and wolverines, although bears will also kill adult sheep. In addition lambs are at risk from foxes and raptors. Domestic and feral dogs also prey on sheep. The sheep is largely defenseless against predator attacks, their main antipredator strategies being flocking, and flight to cover or escape terrain where they can successfully evade predators. The importance of escape terrain in antipredator responses is demonstrated by a study that found juvenile

Bighorn that encountered a predator in the open had a mortality rate of 2.4 times that of encounters in escape terrain (Bleich 1999). Whilst in the escape terrain, sheep spend more time feeding and engaging in social interaction, and less time alert, than they do when out in the open (Jansen et al. 2006). Flight is, therefore, the main response to a predator when in the open although Bighorn females will stand their ground and act aggressively when encountering a predator on escape terrain. Mountain sheep will defend their young if attacked by small predators, such as crows and foxes, by standing over the lamb and have successfully driven off or killed eagles with their horns when protecting their lambs (Geist 1971). Lambs and juveniles are most likely to be killed by predators, possibly because lambs are less likely to survive an encounter than adults (Bleich 1999).

2.2.2 Habitat Preferences

Sheep have colonised some of the harshest and most severe environments on earth. They are found in mountainous regions across the globe, existing in the Himalayas and Rocky Mountains, where the temperature extremes between summer and winter can exceed 40°C. Wild sheep are found both in hot and arid desert environments and live inside the Arctic Circle in Canada, Alaska and Siberia, and in the sub Antarctic. They are, therefore, able to cope with thermal extremes, poor quality forage and limited availability of water.

Because of the importance of escape terrain for predator defence, this is an important feature of the wild sheep preferred environment, in addition to suitable forage and a water supply. The area and type of escape terrain available appear to act as limiting factors for population growth, suggesting that escape terrain is the dominant feature of habitat selection (McKinney et al. 2003). However, in desert populations, precipitation also regulates population size, primarily by affecting lamb production and survival (Bender & Weisenberger 2005), and influences home range size (Oehler et al. 2003). Bighorn and Dall's sheep are found in remote mountain and desert regions, and prefer open forest, shrub and grasslands whilst closed forests are avoided. European mouflon select meadows and broom moorlands and open woodland on the mainland, or steep wooded mountains on Corsica and Sardinia. Elevations above 1000 m appear to be preferred although sheep also avoid deep snow and remain below the permanent snow line on mountains. Sheep are rarely more than 200 m from escape terrain and show distinct preferences for rough, steep slopes with flat, smooth ground being avoided. Desert-living sheep are generally within 400 m of a water source, and select waterholes in or near escape terrain, with poor cover and good visibility. Island-dwelling Soay sheep live at much lower elevations, but make use of steep slopes and cleits (abandoned, man-made drystane storage houses offering a secluded shelter) for camping and as escape terrain. Preferred camping grounds (night habitat used mainly for resting) of mountain sheep are elevated rocky outcrops with good visibility where they can see long distances but are hard to spot themselves as they blend into the surroundings. These environmental preferences are expressed by a diurnal movement of sheep rapidly down from the camping grounds to graze in the morning, and then a slow movement back up to the camp grounds at night.

Wild sheep are primarily grazers of grasses and forbs, the absence of upper front teeth means that they can graze closer to the ground than other ungulate species. However, sheep will also browse on a variety of shrubs and, for desert-living sheep, browse species, including cactus, are their main food source. Dall's sheep living in Alaska and Northern Canada and mouflon in the sub-Antarctic also feed on lichens in winter. Mountain Bighorn have an attraction for salt, particularly in spring, and the distance to a suitable salt lick can be an important factor affecting habitat use (Shannon et al. 1975).

Wild sheep are not territorial, as resources are not defended, however they establish a home range that is utilised throughout their life. Home ranges are common areas shared by a number of animals where group members recognise and avoid sheep from adjacent groups. Home ranges can be large (in excess of 10 km^2 for Bighorns, Krausman et al. 1989) and include the main resources required by sheep (feed, water and escape terrain).

2.2.2.1 Seasonal and Weather Effects on Habitat Use

Most wild sheep undergo seasonal migrations about their home range, and in winter restrict themselves a smaller area than in summer. In winter, for example, Bighorn feed on the middle and lower slopes of their range, moving up to the higher elevations in the summer. In general the seasonal movement of sheep about the home range appears to be dictated by growing seasons of forage species but is also influenced by the weather. Animals are willing to move further from water sources in spring, presumably as they are able to meet some of their moisture requirements from forage. Warmer south and south-western slopes are preferred during the winter. Mouflon and Soay avoid open habitats in winter when wind speeds are strong, seeking shelter in more enclosed habitats (Grubb & Jewell 1966; Cransac & Hewison 1997). Grubb & Jewell (1966) observed that shelter-seeking in poor weather was more frequent in Soays when they were physically fit, suggesting responses to bad weather may be traded-off against the need to acquire good quality forage of animals in poor condition. Low barometric pressure in winter also limits use of open habitat (Tilton & Willard 1982). Snow depth influences winter habitat choice, deep snows both impede escape from predators and reduce forage availability and thus areas of deep snow are avoided. Snow cover results in habitat shifts from open sites to shrub cover (Goodson et al. 1991) and an increase in time spent foraging, although sheep cannot completely compensate for the decreased availability of forage with snow. Stone's sheep (a subspecies of Dall's) will dig for forage in the snow until snow cover reaches 32 cm but cease to work for food at snow depths greater than this (Seip & Bunnell 1985).

Spring and autumn are marked by considerable locomotor activity and movement about the range (Bon et al. 1993). In the summer north and northwestern upper slopes are used which provided a cooler and moister environment for Bighorns living in arid areas (Gionfriddo & Krausman 1986). Likewise, European mouflon prefer to use cooler open habitat with strong winds in summer. Availability of shade is also particularly important to mountain sheep during hot, dry summers. For desert living sheep in the summer the distance to a water source is an important component of habitat use (Payer & Coblentz 1997). During drought years the range used by Bighorns is much smaller than during relatively wetter years (McKinney et al. 2001). Desert-living Bighorns also adjust their diurnal activity patterns during the summer, when water sources are scarce and temperatures high, resting for most of the day and feeding in the evening and before dawn (becoming crepuscular). They feed particularly on cacti and succulents, to increase their water intake from plants. Soay sheep, living in a more temperate climate, respond to warm sunny weather by ranging more widely and using the upper slopes of their range more frequently. The primitive, feral island-living North Ronaldsay sheep, where the main diet is seaweed, also adjust their diurnal rhythm in response to environmental stimuli, in this case the tides. North Ronaldsay sheep are most active in the four hours preceding low tide, when the algae are exposed (Paterson & Coleman 1982). Wild and feral sheep therefore appear to use habitats both to access the best available forage and to minimise their need to expend energy on thermoregulatory processes.

2.2.2.2 Age and Sex Differences in Habitat Use

Sheep are sexually dimorphic (with males on average 40–50% larger than females) and show distinct patterns of sexual segregation. Ewes and juveniles form matrilineal groups led by the older ewes, followed by her daughters and their offspring. Males move away from the ewe and juvenile groups, as they reach the age of one or two years, and join smaller bachelor groups of rams of a similar age and body size. Males occupy a much larger home range than female groups, their range overlapping with or encompassing the home range of the female groups. Males are generally more exploratory than females with young males making short duration visits to remote parts of the home range in spring and summer (Krausman et al. 1989). As animals become older their spatial patterns and use of the home range becomes more fixed, perhaps through increasing knowledge of the environment.

Ram groups graze further from the slopes and spend more time in the open than ewes (Woolf et al. 1970; Berger 1991; Bleich et al. 1997; Corti & Shackleton 2002). As males are larger than females their predation risks are less than for females when occupying the same habitat. Male sheep are also less likely to flee and are less vigilant than ewes when in the open (Schaller 1977; Bleich 1999; Laundré et al. 2001). Adult male sheep are less responsive than ewes to the presence of potential predators (man): although ewes decrease resting time and increase foraging when disturbed the behaviour of rams is unchanged (Loehr et al. 2005). Forage is often more abundant out in the open plains than on the slopes meaning that the reduced threat of predation on rams allows males to obtain better quality diets than ewes (Bleich et al. 1997; Mooring et al. 2003). As male reproductive success is dependent on body mass and growth, this means that males trade-off increased predation risks against maximising reproductive success.

Ewes do not maintain the same pattern of habitat use throughout their reproductive cycle. Ewes forage on the better grazing during the high nutritional demands of their pregnancy even if this means incurring greater risk of predation. By contrast at the onset of birth Bighorn, Dall's, mouflon and Soay sheep withdraw to rocky and secluded parts of their home range where predation risks are minimal (Grubb & Jewell 1966; Geist 1971; Bon et al. 1995; Rachlow & Bowyer 1998). Shelter, absence of windchill, dry and southern aspects are important features of lambing sites, particularly in Dall's sheep living in severe Arctic and sub-Arctic environments. For desert-living Bighorn ewes, parturition sites are selected particularly for their elevation and are more rugged than pre-parturition habitat. Parturition sites also have lower visibility than slopes used before and after parturition (Bangs et al. 2005). Ewes and their lambs return to the matriarchal groups some days after the birth but lactating ewes remain closer to escape terrain than non-lactating ewes, are more vigilant and spend less time active in open habitat and more time active in the escape terrain (Berger 1991; Bon et al. 1995; Bleich et al. 1997; Walker et al. 2006). Thus it appears that wild ewes use less vulnerable areas during lactation, presumably to minimise predation on their young, even though this means utilising areas of poorer forage. However, in years when forage abundance and quality are low, maternal ewes also select habitats on the basis of forage to meet the energetic expenses of lactation and trade-off some of the safety features of habitats with poorer forage (Rachlow & Bowyer 1998).

2.2.3 Morphological Adaptations to Environments

Wild sheep vary in size from the 'giant' argali, weighing up to 180 kg, to the mouflon weighing 50–60 kg. Feral Soay sheep are smaller at approximately 25 kg. All species of wild sheep share a number of similarities in appearance: they vary in colour from pale brown to dark reddish brown, their belly, face and rump may be paler in colour, and males may also have a paler saddle. The exceptions are Arctic-living Dall's, which are predominantly white. In all species males carry spectacular horns that usually curl round or corkscrew and, in argali, can reach nearly 2 m in length. Mature males are often aged by the degree of curvature of their horns as horn grows throughout their lives. Females have shorter and straighter horns or can be polled (Soay and mouflon). Rams of several species (mouflon, argali, Soay) grow a neck ruff of longer hair and the argali also develops a pronounced dorsal crest in winter. Mortality is high in young animals up to two years old but animals that survive beyond this age can live to be at least 10 years old, with life spans of nearly 20 years reported for many wild sheep species (Hoefs 1991).

Wild sheep are relatively long-legged, fast and agile over rough slopes to escape predators, and well adapted to mountain and rough terrain. They have cloven feet that enhance surefootedness and Bighorn have elastic padded soles for improved agility on steep slopes. Young are mature at birth and on their feet quickly, so that they are able to follow their mothers over steep slopes within hours of birth.

Wild sheep and Soay have a double-layered coat consisting of outer guard or kemp hairs over a short, fine crimped undercoat. As an adaptation to their Arctic habitat where temperatures remain below zero, Dall's sheep have hollow hairs that allow better insulation. Growth of the wool undercoat occurs in late summer, reaching a maximum in autumn, and ceases in winter (Ryder 1973). The wool undercoat provides insulation during the winter, the coat is then shed in the spring for the warmer summer months. Despite the harshness of winter weather in many sheep habitats, and its variability from one year to the next, no relationship has been found between snowfall, winter temperatures and lamb survival over winter (Portier et al. 1998). This suggests that the adaptations of sheep to harsh environments are sufficient to ensure survival through the winter.

Male horns act as rank symbols in contests to determine dominance and breeding success. Thus large horn sizes are favoured. However, the highly vascularised bone core of horn may act as source of heat loss in cold temperatures. Sheep adapted to colder climates have a smaller bone core and a thicker keratin sheath on their horns, compared to tropical species, to reduce heat loss (Picard et al. 1996). Desert Bighorn, for example, have wider and flatter horns than mountain Bighorn, which increase heat loss in the desert adapted subspecies.

2.2.4 Physiological Adaptations to Environment

Wild sheep face various physiological challenges from the environments that they inhabit. Temperature fluctuations mean that animals may face both very hot temperatures in summer and very low temperatures in winter. Many of the environments inhabited by sheep are arid, thus sheep must also cope with periods when water is hard to come by. The severity of the habitat means that forage availability and nutrient quality are very variable throughout the year. As described above, wild sheep use behavioural adaptations to counter some of these challenges: selecting shade or shelter, varying elevation in response to plant growth cycles and selecting sunny or sheltered aspects depending on season and weather. These are supported by a number of physiological adaptations that also help the sheep to survive in hostile environments.

Sheep adapted to desert environments will drink every day if water is available but can exist for long periods without drinking. Although ewes and lambs drink nearly every day, Bighorn rams may drink only once a week. This is reflected in their habitat use as desert-living Bighorn ewes forage much closer to sources of free water than ram groups (Mooring et al. 2003). Desert-adapted sheep show a decrease in appetite and an increase in feed utilization when water is restricted (Silanikove 1992). Bighorns can lose more than 20% of their bodyweight when water is scarce and tolerate a loss of 48% in plasma volume (Turner 1979). Dehydration weight loss is reversed within an hour of drinking and plasma volume increases in less than four hours. In common with other desert-adapted species, the Bighorn also produces very concentrated urine such that water is conserved as much as possible.

Just as the desert-adapted sheep have a number of mechanisms to allow them to cope with hot weather and restricted water supplies so the species which inhabit very cold Arctic and sub-Antarctic environments have adaptations to deal with low temperatures. Sheep are particularly resistant to cold weather, and winter weather may not be an important factor in survival (Jorgenson et al. 1997). In Bighorn, for example, the thermoneutral range in winter extends to at least −20°C, with increases in metabolic rate seen only below this temperature (Chappel & Hudson 1978). Even newborn lambs are able to maintain body temperature in air temperatures of well below freezing provided the lamb is dry (McCutcheon et al. 1983), and Soay lambs survive a cold, dry winter better than a warm, wet one (Coulson et al. 2001). Ability to survive the winter does, however, seem to be influenced by the condition of the sheep at the onset of winter. Therefore, factors that can impact on summer forage availability (such as population density and spring weather) also affect ability to cope over the winter.

Ungulates living in the harsh environments of extreme latitudes need to cope with considerable seasonal variation in forage availability. Limited availability usually coincides with other environmental challenges. In sheep, as in other ruminants, appetite and metabolic rate both naturally decline in autumn (Kay 1979; Argo et al. 1999) in line with a decrease in food availability and time spent foraging (Seip & Bunnell 1985). This mechanism is entrained by the change in day length, and may also be influenced by nutrient supply, and is accompanied by changes in endocrine responses to feeding and photoperiod (Argo et al. 1999; Rhind et al. 2001). These seasonal influences accompany a reduction in body weight gain of sheep and in the growth of the coat and horns, which cease growing in the winter and resume in the spring.

2.2.5 Reproductive Adaptations

Environmental pressures, particularly on Arctic and sub-Antarctic living sheep, means that these animals grow slowly and reach sexual maturity relatively late. The age at which sexual maturity is reached is dependent on the condition of the animal. Sexual maturity for ewes is reached between 18 and 36 months of age, males do not begin to participate in the rut until they are more than two years old, and may not achieve breeding success until 4 or 5 years of age. Two-year old ewes are less productive than older animals and produce lambs later in the season (Bon et al. 1993).

Sheep show seasonal breeding behaviour, entering the rut usually in November to December when ram flocks migrate back to the home ranges of the ewe and juvenile bands. The gestation period of wild sheep lasts for 150 to 180 days and lambs are born in mid to late spring. The exceptions are the desert Bighorn, where the rut can last for nine months of the year peaking in late summer. Lambs are then born throughout the year, although reaching a peak, corresponding to the rut peak, in the cooler months of late winter and early spring. Likewise, mouflon living on the sub-Antarctic Kerguelen Archipelago reproduce twice a year in summer and winter lambing seasons (Reale & Bousses 1999). Seasonal breeding ensures that mating occurs when males and females are normally in peak condition, following summer grazing, and that lambs are born as the spring forage availability increases. Synchronous breeding behaviour, where all ewes give birth within a few weeks, reduces predation on newborn lambs as vulnerable lambs are only present for a relatively short period of the year. This is particularly important in 'following species' (where the lamb accompanies the dam from birth, also see Chapter 3) where the young lambs and their mothers join together as a lactating flock (Rutberg 1987).

Weaning of lambs occurs on average at about six months, with some variation due to ewe condition, age, social status, parasite burden and timing of births within the season. Although female lambs will remain in the social group with their mother, the ewe and lamb no longer preferentially associate with one another. Under some circumstances, however, where a ewe has lost her newborn lamb, she may continue to associate with her yearling offspring, although this is not seen in non-lactating ewes or ewes with a new lamb (L'Heureux et al. 1995). In Arctic and sub-Antarctic conditions, where time to rear lambs is short, in years of greater nutritional constraint births occur later in the season and less synchronously. Dall's ewes adjust their maternal behaviour under these conditions, nursing lambs for longer bouts after parturition but weaning them earlier (Rachlow & Bowyer 1994). Mouflon, Soay and Bighorn all respond to nutritional stress by reducing maternal care and favouring their own body mass maintenance over the development of their lambs (Festa-Bianchet & Jorgenson 1998; Reale & Bousses 1995; Robertson et al. 1992). Lambs deal with earlier weaning by grazing earlier and playing less than under conditions of better nutrition (Reale et al. 1999; Berger 1979).

2.3 Domestication and Adaptation

The previous section summarised some of the adaptations used by wild sheep to survive and reproduce under particularly harsh environmental conditions. They are able to achieve this through a combination of behavioural, physical and physiological adaptations. Examples of behavioural adaptations are shifts in habitat preferences, use of shelter and shade, changes in feed preferences and diurnal rhythms, and altering maternal behaviours dependent on environmental constraints. For these adaptations to have relevance to domestic sheep welfare, however, we need first to consider how the process of domestication has altered the responses of sheep and whether these adaptations are still functioning in the domestic animal. Secondly, sheep have been selected for particular production traits (wool, meat, dairy) and some breeds have been subject to more selection than others. It may be, therefore, that breeds also differ in their potential ability to cope under different environmental conditions.

2.3.1 Process of Domestication

Domestication is the process whereby wild animals are brought under the management and control of humans. This involves management of most aspects of their lives and, through control of breeding and mate selection, man is also able to shape and change the animal under domestication. However, some of the apparently genetic changes in domestic animals may also be a consequence of the changed environment. Thus domestication can be defined as the process by which animals become adapted to man and captivity through genetic change and environmentally-induced development events (Price 1984).

Particular groups of animals can be considered as 'pre-adapted' to domestication (Price 2002), that is the wild progenitors of the domestic animal display a collection of biological and behavioural traits that facilitate a life in captivity. These include many traits that are displayed by wild sheep as shown in Table 2.1. Thus we might consider that a number of the behavioural characteristics of wild sheep are likely to be preserved in the domestic animal as they are traits already favoured for domestication. In addition, unlike many other farmed livestock species, sheep are rarely housed throughout the year. Thus some aspects of environmental adaptation may be preserved in the domesticated animal as these traits are still required for the survival of the animal. In other aspects of their lives, however, even extensively managed sheep are subjected to a number of manipulations that differ from the wild sheep (Table 2.2).

2.3.2 Effect of Domestication on Physical Attributes of Sheep

Domestication, and the husbandry practices connected with it, is generally associated with a number of morphological and physiological changes in the animal which may be the result of unconscious selection (Clutton-Brock 1992; Zohary et al. 1998). Whether conscious or unconscious, in general domestication is associated with: (i) a reduction in body size (although this does not appear to be applicable for sheep); (ii) increased diversity of outward appearance (coat colour, fibre diversity); (iii) increased fat storage under the skin rather than around organs; (iv) a reduction in the relative size of the brain and sense organs; (v) shortening of the jaws and facial regions and decreased tooth size; and (vi) increased diversity of horns and polling in females (Clutton-Brock 1987). Many of these changes may well result from relaxation of certain selection pressures, in particular a reduction in the threat of predation (e.g. diversity in coat colour, polling), culling of certain types of animals and alterations in the diet of animals with domestication (Zohary et al. 1998).

Domestic sheep are somewhat smaller than the argali, but otherwise of similar size to most wild sheep and most European breeds are larger than mouflon or Soay. In other physical characteristics, however, they have been affected by domestication, most notably by selection for growth of the wooly undercoat (except in the hair sheep breeds, see below). This has resulted in sheep with a thicker wool coat

Table 2.1 Favourable characteristics of sheep leading to their pre-adaptation for domestication

Characteristics favouring domestication	Characteristics of wild sheep
Large gregarious social groups	Highly social, flocking and following behaviours, attraction to similarly aged and sized conspecifics
Dominance hierarchy	Matrilineal structure of ewe flocks; age and size related dominance in ram groups, following behaviours of juveniles
Males affiliate with social group	Overlapping home ranges of rams and ewes, males with females for part of the year
Non-aggressive within and between species	Low levels of aggression within social group, relying on stable social structure and rank symbols for dominance; non-aggressive to other species
Promiscuous mating structure	Rams serve several ewes, no pair-bonding
Male initiated sexual behaviours	Rams court ewes although ewes will occasionally initiate sexual contacts
Sexual signals	Sheep predominately use postural and movement signals indicating willingness to mate, rather than pheromonal or other marking signals
Critical period for parent:offspring attachment	Ewes form a rapid and exclusive attachment to their own young at birth
Precocial development of the young	Lambs are quickly able to stand and follow their mothers after birth
Young easily separated from the mother	Lambs are relatively 'hardy' allowing them to survive early weaning, and quickly transfer social attachments to other sheep
Short flight distance to man	Flight distance of sheep relatively short compared to other wild ungulates that have not been domesticated e.g. deer
Non-aggressive to humans	Sheep relatively placid in the presence of humans
Readily habituated and adaptable to a range of habitats	Sheep are highly adaptable to diverse habitats: mountains, deserts, islands etc.
Limited ability to escape	Although agile in some terrain (particularly on slopes) sheep are slower than plains ungulates (e.g. antelope, gazelle) and less agile than mountain goats
Catholic eating habits	Sheep are generalist feeders and able to exploit a wide range of forage species

Adapted after Hale (1962).

than wild sheep and a loss or decrease in diameter of the kemp hairs (Ryder 1991). Selection for animals with white wool rather than the wild pattern of brown with a white belly has also occurred, perhaps with the advent of dying fleeces. Animals have also been selected not to moult their wool, so that it can be harvested in all animals simultaneously, thus wool shedding is only present in the more primitive sheep breeds. Polling is also common, in both sexes, in many sheep breeds, although in general most hill and mountain breeds retain horns in both rams and ewes.

One expected impact of domestication on the lives of animals would be improved survival and longevity. However, survival to 10 years, and even to 20 years has been reported in wild sheep (see Section 2.2.3 above), whereas domestic ewes are usually

Table 2.2 Comparison of the lifestyles of wild and extensively managed sheep

Features of extensively farmed sheep, compared to wild sheep
Population and groups structure
Age composition distorted (old ewes removed, generally kept in peer groups)
Sex composition distorted (juvenile males removed or kept separate from ewe groups)
Kin structure disrupted or non-existent (e.g. ewe-lambs often separated from breeding ewes)
Movement
Fences limiting freedom of movement
Seasonal migrations prevented
Home range and diurnal movements may be disrupted
Predation and Disease
Less subject to natural predation, but selectively culled by man
May be both more and less exposure to disease pathogens, vaccination programmes
More veterinary care
Food Availability
Naturally available feed supplemented (e.g. concentrates)
Less varied in composition (e.g. artificial swards)
Less seasonal variability in food supply
Reproduction
Reduced or absence of mate choice
Controlled mating and birth season
Veterinary interventions at lambing time
Parental care
Human assistance with obstetrics and early ewe-lamb interactions may be provided
Early weaning of lambs before natural weaning
Separation of ewe and lambs before full range of cultural transmission has been transmitted
Shelter and shade
May be supplemented or less available
Contact with man
Handling (management, veterinary care, culling, etc.)
Natural selection
Reduced by husbandry leading to relaxation of some selection pressures
Supplemented (e.g. selection for 'easy-care' traits; for weather-resistant fleece properties, for disease or parasite resistance)

After Deag (1996).

culled when it is considered that their reproductive performance is less than optimal (which may occur as early as 5–6 years, Mysterud et al. 2002).

2.3.3 Effects of Domestication on Behaviour of Sheep

Several authors suggest that domestication in all species causes behavioural changes associated with a decline in environmental responsiveness (Hemmer 1990; Price 1998; Zohary et al. 1998; Künzl & Sachser 1999). In general, domesticated animals show a reduction in aggressiveness, attentiveness and flight behaviours and an increase in sexual and courtship behaviours in comparison to their wild ancestors. These behavioural differences are, however, considered to be due to changes in the frequency of expression of behaviour patterns based on a shift in threshold (Ratner & Boice 1978; Price 1984). The loss of behavioural elements or the addition of novel behaviour patterns is not believed to occur during the normal process of domestication. In particular, Boissy (1998) argues that as fear and anxiety-related behaviours have adaptive value in promoting survival in domestic species (particularly in a species that is generally extensively managed), the anti-predator strategies that evolved in wild sheep will persist in domestic animals. Furthermore, the social behaviours of wild sheep will also be retained in the domestic sheep as it is these very features of the Genus that promoted their domestication in the first place (Clutton-Brock 1987).

Several authors report that domesticated species show a reduced alertness and attentiveness to the environment in comparison to wild species and attenuated flight distances (Price 1984; Hemmer 1990). This is accompanied by functional alterations in the adrenal glands (Hemmer 1990) and the reactivity of both the hypothalamic-pituitary-adrenal axis and the sympathetic-adrenomedullary system (neuroendocrine systems that regulate the animal's responses to stressful events; see Chapter 1) are reduced in domestic animals (Künzl & Sachser 1999). The threshold for stress and fear-associated behaviours, such as flight, appear to be elevated in domestic species. This may be due to artificial selection by man for docility and ease of handling in domestic species, or an adaptation of the species being domesticated to cope with the environments under which they are being kept. Although the threshold for these types of behaviours (e.g. fear responses) may be elevated in domestic sheep there is no evidence that these behaviours are not expressed once that threshold has been reached. Hemmer (1990) compared the behavioural responses of mouflon, Soay and domestic sheep (Texel). He reported that there was a gradation in all behavioural characteristics from mouflon via the more primitive Soay to the domesticated woolly breeds. Domestic sheep aggregate into larger flocks than mouflon, spend less time in rapid locomotion and more time standing inactive and have a reduced flight distance (Hemmer 1990). Soay sheep were intermediate between mouflon and Texels. Thus the wild, feral and domestic sheep showed similar behavioural repertoires but differed in the frequency that the behaviours were expressed.

Domestic animals are often kept in larger groups and/or more crowded conditions than they would experience in the wild. It would, therefore, be adaptive for them to perform more sociopositive and less aggressive behaviours under these conditions. For example, the vocal behaviour of domestic sheep is increased in comparison to wild sheep (Kiley 1972). This may have arisen because of the need to have more complex social signals in larger groups (as suggested by Berger 1979, in Bighorn sheep) and because the selection pressure against vocalisation, the increased risk of predation, is reduced in domestic sheep (Price 1984). Domestic sheep are vocal when socially isolated but tend to have inhibited vocalisations in other situations, for example in the presence of a tethered dog (Torres-Hernandez & Hohenboken 1979), which mirrors the behaviour of wild sheep in the presence of a predator (Kiley 1972).

2.3.4 Effect of Domestication on Habitat Use in Sheep

In many situations domestic sheep are not offered a choice of habitat and may be kept in rather small fenced paddocks. However, several studies have investigated the habitat use of sheep when offered a much greater area in which to disperse, particularly in the temperate hills of Scotland, ranges in Norway or arid regions in Australia. These studies have shown that hill and mountain sheep (e.g. Scottish Blackface or North and South country Cheviots) do show home range behaviour when the opportunity arises. These home ranges vary in size depending on the habitat but average 40–60 hectares in the summer. Furthermore the changes in home range usage by sheep reflects that seen in wild sheep if there are no management interventions. For example, there is a reduction in dispersal and home range size in the winter as sheep move closer together and migrate to a smaller area of the home range (Hunter & Milner 1963; Lawrence & Wood-Gush 1988). Food preferences also show seasonal shifts reflecting differences in availability and nutrient content of plant species, and in dry regions, the water content of plants (Hunter 1962; Griffiths 1970; Lawrence & Wood-Gush 1988; Kronberg & Malechek 1997). The formation of home ranges by Merino sheep in Australia is affected by environmental features: no home range behaviour is seen in sheep kept on treeless plains or in paddocks of less than 40 hectares, but sheep do form subgroups in hilly paddocks (Lynch 1967; Squires 1974). Domestic sheep segregate by age and sex, as seen in wild sheep, where there are no specific interventions to disrupt this, such as removal of males (Arnold & Pahl 1967). Home ranges tend to be composed of related animals if daughters are left in the social group with their mothers (Hunter 1964; Lawrence 1990). However this seems to reflect the tendency of ewes to remain in the region in which they were born, and may be more strongly related to the need to be part of a social group than familial ties. For example, in experiments where unrelated animals were kept as a group before introduction to the range, or where animals were removed from their home range for some months and then returned, the introduced animals formed a new home range together (Hunter & Davies 1963).

Domestic sheep show a similar diurnal rhythm of behaviour to that expressed by wild sheep. Sheep generally camp in the hills if available, or on elevated ground, and move down to the lower regions at dawn to graze. In temperate climates sheep graze in the morning, rest and ruminate at midday, graze again in the evening and move uphill to their campgrounds. A single nighttime grazing bout in the vicinity of the bedding area often occurs (Arnold 1984). Total time spent grazing ranges between 8 and 12 hours each day, depending on food requirements and forage availability and quality (Iason et al. 1999). For sheep grazing in arid environments movement to and from a water source may mean that sheep walk for up to 16 km each day, although preferred grazing is often less than a km from a water source (Squires 1974). In hot weather sheep spend more time in the shade and move their grazing patterns such that most grazing occurs in the evening and at night, in a similar manner to that seen in the desert-living Bighorn described above. Sheep tend to graze into the wind in summer, particularly at high temperatures (Scott & Sutherland 1981). In cold and wet weather activity declines. In winter, sheep will seek shelter when wind speeds increase, although shelter seeking will be increased when the fleece is short.

As with wild sheep, maternal domestic ewes withdraw from the social group at parturition and seek out secluded and rocky parts of the home range in which to give birth (Hewson & Wilson 1979; Hewson & Verkaik 1981). This behaviour, as with wild sheep, is assumed to have an antipredator function. When given the opportunity, domestic lactating ewes with lambs also form a separate flock from non-lactating ewes and seem to use different parts of the home range (Squires 1975; our recent observations), showing similar behavioural response to wild ewes. Ewes show more attentive or vigilant postures than younger animals, as has also been seen in wild sheep (Risenhoover & Bailey 1985). The environment has an impact on the behaviour of sheep, for example, the frequency of vigilance postures in Merinos is higher in barren paddocks than in environments with greater topographical complexity (Stolba et al. 1990). This response can again be interpreted as an antipredator response in domestic ewes, which may be analogous to the greater vigilance of wild sheep when further from escape terrain (Risenhoover & Bailey 1985; Frid 1997).

As a whole, the data on domestic sheep suggest that they adopt broadly similar rules in their use of habitat, and that similar habitat requirements are common to both wild and domestic sheep. It seems likely, therefore, that information about the environmental responses of wild sheep will be relevant to assessing the requirements and preferences of domestic sheep. An important additional factor to consider in domestic sheep is the existence of many different breeds often selected and bred to produce different products or to thrive under local environmental conditions. The differences between different breeds will therefore be considered in the next section.

2.4 Breed Differences in Environmental Adaptation

There are in excess of 850 breeds of sheep worldwide (Ollivier et al. 1994), although exact numbers may vary as breed definitions change and new strains are developed. Sheep breeds can be broadly divided under geographical/environmental classes (e.g. Squires 1975) where sheep breeds are classified as:

(a) Temperate – a wide variety of sheep breeds from mountain, longwool and down breeds to Merinos found in Europe, North and South America, Australia and New Zealand;
(b) Northern desert sheep found in the Mediterranean border of the Sahara, Syria, Iraq and parts of Afghanistan
(c) Southern desert sheep of sub-Saharan Africa and India.

Alternatively breeds are classified by morphology (essentially 'tail-type' and fleece quality, Mason 1991, 2002). Here breeds are divided into thin-tail (e.g. most European temperate breeds), fat-tail, fat-rump, short and long tail and by hair, coarse, medium and fine-wool types. Temperate, thin-tail sheep bred mainly for meat and wool are the predominant type of sheep breeds in the world. Temperate breeds are moderate in size, short limbed and compact with thick coats. Northern desert sheep are less compact, with thinner necks, longer legs and markedly longer ears and are often fat-tailed, (e.g Awassi). They have woolly coats, which are coarser and less dense than temperate breeds. Southern desert sheep have elongated extremities, long ears and tails and are hair sheep (e.g. Djallonké). From the above description it is clear that temperate breeds, with the better insulation of a dense wool coat and a short compact conformation will be better adapted to colder temperatures and heat conservation, whereas the desert sheep with thin coats and longer extremities better at dealing with hot temperatures and heat dissipation. Within the temperate breeds (where most research has been carried out) there is much variation between different breeds of animals in their distribution and potential for adaptation. It is, however, difficult to prove that breeds are adapted to their environment (Ryder 1983), as confounds between genotype and environment occur. Even differences in morphology, as described above, can be influenced by the temperature of the rearing environment, as elegantly demonstrated in pigs (Dauncey et al. 1983). In these studies piglets from the same litter were reared at different temperatures and considerable differences in morphology and appearance were found: the high temperature piglets had long limbs, noses and ears and a fine silky coat, whereas the cold pigs were compact and squat with short legs, and extremities. Some attempts have been made to accurately identify these effects, particularly behavioural traits, in sheep breeds (e.g. Key & MacIver 1980; Dwyer & Lawrence 2000), but potential environmental effects on genotype are an important consideration for the following sections. These will describe some of the apparent adaptations of breed to environment to illustrate some of the differences that may affect ability to cope under different conditions. Whether adaptation occurs by genetic or environmental means may not, of course, have an influence on the welfare of the animal placed in an inappropriate environment to which it is not adapted, but may be important in how this situation can be improved or avoided.

2.4.1 Physiological Adaptations of Different Breeds

The ability of different breeds to cope, physiologically, with environmental constraints has been investigated in a number of breeds, particularly the temperate breeds, although increasingly investigations into the adaptations of desert sheep have been carried out. Environments can vary by temperature, rainfall, and humidity as well as geographical conditions such as soil, altitude and management, which all can have an impact on the sheep living there and the feed types available to them. In an extensive review of the adaptations of sheep breeds, Terrill & Slee (1991) showed that 90% of sheep breeds were adapted for dry or medium humidity with only 1% of sheep breeds adapted for wet climates. The majority of breeds were also adapted for medium temperatures, although 20% were adapted for hot temperatures, particularly the hot and dry environments of India and the Middle East.

2.4.1.1 Adaptations to Hot and Dry Environments

Sheep have rectal temperatures ranging from 38.3 to 39.9°C with an average of 39.1°C. A rise in ambient temperature brings about an increase in heart rate, respiration rate, panting (to increase evaporation from the respiratory tract) and sweating, accompanied by reduced feed intake, and reduced water loss in faeces and urine. Panting and loss from the respiratory tract seem to be the main evaporatory heat loss mechanisms in sheep (Silanikove 2000). Plasma cortisol increases following acute heat exposure whereas a decrease in thyroid hormone activity occurs with chronic exposure to high temperatures resulting in a decreased metabolic rate (and hence heat production).

Many studies have shown that breeds native to India, Africa and the Middle East (e.g. Barki, Malpura, Awassi, Chokla and Barbados Blackbelly) are apparently better adapted to hot and dry climates than imported temperate breeds (e.g. Suffolk, Merino, Rambouillet, Texel and Dorset). These studies show that exposure to heat or sun causes a greater increase in rectal temperatures and respiration rates in temperate breeds than indigenous breeds (Ross et al. 1985; Gupta & Acharya 1987; Maurya et al. 1998). Imported breeds also sweat more than do the native breeds (Rai et al. 1979) and have higher plasma thyroxine, indicative of a higher metabolic rate (Ross et al. 1985). Additionally, the indigenous breeds have better production under high temperature: a higher ovulation rate, increased feed intake and a greater weight gain. During exercise in hot temperatures Awassi sheep maintain lower heart rates and respiration rates than do Awassi sheep crossed with Texels or Finns (AbiSaab & Sleiman 1995). A comparison of Egyptian Barki sheep with temperate Suffolks in their response to water deprivation demonstrates that Barkis are better able to conserve water than are Suffolks (Ismail et al. 1996). Egyptian Rahmani sheep have a smaller lung volume and respiratory surface than Merinos, but greater dead space (Shafie & Abdelghany 1978), which seems to be an adaptation for heat dissipation in the African breed. Thus sheep breeds adapted to the hot and dry environments of Africa and Asia are better able to cope with elevated temperatures and lack of water than are temperate breeds of sheep. Within the temperate breeds Merino and Rambouillet (wool) sheep have lower body temperatures and respiration rates than Southdown and Hampshire (meat) breeds on exposure to hot weather (Miller & Monge 1946). This may reflect the environments in which these breeds were originally created: the warmer and drier Mediterranean environment in Spain, where the Merino originated, in comparison to the cooler and wetter climate in the UK. These data suggest that, even within the temperate breeds, some are better adapted to hot weather than others.

2.4.1.2 Adaptation to Cold Climates

Sheep are more likely to experience exposure to cold conditions than most other livestock species. Because of their good insulation they are generally considered well adapted to cope with cold climates. The lower critical temperature, that is the temperature at which the sheep needs to increase heat production to maintain core body temperature, can be less than 0°C in fully fleeced adult sheep, although it is considerably higher in shorn animals or newborn lambs (Terrill & Slee 1991). Peak heat production can normally be maintained for several hours, and temperatures of −60°C can be resisted for short periods by shorn adult sheep in dry and wind-free conditions (Alexander 1979). Sheep generate heat predominantly by shivering, although non-shivering thermogenesis through metabolism of brown fat is a significant source of heat production in newborn lambs. Heat loss can also be reduced in cold weather by decreasing blood flow to the extremities.

Breed differences exist in the ability of sheep both to produce heat, and to reduce heat loss. The cold resistance, or the ability to withstand hypothermia, of shorn adult sheep demonstrates significant variation with Scottish Blackface and Tasmanian Merinos representing two of the most and least resistant breeds, respectively (Terrill & Slee 1991). Part of this difference is due to greater peripheral vasoconstriction in Blackface sheep on cooling compared to Merinos (Slee 1968), thus reducing heat loss. Similarly, in unshorn sheep, heat loss after exposure to cold, rain and wind, and corresponding metabolic responses, differs between breeds (Blaxter et al. 1966). The dense fleeces of some lowland breeds (e.g. Hampshire) provide the best resistance to wind, whereas the open fleeces of hill breeds (e.g. Scottish Blackface) are most resistant to rain. The cold resistance of new-born lambs of various breeds has also been assessed by monitoring the time taken for the lambs' rectal temperature to fall to 35°C when immersed in a cooling water bath (Samson & Slee 1981). The lambs of hill breeds (Welsh Mountain, Scottish Blackface, Cheviot) are significantly more resistant to cold than other breeds (Finnish Landrace, Merino, Southdown). When adjusted for differences in body weight, hill and feral (e.g. Soay) breeds had the highest weight-specific cold resistance. Skin thickness and birth coat depth are important components of cold resistance in the neonatal lamb and these are greatest in the hill and feral breeds in comparison to lowland breeds (Samson & Slee 1981). Differences in peripheral vasoconstriction, as seen in adults, also form part of breed differences in cold resistance of lambs, with Drysdale and Romney lambs in New Zealand better able to conserve heat with mild cold exposure than Merino lambs (McCutcheon et al. 1983). Studies with neonatal Scottish Blackface and Suffolk lambs suggest that, in addition to better heat conservation, the ability to generate heat may also be greater in hill breeds, which have higher thyroid hormone concentrations, important for non-shivering thermogenesis (Dwyer & Morgan 2006).

Studies in Merino and Scottish Blackface lambs suggest that cold resistance is a heritable trait in the lamb (Slee & Stott 1986; Slee et al. 1991). However, previous exposure to cold has a remarkable ability to increase cold resistance in adult sheep by as much as 50% and results in increased heart rates, heat production capability and resting metabolic rate (Webster et al. 1969; Webster 1983). Cold exposure of ewes in late pregnancy also improves cold resistance of newborn lambs by increasing their capacity for non-shivering thermogenesis (Stott & Slee 1985), as does rearing in a cold environment. Thus, although breed differences in adaptation to cold exist, these can be modulated to some extent by exposure to a cold environment.

2.4.1.3 Seasonality

Sheep show seasonality in their reproductive cycles, voluntary feed intake, fat metabolism, and pelage and horn growth. Seasonality allows sheep to cope with fluctuating food supplies and climatic variability. Thus reproductive effort, for example, can be timed to coincide with the growing season of plants. Ewes of temperate breeds begin to show oestrus cycles in late summer and autumn ensuring that lambs are born in spring as new grass begins to grow. Tropical and sub-tropical breeds, where the environment may be more affected by cycles of rain and drought, show oestrus and are able to conceive throughout the year (de Combellas 1980; Aboul-Naga et al. 1991). Within the tropical sheep breeds, some show short periods of anoestrus of 2–3 months (e.g. Awassi, Ossimi) whereas other breeds do not show anoestrus (e.g. Rahmani). Even within the temperate breeds the onset of estrus cycles and the duration of the breeding season varies. Various studies have shown that the breeding season is very short for some northern latitude breeds (e.g. Scottish Blackface) but longer for Galway, Suffolk and Finnish Landrace and their crosses (Quirke et al. 1986). Likewise, Mediterranean and equatorial breeds such as Rambouillet and Criollo have a longer breeding season than more temperate Romney, Corriedale or Suffolk ewes, when kept at the same latitude. For all seasonal breeds the endogenous rhythm of reproductive activity is synchronised by photoperiod, breed differences in timing seem to reflect differences in how these are synchronised (O'Callaghan et al. 1992).

Rams also show seasonality in testis size and libido, which varies with breed (Lincoln et al. 1990; Xu et al. 1991). Seasonal changes are most pronounced in wild rams (mouflon) and occur later than in feral and domestic breeds (Soay, Shetland, Herdwick, Scottish Blackface, Norfolk and Wiltshire Horn). For the Southern breeds (Portland and Merinos), the onset of testis activity is even earlier (Lincoln et al. 1990). Defined seasonal cycles of horn growth also occur in spring, which coincides with changes in plasma prolactin and a resurgence of pelage growth, in mouflon, Soay, Wiltshire Horn, Herdwick and Shetland sheep (Lincoln 1990). Selection for fleece characteristics in other breeds, such as Merinos, has changed the seasonal cycle of wool growth and prolactin secretion, and pelage growth continues all year without the moult seen in wild, feral and less domesticated breeds.

2.4.1.4 Adaptation to Feed Types and Availability

In addition to the seasonal changes mentioned above, temperate sheep show seasonal declines in appetite, voluntary food intake, metabolic rate and bodyweight gain in winter. Changes in metabolic rate appear to precede appetite shifts (Argo et al. 1999) and this change is considered an adaptation to food scarcity. Breed differences in this seasonal shift are also known to occur: as with other seasonal changes Soay, Scottish Blackface and Shetland sheep have a greater seasonal variation in voluntary feed intake than Dorset Horns. A more extreme adaptation to fluctuating food availability are the fat-tails and fat-rumps of sheep breeds adapted to arid environments. During favourable periods fat stores accumulate to be utilised during periods of undernutrition. The fat depots in the tail of fat-tailed sheep contain small adipocytes, which are very sensitive to the fat metabolising effects of catecholamines (Chilliard et al. 2000), making this source of energy rapidly available to the sheep when food is limited. Storing fat in the tail or rump, rather than back fat, in these animals may also help them to cope with the hot temperatures experienced in these environments.

Sheep are able to exploit a number of food resources, including cacti in the desert, lichens, tree leaves and fruit (Hadjiesterkotis 1996). Perhaps the most striking adaptation to food availability is the ability of the North Ronaldsay sheep in Orkney, Scotland, to survive on a diet largely of seaweed (Paterson & Coleman 1982). These sheep are feral and prevented from accessing inland pastures by a sea wall, which confines the sheep to foraging on the foreshore (Tribe & Tribe 1949). Seaweed has few of the structural components of land plants (e.g. only 4% cellulose compared to about 40% in land plants; Greenwood et al. 1983), and is low in copper, but high in salt, arsenic, and zinc. North Ronaldsay sheep are well adapted to their copper-impoverished diet, appear to be able to meet their energy intake needs from seaweed (Hansen et al. 2003) and are able to absorb copper more efficiently from the diet than other breeds. However, when North Ronaldsay sheep are exposed to pastures with normal levels of copper, or even marginally deficient, they can develop copper toxicity with elevated copper in the liver and plasma (Wiener et al. 1977). Most North Ronaldsay sheep (95%) also display a low-potassium phenotype with high erythrocyte sodium (Hall et al. 1975), which may be an adaptation to their high salt intake to maintain the electrolyte gradient across erythrocyte membranes.

2.4.2 Behavioural Adaptations of Different Breeds

The effect of breed and breed crosses on a number of behaviour patterns, predominantly social and maternal behaviours, has received considerable attention. The influence of breed on maternal behaviour and neonatal vigour will be discussed in the following Chapter and will not be elaborated on here. More recently there has also been interest in how different genotypes respond to fear-eliciting and stressful stimuli such as transport or predator stimuli.

2.4.2.1 Social Behaviours

Although flocking remains an integral part of the behavioural response of sheep, there is considerable evidence that the formation of home ranges and propensity to sub-group are variable amongst sheep breeds, as described above. The preferred size of the social group is also apparently affected by breed: lowland, highly selected breeds, such as Clun Forest or Suffolk, aggregate into larger sub groups than hill breeds, such as Scottish Blackface or Dalesbred (Winfield & Mullaney 1973; Shillito-Walser & Hague 1981; Dwyer & Lawrence 1999). This is also accompanied by a tolerance for greater crowding, as closer nearest neighbour distances are maintained in lowland breeds.

However, as described above for home range behaviours and sub-grouping (Section 2.3.4), the potential impact of environment on these behaviours may be partly responsible for the apparent breed responses. Animals in a hill environment, for example, where grazing sites may be patchy and nutrient poor, may need to spend more time grazing and cover a wider area, than lowland animals in a relatively homogeneous environment. When examined in different environments, Blackface and Suffolk ewes show alterations in social behaviour depending on environment, although breed differences persist (Dwyer & Lawrence 1999). Whereas the Blackface ewes respond to a larger, more complex environment by increasing nearest neighbour distances, the Suffolk ewes aggregate into larger subgroups. Research in wild sheep suggest that social group size is also influenced by environment (Frid 1997) which may be related to perceived riskiness as group size, environment and vigilance responses interact. Other studies using cross-fostering or embryo-transfer (Key & MacIver 1980; Dwyer & Lawrence 2000) suggest that aspects of social behaviour, including preferred proximity, may be acquired through maternal transmission rather than genetic influences.

2.4.2.2 Responses to Predators

As described above (Section 2.2.1) the main defense of sheep against predators is flight. Increased vigilance, particularly of lactating mothers, and a close ewe-lamb spatial distance are also reported to be antipredator strategies to reduce predation particularly on vulnerable lambs (Hewson & Verkaik 1981; Warren et al. 2001; Wolff & van Horn 2003). Few studies have addressed breed differences in response to predators yet these may be important in free-ranging sheep in areas of high predation. Comparison of the behaviour of Norwegian breeds on exposure to stuffed predator stimuli demonstrates that the most primitive and least selected breed (Old Norwegian) has the greatest flight distance, the longest recovery time and are least likely to bleat (Hansen et al. 2001). The most selected and heaviest breeds (Suffolk, Steigar, Dala) are the least responsive but most vocal, whereas the medium light breeds selected for some improved carcass qualities (Spælsheep, Norwegian fur sheep) are intermediate. These behavioural differences may explain the greater than expected wolverine predation on Dala sheep on Norwegian summer ranges (Landa et al. 1999), whereas losses of Old Norwegian and Norwegian fur sheep are lower than expected. Other studies show breed differences in the response to the presence of dogs: primitive Soay sheep are reported to scatter (Boyd et al. 1964), reassembling as a flock later, whereas more domestic breeds (Merino, Wiltshire Horn) form into a tight flock immediately (Winfield & Mullaney 1973).

Studies with hill (Scottish Blackface) and lowland (Suffolk) ewes with lambs demonstrate an increased frequency of vigilance postures in Blackface ewes compared to Suffolks, particularly in early lactation (Pickup & Dwyer 2002). When exposed to a dog Blackface ewes also make more vigilance postures, and are more active, than are Suffolk ewes. Blackface ewes maintain consistently closer spatial relationships with their lambs than Suffolk ewes throughout lactation (Dwyer & Lawrence 1999; 2000), suggesting that, similar to the Norwegian study, the hill ewes show greater antipredator responses than lowland animals.

2.4.2.3 Other Fear Responses

Breed influences on fearfulness have been investigated by tests measuring sheep responses to surprise effects, the presence of a human or novel object, exposure to an open-field or an unfamiliar environment (Romeyer & Bouissou 1992), and feeding behaviour in the presence of a human intruder (Le Neindre et al. 1993; Lankin 1997). Taken together these data suggest that less selected and specialised breeds of sheep (e.g. Romanov, Karakul) are more fearful than more specialised breeds (e.g. Ile-de-France, Merino, East Friesian). Fearfulness was shown by a higher incidence of withdrawal from humans, immobilisations, low pitched bleats, escape attempts and unwillingness to interact with novel objects. In other studies, Scottish Blackface lambs are found to have higher heart rates and plasma cortisol following an open-field test than lambs of a more highly selected meat breed, the Texel (Goddard et al. 2000). Lowland genotypes are also reported to have a weaker cortisol response to transportation than hill or upland breeds (Hall et al. 1998). In summary, these studies, together with the responses to predators, suggest that the less-selected hill and upland breeds or more primitive breeds have a greater reactivity to the same stressor and take longer to recover than do more selected and specialised lowland breeds.

2.4.2.4 Shading, Sheltering and Grazing Behaviours

Total amount of time spent grazing has been shown to be influenced by breed, as is the timing of diurnal shifts between grazing and ruminating in breeds grazing in a Mediterranean climate (Dudzinski & Arnold 1979). Suffolk sheep in this study grazed for longer than other breeds (Southdown, Border Leicester, Dorset Horn, Romney, Cheviot) and had a distinctly different grazing pattern to other breeds. Environmental variables, particularly relative humidity, influence diurnal grazing patterns although breeds differ in their sensitivity to environment: Border Leicesters are most and Dorset Horns least responsive to these influences. In separate experiments Cheviot and Border Leicester sheep were shown to respond to increases in temperature by shifting to more nighttime grazing at temperatures where the grazing patterns of Southdown or Merino breeds were unaffected (Daws & Squires 1974; Dudzinski & Arnold 1979). In the arid, sub-tropical environment of Egypt, local Ossimi sheep and imported Texel and Merinos were shown to have a characteristic breed pattern of time spent grazing in the sun and time in the shade (Sharafeldin & Shafie 1965). The Ossimi sheep walked more rapidly to pasture, showed fewer signs of fatigue and grazed longer in the sun than other breeds. Ossimi ewes were also the only breed that did not necessarily seek shade to rest and ruminate. In studies where Merino and Border Leicester sheep were required to walk varying distances between feed and water, the Merinos quickly moved from drinking twice a day to once a day when the distance increased (Squires & Wilson 1971) whereas Border Leicester sheep maintained the twice daily drinking pattern for longer.

Adult ewes in full fleece seek shelter only in climatic conditions (wet, cold and windy weather) when they are outside their thermoneutral zone (Lynch & Alexander 1976; Alexander et al. 1979) which may occur infrequently in temperate sheep (Duncan et al. 2001). Sheep are, however, attracted to use trees for shelter on windy days, although they tend to move away from them in rain (Sibbald et al. 1996). Shorn ewes make greater use of shelter than full-fleeced ewes (Alexander & Lynch 1976) as do breeds, such as the Lacaune, which have relatively thin fleeces (Lecrivain & Janeau 1987). However, neither Merino nor Corriedale ewes (both woolly breeds) seek shelter unless wind speeds exceed 32 km/h with rain. Breed differences in sheltering behaviour, therefore, probably reflect mainly fleece characteristics rather than differences in the desire to seek shelter.

2.5 The Environment and Welfare

Movement of sheep from one environment to another, whether this is from pasture to indoors, from a valley to the mountains, from a cold to a hot climate, can cause disturbance and stress. This may be expressed by inactivity, apathy and a decrease in eating or drinking behaviours or even a refusal to do so. Animals may lose weight and condition and suffer from high levels of parasitism, particularly if moved to humid pastures. Acclimatization and adjustments to these environments may occur over a period of days or weeks but, as described above, some breeds or types of sheep will experience greater stress than others in different environments. In this section we will discuss potential welfare issues arising from the environment in which sheep are kept. This section will consider welfare issues pertaining particularly to extensive systems, as well as welfare problems that might occur in more confined environments.

2.5.1 Exposure to Thermal Extremes

In the UK the most common environmental stressor the sheep is likely to face will be cold (often increased by precipitation and wind-chill), although housed sheep in full fleece, as well as sheep without shade in Australia or Africa, may be more likely to suffer from heat stress. Sheep are well adapted to cope with both extremes, and ruminants are known to have a wide thermoneutral range (Webster 1983). Sheep are, therefore, able to adapt physiologically and behaviourally to regulate heat loss and to cope with thermal extremes. Provision of shelter and shade are important for protection from solar radiation and precipitation. For example, with shade, sheep are able to maintain body temperature in ambient temperatures of up to 50°C (Johnson 1987). During cold exposure sheep increase feed intake (Kennedy 1985), flock more closely together and make use of shelter, particularly if they are likely to be more susceptible to hypothermia (e.g. lambs, lactating ewes and shorn sheep, Miller 1968; Alexander et al. 1979; Pollard et al. 1999). Thus behavioural mechanisms appear to be important for dealing with thermal extremes in sheep. Traditional hill sheep farming practices facilitate these behaviours, since they allow for the formation of home ranges. In these ranges, ewes of the same social group restrict themselves to particular areas where they become familiar with the location of resources such as food, water and shelter (Hunter & Davies 1963; Hewson & Wilson 1979; Lawrence & Wood-Gush 1988). Sheep may experience distress under thermal extremes if confined within open and exposed pastures lacking in shelter or shade. In both hot and cold extremes, however, these are likely to be part of a compound stressor (e.g. hypo/hyperthermia with undernutrition and/or dehydration) which can reduce the ability of the sheep to cope with the situation. We might expect that difficulties in coping, where the animal is able to express species typical behaviour, should result in both changes in the animal's biological functioning, and in its negative subjective state or feelings and thus we would expect welfare to be impaired.

2.5.2 Undernutrition

Food availability, and consequent malnutrition, can be a serious problem particularly for pregnant ewes, which are carrying lambs through the winter when food is most likely to be scarce. Pregnant hill sheep have been shown to sustain losses of up to 20% of their pre-pregnancy body weight (Thomson & Thomson 1949) and may lose 85% of their subcutaneous fat during pregnancy and lactation (Russel et al. 1968). Surveys of sheep production in hill flocks demonstrate that lack of supplementary feed during pregnancy results in the deaths of about one third of neonatal lambs and 11% of ewes annually (Orr & Fraser 1932). Sheep generally graze for about 8 h a day, but can increase this to up to 13 h when food is limited (Lynch et al. 1992). An important constraint on the time budget of ruminants is the need to find time to ruminate, thus sheep cannot increase intake maximally to compensate for low food availability. The rumen acts to buffer the sheep from the effects of food and water deprivation and, although food deprivation increases the motivation to feed, whether the rumen also protects the ruminant from the sensations of hunger and thirst is unclear. It seems reasonable to assume, however, that an animal losing both weight and condition, whilst making futile attempts to find food, is not in good welfare, particularly when this semi-starvation can end in death. Moreover, lamb survival is greatly affected by the nutritional intake of the ewe in pregnancy (Waterhouse 1996). The consequences of maternal undernutrition, therefore, will cause distress and suffering to the lamb and ewe as a result of poor nutrition, even in the absence of distress from food deprivation *per se*.

The welfare codes of many countries state (in some form or other) that sheep should be fed to maintain full health and vigour. Most codes also point out that nutritional requirements can vary depending on stage of growth, reproductive status, whether recently clipped and for rams entering the breeding season (when they are likely to lose weight and condition). In addition, the effect of dental disease and tooth loss on the ability of sheep to feed themselves adequately in some forage conditions should be recognised. Both New Zealand and UK welfare recommendations state that sheep should not be food deprived for longer than 24 h and (in New Zealand) must not be deprived of feed for longer than 48 h. Whilst this is relatively easy to adhere to in housed sheep, heavy snowfall (for example) may make feed largely inaccessible to extensive sheep for periods longer than 48 h. UK Welfare codes appear to recognise this as they suggest that 'A condition score[1] in a significant number of the flock of ... 1.5 for those [sheep] on the hill can indicate inadequate management ...'. However, lowland sheep should be maintained at condition scores above 2 (or at condition score 3 in the New Zealand recommendations). Thus, whilst recognising some of the difficulties associated with feeding extensive sheep, the potential for these animals to spend part of the year in poorer welfare conditions than lowland or more intensively farmed sheep is implicit.

2.5.3 Predation

A feature of domestication has been the protection of livestock from predators. However, extensive sheep, particularly when managed on unenclosed pastures or ranges without shepherding, are still vulnerable to attack by predators. Both wild and domestic sheep can be relatively easy kills for wild canids or felids and, particularly ewes and juveniles, are largely defenceless other than expressing antipredator behaviours (flight). Lambs and subadult sheep are generally the most vulnerable to predator attacks, adult sheep are reportedly killed preferentially only by bears, and mountain lions. The extent of sheep predation is very variable, depending on the types of sheep management and the abundance of potential predators (see Table 2.3). For example, in the UK, foxes are probably the only significant wild predator of sheep, preying on young lambs. Foxes and crows may scavenge dead and moribund ewes and lambs but the only predator likely to attack adult sheep are domestic dogs. In the USA, by contrast, coyotes kill significant numbers of lambs, and coyote attacks account for nearly all lamb deaths after the neonatal period. In Australia lambs are killed by dingoes and feral pigs, in Norway by foxes, bears, lynx, wolverines and raptors, and in Africa sheep are also attacked and killed by baboons and cheetahs. Predation by domestic dogs also occurs in many countries and causes the death of between 1 and 2% of sheep in a study in the USA (Blair & Townsend 1983), sheep losses to dogs seem to be primarily adult ewes rather than lambs. In the UK,

[1] A condition score is an assessment of back fat and muscle over the spine between the last rib and the pelvis, made by manual palpation using the fingertips. The sharpness of the vertical processes of the vertebrae, and the amount of fat and muscle over the horizontal processes are assessed and the animal assigned a score from 0 (emaciated) to 5 (obese). The technique should be used to maintain ewes at a condition score of between 2 and 4. For more information see *Condition scoring of sheep: action on animal welfare* (1994) MAFF Publications PB1875, Stationery Office, UK.

Table 2.3 Published estimates of the incidence of predation on domestic sheep (where ranges are given these reflect the variation on different farms or locations)

Predator	% mortality	Location	Source
Red fox	0.0008–0.26 lambs per ewe*	UK	Moberly et al. 2003
Red fox	0.6–1.5% lambs	Scotland hill farms, UK	White et al. 2000
Red fox, lynx, Golden eagle, wolverine	10.7% lambs	Norway	Warren et al. 2001
Wolverine	5–10% lambs	Snøhetta Plateau, Norway	Landa et al. 1999
Brown bears	7.2% lambs 12.5% ewes	South east Norway	Warren & Mysterud 1995
Lynx	3.8% lambs	Southern Norway	Mysterud & Warren 1991
Lynx	0.14–0.59% all sheep	Jura, France	Stahl et al. 2001
Wolf, domestic dogs	0.35% all sheep	Tuscany, Italy	Ciucci & Boitani 1998
Red fox	0.25–10.25% lambs	Australia	Greentree et al. 2000
Feral pigs	32% lambs	New South Wales, Australia	Plant et al. 1978
Feral pigs	8–36% lambs	New South Wales, Australia	Choquenot et al. 1997
Baboons	2.5% sheep and goats	Zimbabwe	Butler 2000
Coyotes	5.3% lambs	California, USA	Neale et al. 1998
Coyotes	5.8% lambs	Utah, USA	Taylor et al. 1979
Coyotes	6.3% lambs	California, USA	McAdoo & Klebenow 1978
Mountain lions	Up to 84% all sheep	Brazil	Mazzoli et al. 2002
Unspecified	4.5% lambs	Brazil	Oliviera & de Barros 1982
Unspecified	11.5–23.6% lambs	Brazil	Del Camen-Mendez et al. 1982
Unspecified	4.7% lambs	Rio Negro, Argentina	Olaecha et al. 1981
Unspecified	1.82% all sheep	Rajasthan, India	Mathur et al. 1982

*Based on farmer questionnaire and perceived predation.

the National Farmers Union estimates that 24,000 sheep are killed or mutilated by domestic dogs annually, but note that the numbers may be considerably higher as dog attacks often go unreported.

How does predation affect sheep welfare? Except where sheep are confined and unable to escape, extensive sheep are able to express their evolved antipredator behaviours. However, we might expect these natural behaviour responses to be associated with negative emotional states, particularly if these emotional states aid the sheep in displaying antipredator behaviour more efficiently. Animals showing high levels of antipredator behaviour and vigilance also have to trade-off these behaviours against engaging in other maintenance behaviours such as feeding, mating and maternal care. Thus there may be costs to biological functioning in being exposed to predation, even if the animal is not attacked.

Clearly for animals that are subjected to attacks by predators the consequences can be extreme. Except when training cubs, kills by coyotes and dingoes are generally described as quick, bites to the throat damaging the trachea and the major blood vessels in the neck, and coyotes are likely to kill only one or two members of the flock. Attacks are usually on lambs, which rarely survive an encounter with a predator, although adult sheep may survive, as evidenced by those with sustained injuries. The biggest threat to sheep welfare is likely to come from predation by domestic dogs. Dogs are generally inefficient and indiscriminate in their hunting behaviour, rarely killing sheep outright but leaving numbers of injured and mutilated sheep following an attack, and causing considerable stress to the chased but uninjured members of the flock. Thus, for sheep, exposure to predation can be a significant welfare concern that is not present for intensively husbanded, housed animals.

2.5.4 Effects of Pasture Features and Social Group

Animals managed outdoors are often perceived to be in good welfare. Although their welfare may well be generally higher than in confinement agriculture, the environment may not necessarily meet all their species specific requirements. Pastures where sheep are held at high stocking density in relatively flat and featureless environments, without shelter or shade and an escape terrain, do not contain many of the requirements for sheep to be in good welfare. The sheep is better adapted for relatively short flight to rocky terrain than prolonged flight over plains, and maintaining proximity to escape terrain has been shown above to be very important to wild sheep breeds. The behaviours of sheep have been shown to change with different environments, suggesting that some are not optimal. For example, Merinos show sub-grouping and form home ranges only when pastures are large and hilly (Lynch 1967). Vigilance behaviours also decrease when the environment is more complex (Stolba et al. 1990), suggesting that featureless pastures without escape terrain are perceived as more threatening. Time spent on grazing behaviour is reduced in very small groups (less than 3 animals; Penning et al. 1993), perhaps as more time is spent on vigilance, and at space allowances of 50 m^2 per head in comparison to 200 m^2 (Sibbald et al. 2000). Sheep that are with familiar social companions graze more and are less vigilant and vocal than with unfamiliar animals (Boissy & Dumont 2002). Inter-individual distances also change with familiarity with social companions and with different environments (Dwyer & Lawrence 1999; Boissy & Dumont 2002).

Although group size and social relationships are important parts of the sheep environment, the social group can also act as a source of stress and welfare compromise, particularly when resources are limited. Agonistic encounters and the increased importance of a social hierarchy may appear when sheep are crowded together and resources are limited (McBride et al. 1967). An increase in aggression is associated with sudden environmental change, lack of space, a large social group size and when food or feeder space may be restricted (Arnold & Maller 1974;

Kiley-Worthington 1977; Done-Currie et al. 1984). Many expressions of dominance in sheep are not necessarily associated with overtly aggressive behaviours. Sheep maintain social hierarchy through subtle behaviours associated with head movement and eye contact. Dominant sheep may displace subordinates from the feed troughs and from preferred lying positions by resting their chins on the backs of subordinate sheep, or by pawing (Done-Currie et al. 1984). There are a number of consequences for the subordinate sheep. When feeder space is limited the number of displacements or disturbances from the trough increases (Arnold & Maller 1974) and a progressively greater proportion of sheep cease to compete for food becoming non-feeders. These animals are usually the very young or older sheep which are subordinate (McBride et al. 1967). Sheep may also be displaced from preferred feed patches when grazing heterogeneous pastures (Sibbald & Hooper 2003). Subordinate animals are, therefore, likely to have a low feed intake, they are usually at the tail of movement order and may eat the poor quality or contaminated forage, leading to high worm burdens (Lynch & Alexander 1973). Subordinate sheep may also be displaced from shelter or shade during conditions of thermal extremes if space is limited (Sherwin & Johnson 1987; Deag 1996). Subordinate sheep may, therefore, be chronically stressed, particularly when resources are limited and competition is great.

2.5.5 Movement Between Environments

Welfare issues surrounding sheep housing will be discussed in more detail in later Chapters so will not be expanded upon here. This section will deal specifically with the welfare issues in the movement of sheep from one environment to another. Sheep transferred indoors to single pens from a pasture appear to go through a period of behavioural inhibition or withdrawal, with increased time spent lying, for the first 2–3 weeks of confinement (Done-Currie et al. 1984; Fordham et al. 1991). Newly confined sheep also show a lack of environmental awareness and are inattentive to activities occurring around them (Done-Currie et al. 1984). Thereafter there is an increase in behavioural activity although this may be associated with performance of different parts of the ethogram, or the performance of stereotypical behaviours (Done-Currie et al. 1984; Marsden & Wood-Gush 1986; Fordham et al. 1991). Comparison of newly confined sheep with sheep that had been housed in single pens for 6 months demonstrated that newly housed sheep drink considerably less, and ruminate more, than sheep confined long term (Done-Currie et al. 1984). It is not clear, however, whether these reflect acute stress responses in the newly confined sheep, or chronic stress in the long-term confined sheep. These data do, however, demonstrate the temporal change in behaviours seen over a period of confinement. Confinement disrupts feeding patterns, leading to either over-eating or a refusal to eat. Differences in the circadian rhythm of behaviour or activity of animals in confinement have also been reported: the onset and decline in activity patterns are more abrupt in confined animals than sheep at pasture (Tobler et al. 1991). Waves of activity lasting for a few minutes interspersed with inactivity occurring throughout the day are also reported in groups of individually housed sheep (Done-Currie et al. 1984; Fordham et al. 1991). These responses are not seen in pastured sheep that show more pronounced diurnal rhythms of activity (Tobler et al. 1991), as described above.

Moving ewes to different pens and social groups results in a higher degree of activity and aggressive encounters when compared to control ewes maintained in a consistent physical and social environment (Sevi et al. 2001). Moved and regrouped animals also have lower immune responses and reduced milk production compared to controls. Movement to a new environment is linked to unwillingness to eat novel foods. At its most extreme sheep may refuse to eat altogether, leading to prolonged inappetance and eventual death, as can occur in long distance sea voyages where sheep are exposed to both novel environments and novel food. In an unfamiliar environment sheep are more likely to eat familiar food, even if their previous experiences of this feed are aversive, but less likely to eat novel foods (Burritt & Provenza 1997). This might mean that sheep are more likely to eat familiar plants containing toxins in an unfamiliar environment, even if they have previously learned to avoid them in another situation. Similarly, animals that are familiar with a particular environment have longer grazing bouts and ingest more food than sheep that are unfamiliar with the location (Ramos & Tennessen 1992).

These data suggest that sheep find environmental change, whether physical, social or both, to be stressful. In the wild sheep studies described above, ewes tend to live within consistent social groups, and generally also occupy the same home range throughout their lives. The importance of familiarity with the physical environment, and the social group for survival suggests that changes in either of these will have at least a short-term negative impact on sheep welfare.

2.6 Conclusions

The wild ancestors of domestic sheep have been able to exploit a diverse range of extreme environments, from deserts to lands within the Arctic Circle, and are well-adapted to cope with thermal extremes and poor quality diets. They are able to achieve this through a combination of physical, physiological and behavioural adaptations. Domestication has altered some of these characteristics, and led to the development of breeds that may be more specialised for particular environments. Thus the fact that some breeds and wild sheep can survive in a desert habitat, for example, cannot be taken to mean that all sheep can thrive with little water. Nevertheless, sheep are able to cope with environmental extremes that other domestic species are not. However, their ability to do so will be seriously impaired if they are not given the opportunity to display the behavioural adaptations that form part of their coping strategy.

An extensive environment is often considered as synonymous with good welfare. However, merely being outdoors with the freedom to express some behaviours does not necessarily mean that the environment will meet all the requirements and needs of sheep. In the behaviour and life of wild sheep, for example, the presence of an escape terrain is an important feature of the environment for the expression of antipredator behaviours, and ewes in particular maintain a close spatial relationship to this habitat. However, domestic sheep are often kept in featureless paddocks without access to anything similar to an escape terrain, such as slopes or cover. Concern about the welfare of other species kept in greater confinement than sheep, has led to studies of their environmental requirements (requirements for perches, nest boxes and substrates in which to dust bathe in chickens for example). These questions have, however, never been asked of extensive species, the assumption being that lack of confinement means that the animal's requirements have been met. An extensive environment contains greater diversity than intensive systems. Although this variety may be positive in many ways, variations in food availability and quality, climate and the presence of predators can have a serious impact on the welfare of the sheep.

References

AbiSaab, S. & Sleiman, F. T. (1995) Physiological responses to stress of filial crosses compared to local Awassi sheep. *Small Ruminant Research*, 16, 55–59.

Aboul-Naga, A. M., Mansour, H., Aboul-Ela, M. B., Mousa, M. T., Hassan, F. & El-Hommosi, F. (1991) Oestrous activity in three subtropical sheep breeds in Upper Egypt and response to long-day light treatment. *Journal of Agricultural Science,* 116, 139–143.

Alexander, G. (1979) Cold thermogenesis. (Ed. D. Robertshaw) *International Review of Environmental Physiology,* 20, 43–155.

Alexander, G. & Lynch, J. J. (1976) Phalaris windbreaks for shorn and fleeced lambing ewes. *Proceedings of the Australian Society of Animal Production*, 11, 161–164.

Alexander, G., Lynch, J. J. & Mottershead, B. E. (1979) Use of shelter and selection of lambing sites by shorn and unshorn ewes in paddocks with closely or widely spaced shelters. *Applied Animal Ethology*, 5, 51–69.

Argo, C. M., Smith, J. S. & Kay, R. N. B. (1999) Seasonal changes of metabolism and appetite in Soay rams. *Animal Science*, 69, 191–202.

Arnold, G. W. (1982). Some factors affecting the grazing behaviour of sheep in winter in New South Wales. *Applied Animal Ethology*, 8, 119–125.

Arnold, G. W. & Maller, R. A. (1974) Some aspects of competition between sheep for supplementary feed. *Animal Production,* 19, 309–319.

Arnold, G. W. & Pahl, P. J. (1967) Sub-grouping in sheep flocks. *Proceedings of the Ecological. Society of Australia,* 2, 183–189.

Bangs, P. D., Krausman, P. D., Kunkel, K. E. & Parsons, Z. D. (2005) Habitat use by desert bighorn sheep during lambing. *European Journal of Wildlife Research*, 51, 178–184.

Bender, L. C. & Weisenberger, M. E. (2005) Precipitation, density and population dynamics in desert bighorn sheep on San Andres National Wildlife Refuge, New Mexico, *Wildlife Society Bulletin*, 33, 956–964.

Berger, J. (1979) Social ontogeny and behavioural diversity: consequences for Bighorn sheep *Ovis canadensis* inhabiting desert and mountain environments. *Journal of Zoology, London,* 188, 251–266.

Berger, J. (1991) Pregnancy incentives, predation constraints and habitat shifts: experimental and field evidence for wild bighorn sheep. *Animal Behaviour,* 41, 61–77.

Blair, B. J. & Townsend, T. W. (1983) Dog predation of domestic sheep in Ohio. *Journal of Range Management,* 36, 527–528.

Blaxter, K. L., Clapperton, J. L. & Wainman, F. W. (1966) The extent of differences between six British breeds of sheep in their metabolism, feed intake and utilisation and resistance to climatic stress. *British Journal of Nutrition*, 20, 283–294.

Bleich, V. C. (1999) Mountain sheep and coyotes: Patterns of predator evasion in a mountain ungulate. *Journal of Mammalogy,* 80, 283–289.
Bleich, V. C., Bowyer, R. T. & Wehausen, J. D. (1997) Sexual segregation in mountain sheep: Resources or predation? *Wildlife Monographs,* 134, 3–50.
Boissy, A. (1998) Fear and fearfulness in determining behavior. In: *Genetics and the Behavior of Domestic Animals* (Ed. T Grandin), pp. 67–111. Academic Press, San Diego, USA.
Boissy, A. & Dumont, B. (2002) Interactions between social and feeding motivations on the grazing behaviour of herbivores: sheep more easily split into subgroups with familiar peers. *Applied Animal Behaviour Science,* 79, 233–245.
Bon, R., Badia, J., Maublanc, M. L. & Recarte, J. M. (1993) Social grouping dynamics of mouflon (*Ovis ammon*) during rut. *International Journal of Mammalian Biology,* 58, 294–301.
Bon, R., Joachim, J. & Maublanc, M. L. (1995) Do lambs affect feeding habitat use by lactating female mouflons in spring in areas free of predators? *Journal of Zoology,* 235, 43–51.
Boyd, J. M., Doney, J. M., Gunn, R. G. & Jewell, P. A. (1964) The Soay sheep of the island of Hirta, St. Kilda: a feral population. *Proceedings of the Zoological Society, London,* 142, 129–163.
Burritt, E. A. & Provenza, F. D. (1997) Effect of an unfamiliar location on the consumption of novel and familiar foods by sheep. *Applied Animal Behaviour Science*, 54, 317–325.
Butler, J. R. A. (2000) The economic costs of wildlife predation on livestock in Gokwe communal land, Zimbabwe. *African Journal of Ecology*, 38, 23–30.
Carmen Mendez, M. del, Riet-Correa, F., Ribeiro, J., Selaive, A. & Schild, A. L. (1982) Perinatal lamb mortality in the districts of Bage, Pelotos and Santa Victoria do Palmar, Rio Grande do Sul. *Pesquisa Veterinaria Brasileira*, 2, 69–76.
Chappel, R. W. & Hudson, R. J. (1978) Winter bioenergetics of Rocky Mountain bighorn sheep. *Canadian Journal of Zoology*, 56, 2388–2393.
Chilliard, Y., Ferlay, A., Faulconnier, Y., Bonnet, M., Rouel, J. & Bocquier, F. (2000) Adipose tissue metabolism and its role in adaptations to undernutrition in ruminants. *Proceedings of the Nutrition Society*, 59, 127–134.
Choquenot, D., Lukins, B. & Curran, G. (1997) Assessing lamb predation by feral pigs in Australia's semi-arid rangelands. *Journal of Applied Ecology,* 34, 1445–1454.
Ciucci, P. & Boitani, L. (1998) Wolf and dog depredation on livestock in central Italy. *Wildlife Society Bulletin,* 26, 504–514.
Clutton-Brock, J. (1987) *The Natural History of Domesticated Mammals.* Cambridge University Press and British Museum (Natural History).
Clutton-Brock, J. (1992) The process of domestication. *Mammal Review,* 22, 79–85.
Combellas, J. de (1980) Production and reproduction parameters of tropical sheep breeds in improved production systems. *Tropical Animal Production*, 5, 266–272.
Corti, P. & Shackleton, D. M. (2002) Relationship between predation-risk factors and sexual segregation in Dall's sheep (Ovis dalli dalli). *Canadian Journal of Zoology*, 80, 2108–2117.
Coulson, T., Catchpole, E. A., Albon, S. D., Morgan, B. J. T., Pemberton, J. M., Clutton-Brock, T. H., Crawley, M. J. & Grenfell, B. T. (2001) Age, sex, density, winter, weather, and population crashes in soay sheep. *Science* 292, 1528–1531.
Cransac, N. & Hewison, A. J. M. (1997) Seasonal use and selection of habitat by Mouflon (*Ovis gmelini*): Comparison of the sexes. *Behavioural Processes,* 41, 57–67.
Dauncey, M. J., Ingram, D. L., Walters, D. E. & Legge, K. F. (1983) Evaluation of the effects of environmental temperature and nutrition on growth and development. *Journal of Agricultural Science*, 101, 291–299.
Daws, G. T. & Squires, V. R. (1974) Observations on the effects of temperature and distance to water on the behaviour of Merino and Border Leicester sheep. *Journal of Agricultural Science,* 82, 383–390.
Deag, J. M. (1996) Behavioural ecology and the welfare of extensively farmed animals. *Applied Animal Behaviour Science,* 49, 9–22.
Done-Currie, J. R., Hecker, J. F. & Wodzicka-Tomaszewska, M. (1984) Behaviour of sheep transferred from pasture to an animal house. *Applied Animal Behaviour Science,* 12, 121–130.

Dudzinski, M. L. & Arnold, G. W. (1979) Factors influencing the grazing behaviour of sheep in a Mediterranean climate. *Applied Animal Ethology,* 5, 125–144.

Duncan, A. J., Mayes, R. W., Young, S. A., Lamb, C. S. & MacEachern, P. (2001) Choice of foraging patches by hill sheep given different opportunities to seek shelter and food. *Animal Science*, 73, 563–570.

Dwyer, C. M. & Lawrence, A. B. (1999) Ewe-ewe and ewe-lamb behaviour in a hill and a lowland breed of sheep: A study using embryo transfer. *Applied Animal Behaviour Science,* 61, 319–334.

Dwyer, C. M. & Lawrence, A. B. (2000) Effects of maternal genotype and behaviour on the behavioural development of their offspring in sheep. *Behaviour*, 137, 1629–1654.

Dwyer, C. M. & Morgan, C. A. (2006) Maintenance of body temperature in the neonatal lamb: effects of breed, birth weight and litter size. *Journal of Animal Science,* 84, 1093–1011.

Festa-Bianchet, M. & Jorgenson, J. T. (1998) Selfish mothers: Reproductive expenditure and resource availability in bighorn ewes. *Behavioural Ecology*, 9, 144–150.

Fordham, D. P., Al-Gahtani, S., Durotoye, L. A. & Rodway, R. G. (1991) Changes in plasma cortisol and β-endorphin concentrations and behaviour in sheep subjected to a change in environment. *Animal Production,* 52, 287–296.

Frid, A. (1997) Vigilance by female Dall's sheep: interactions between predation risk factors. *Animal Behaviour*, 53, 799–808.

Geist, V. (1971) *Mountain Sheep: A Study in Behaviour and Evolution.* University of Chicago Press, Chicago and London.

Gionfriddo, J. P. & Krausman, P. R. (1986) Summer habitat use by mountain sheep. *Journal of Wildlife Management,* 50, 331–336.

Goddard, P. J., Fawcett, A. R., Macdonald, A. J. & Reid, H. W. (2000) The behavioural, physiological and immunological responses of lambs from two rearing systems and two genotypes to exposure to humans *Applied Animal Behaviour Science,* 66, 305–321.

Goodson, N. J., Stevens, D. R. & Bailey, J. A. (1991) Effects of snow on foraging ecology and nutrition of bighorn sheep. *Journal of Wildlife Management*, 55, 214–222.

Greentree, C., Saunders, G., Mcleod, L. & Hone, J. (2000) Lamb predation and fox control in south-eastern *Australia. Journal of Applied Ecology*, 37, 935–943.

Greenwood, Y., Hall, F. J., Orpin, C. G., & Paterson, I. W. (1983) Microbiology of seaweed digestion in Orkney sheep. *Journal of Physiology*, 343, 121P.

Griffiths, J. G. (1970) A note on patterns of behaviour and grazing in hill sheep. *Animal Production*, 12, 521–524.

Grubb, P. & Jewell, P. A. (1966) Social grouping and home range in feral Soay sheep. *Symposia of the Zoological Society of London,* 18, 179–210.

Gupta, U. D. & Acharya, R. M. (1987) Heat tolerance in different genetic groups of sheep in semi-arid conditions. *Indian Journal of Animal Sciences*, 57, 1314–1318.

Hadjisterkotis, E. (1996) Food habits of the Cyprus mouflon (Ovis gmelini ophion). *Zeitschrift fur Jagdwissenschaft*, 42, 256–263.

Hale, E. B. (1962) Domestication and the evolution of behavour. In: *The Behaviour of Domestic Animals* (Ed. E. S. E. Hafez), Balliere Tindall, London.

Hall, S. J. G., Hall, R. J. & Hall, J. G. (1975) Polymorphism of erythrocyte potassium concentration in seaweed-eating sheep. *Nature*, 255, 62.

Hall, S. J. G., Kirkpatrick, S. M. & Broom, D. M. (1998) Behavioural and physiological responses of sheep of different breeds to supplementary feeding, social mixing and taming, in the context of transport. *Animal Science,* 67, 475–483.

Hansen, I., Christiansen, F., Hansen, H. S., Braastad, B. & Bakken, M. (2001) Variation in behavioural responses of ewes towards predator-related stimuli. *Applied Animal Behaviour Science,* 70, 227–237.

Hansen, H. R., Hector, B. L. & Feldman, J. (2003) A qualitative and quantitative evaluation of the seaweed diet of North Ronaldsay sheep. *Animal Feed Science and Technology*, 105, 21–28.

Hemmer, H. (1990) *Domestication: The Decline in Environmental Appreciation.* Cambridge University Press, Cambridge.

Hewson, R. & Verkaik, A. J. (1981) Body condition and ranging behaviour of blackface hill sheep in relation to lamb survival. *Journal of Applied Ecology*, 18, 401–415.

Hewson, R. & Wilson, C. J. (1979) Home range and movements of Scottish Blackface sheep in Lochaber, north-west Scotland. *Journal of Applied Ecology,* 16, 743–751.

Hoefs, M. (1991) A longevity record for Dall sheep, Ovis dalli dalli, Yukon Territory. *Canadian Field-Naturalist*, 105, 397–398.

Hunter, R. F. (1962) Hill sheep and their pasture: A study of sheep grazing in south-east Scotland. *Journal of Ecology*, 50, 651–680.

Hunter, R. F. (1964) Home range behaviour in hill sheep. In: *Grazing in Terrestrial and Marine Environments* (Ed. D. J. Crisp), British Ecological Society Symposium No. 4, pp. 155–171, Blackwell Scientific Publications, Oxford.

Hunter, R. F. & Davies, G. E. (1963) The effect of method of rearing on the social behaviour of Scottish Blackface hoggets. *Animal Production,* 5, 183–194.

Hunter, R. F. & Milner, C. (1963) The behaviour of individual, related and groups of South Country Cheviot hill sheep. *Animal Behaviour*, 11, 507–513.

Iason, G. R., Mantecon, A. R., Sim, D. A., Gonzalez, J., Foreman, F., Bermudez, F. F. & Elston, D. A. (1999) Can sheep compensate for a daily foraging time constraint? *Journal of Animal Ecology,* 68, 87–93.

Ismail, E., Hassan, G. A., Abo-Elezz, Z. & Abd El-Latif, H. (1996) Physiological responses of Barki and Suffolk sheep and their crossbreeds to dehydration. *Egyptian Journal of Animal Production*, 33, 89–101.

Jansen, B. D., Krausman, P. R., Heffelfinger, J. R. & Devos, J. C. (2006) Bighorn sheep selection of landscape features in an active copper mine. *Wildlife Society Bulletin*, 34, 1121–1126.

Johnson, K. G. (1987) Shading behaviour of sheep: preliminary studies of its relation to thermoregulation, feed and water intakes, and metabolic rates. *Australian Journal of Agricultural Research,* 38, 587–596.

Jorgenson, J. T., Festa-Bianchet, M., Gaillard, J. M. & Wishart, W. D. (1997) Effects of age, sex, disease, and density on survival of bighorn sheep. *Ecology*, 78, 1019–1032.

Jugo, B.M. & Vicario, A. (2000) Single-strand conformational polymorphism and sequence polymorphism of Mhc-DRB in Latxa and Karrantzar sheep: implications for Caprinae phylogeny. *Immunogenetics,* 51, 887–897.

Kay, R. N. B. (1979) Seasonal changes of appetite in deer and sheep. *ARC Research Review*, 5, 13–15.

Kennedy, P. M. (1985) Influences of cold exposure on digestion of organic matter, rates of passage of digesta in the gastrointestinal tract, and feeding and rumination behaviour in sheep given four forage diets in the chopped, or ground and pelleted form. *British Journal of Nutrition,* 53, 159–173.

Key, C. & MacIver, R. M. (1980) The effects of maternal influences on sheep: Breed differences in grazing, resting and courtship behaviour. *Applied Animal Ethology,* 6, 33–48.

Kiley, M. (1972) The vocalisations of ungulates, their causation and function. *Zeitschrift fur Tierpsychologie,* 31, 171–222.

Kiley-Worthington, M. (1977) *Behaviour Problems of Farm Animals.* Oriel Press Ltd, Northumberland.

Krausman, P. R., Leopold, B. D., Seegmiller, R. F., & Torres, S. G. (1989) Relationships between desert Bighorn sheep and habitat in Western Arizona. *Wildlife Monographs,* 102, 1–66.

Kronberg, S. L. & Malechek, J. C. (1997) Relationships between nutrition and foraging behavior of free-ranging sheep and goats. *Journal of Animal science*, 75, 1756–1763.

Künzl, C. & Sachser, N. (1999) The behavioural endocrinology of domestication: A comparison between the domestic guinea pig (*Cavia aperea* f. *porcellus*) and its wild ancestor, the Cavy (*Cavia aperea*) *Hormones and Behavior,* 35, 28–37.

Landa, A., Gudvangen, K., Swenson, J. E. & Roskaft, E. (1999) Factors associated with wolverine Gulo gulo predation on domestic sheep. *Journal of Applied Ecology*, 36, 963–973.

Lankin, V. (1997) Factors of diversity of domestic behaviour in sheep. *Genetics, Selection, Evolution*, 29, 73–92.

Laundré, J. W., Hernandez, L. & Altendorf, K. B. (2001) Wolves, elk and bison: re-establishing the 'landscape of fear' in Yellowstone National Park, USA. *Canadian Journal of Zoology*, 79, 1401–1409.

Lawrence, A. B. (1990) Mother-daughter and peer relationships of Scottish hill sheep. *Animal Behaviour*, 39, 481–486.

Lawrence, A. B. & Wood-Gush, D. G. M. (1988) Home-range behaviour and social organization of Scottish Blackface sheep. *Journal of Applied Ecology*, 25, 25–40.

Lécrivain, E. & Janeau, G. (1987) Isolation and shelter-seeking behaviour of ewes lambing in large and shrubby paddocks. *Biology of Behaviour*, 12, 127–148.

Le Neindre, P., Poindron, P., Trillat, G. & Orgeur, P. (1993) Influence of breed on reactivity of sheep to humans. *Génétics, Sélection, Evolution*, 25, 447–458.

L'Heureux, N., Lucherini, M., Festa-Bianchet, M. & Jorgenson, J. T. (1995) Density dependent mother-yearling association in Bighorn sheep. *Animal Behaviour*, 49, 901–910.

Lincoln, G. A. (1990) Correlation with changes in horn and pelage, but not reproduction, of seasonal cycles in the secretion of prolactin in rams of wild, feral and domesticated breeds of sheep. *Journal of Reproduction and Fertility*, 90, 285–296.

Lincoln, G. A., Lincoln, C. E. & McNeilly, A. S. (1990) Seasonal cycles of blood plasma FSH, inhibin and testosterone, and testicular size in rams of wild, feral and domesticated breeds of sheep. *Journal of Reproduction and Fertility*, 88, 623–633.

Loehr, J., Kovanen, M., Carey, J., Hogmander, H., Jurasz, C., Karkkainen, S., Suhonen, J. & Ylonen, H. (2005) Gender- and age-class specific reactions to human disturbance in a sexually dimorphic ungulate. *Canadian Journal of Zoology*, 83, 1602–1607.

Lynch, J. J. (1967) Ranging behaviour of sheep in large areas. *Proceedings of the Ecological Society of Australia*, 2, 167–169.

Lynch, J. J. & Alexander, G. (1973) Animal behaviour and the pastoral industries. In: *Pastoral Industries of Australia* (Eds. G. Alexander & O. B. Williams), pp. 371–400 Sydney University Press.

Lynch, J. J. & Alexander, G. (1976) The effect of gramineous windbreaks on behaviour and lamb mortality among shorn and unshorn Merino sheep during lambing. *Applied Animal Ethology*, 2, 305–325.

Lynch, J. J., Hinch, G. N. & Adams, D. B. (1992) *The Behaviour of Sheep: Biological Principles and Implications for Production.* CAB International, Oxon, UK and CSIRO, Victoria, Australia.

McAdoo, J. K. & Klebenow, D. A. (1978) Predation on range sheep with no predator control. *Journal of Range Management*, 31, 111–114.

McBride, G., Arnold, G. W., Alexander, G. & Lynch, J. J. (1967) Ecological aspects of the behaviour of domestic animals. *Proceedings of the Ecological Society of Australia*, 2, 133–165.

McCutcheon, S. N., Holmes, C. W., McDonald, M. F. & Rae, A. L. (1983) Resistance to cold stress in the newborn lamb. 1. Responses of Romney, Drysdale x Romney, and Merino lambs to components of the thermal environment. *New Zealand Journal of Agricultural Research*, 26, 169–174.

McKinney, T., Boe, S. R. & de Vos, J. C. (2003) GIS-based evaluation of escape terrain and desert bighorn sheep populations in Arizona. *Wildlife Society Bulletin*, 31, 1229–1236.

McKinney, T., Smith, T. W. & Hanna, J. D. (2001) Precipitation and desert bighorn sheep in the Mazatzal Mountains, Arizona. *Southwestern Naturalist*, 46, 345–353.

Marsden, D. & Wood-Gush, D. G. M. (1986) A note on the behaviour of individually-penned sheep regarding their use for research purposes. *Animal Production*, 42, 157–159.

Mason, I. L. (1991). Classification and distribution of sheep breeds. In: *Genetic Resources of Pig, Sheep and Goat* (Ed. K. Majala), pp. 179–193. World Animal Series B8, Elsevier Science Publishers, Amsterdam.

Mason, I. L. (2002) *A World Dictionary of Livestock Breeds, Types and Varieties*. CAB International, Wallingford, UK.

Mathur, P. B., Bhagwan, P. S. K. & Kumar, P. N (1982) Important non-infectious causes of mortality in sheep. *Indian Journal of Animal Health*, 21, 125–129.

Maurya, V. P., Naqvi, S. M. K. & Mittal, J. P. 1998. Reproductive behaviour of Malpura and Bharat Merino ewe lambs under semi-arid conditions. *Indian Journal of Animal Sciences*, 68, 533–535.

Mazzolli, M., Graipel, M. E. & Dunstone, N. (2002) Mountain lion depredation in southern Brazil. *Biological Conservation*, 105, 43–51.

Melinkova, M. N., Grechko, V. V. & Mednikov, B. M. (1995) Investigation of genetic divergence and polymorphism of nuclear DNA in species and populations of domestic and wild sheep. *Genetika*, 31, 1120–1131.

Miller, G. R. (1968) Some responses of hill ewes and lambs to artificial shelter. *Animal Production*, 10, 59–66.

Miller, J. C. & Monge, L. (1946) Body temperature and respiration rate, and their relation to adaptability of sheep. *Journal of Animal Science*, 5, 147–153.

Moberly, R. L., White, P. C. L., Webbon, C. C., Baker, P. J. & Harris, S. (2003) Factors associated with fox (Vulpes vulpes) predation of lambs in Britain. *Wildlife Research*, 30, 219–227.

Mooring, M. S., Fitzpatrick, T. A., Benjamin, J. E., Fraser, I. C., Nishihira, T. T., Reisig, D. D. & Rominger, E. M. (2003) Sexual segregation in desert bighorn sheep (Ovis canadensis mexicana). *Behaviour*, 140, 183–207.

Mysterud, A., Steinheim, G., Yocooz, N. G., Holand, ǿ. & Stenseth, N. C. (2002) Early onset of reproductive senescence in domestic sheep, *Ovis aries*. *Oikos*, 97, 177–183.

Mysterud, I. & Warren, J. T. (1991) Mortality transmitters – new instrument for animal loss research on Norwegian ranges. *Acta Veterinaria Scandinavica*, 32, 415–424.

Neale, J. C. C., Sacks, B. N., Jaeger, M. M. & McCullough, D. R. (1998) A comparison of bobcat and coyote predation on lambs in north-coastal California. *Journal of Wildlife Management*, 62, 700–706.

O'Callaghan, D., Karsch, F. J., Boland, M. P., Hanrahan, J. P. & Roche, J. F. (1992) Variation in the timing of the reproductive season among breeds of sheep in relation to differences in photoperiodic synchronization of an endogenous system. *Journal of Reproduction and Fertility*, 96, 443–452.

Oehler, M. W., Bowyer, R. T. & Bleich, V. C. (2003) Home ranges of female mountain sheep, *Ovis Canadensis nelsoni*: effects of precipitation in a desert ecosystem. *Mammalia*, 67, 385–401.

Olaechea, F. V., Bellati, J. P., Suarez, M., Pueyo, J. M. & Robles, C. (1981) Mortality among newborn lambs in western Rio Negro province, Argentina. *Revista de Medicina Veterinaria, Argentina*, 62, 128–134.

Oliveira, A. de C. & Barros, S. de. B. (1982) Perinatal mortality of sheep in the district of Uruguaiana, Rio Grande do Sul. *Pesquisa Veterinaria Brasileira*, 2, 1–7.

Ollivier, L., Bodo, I. & Simon, D. L. (1994) Current developments in the conservation of domestic animal diversity in Europe. *Proceedings of the 5th World Congress on Genetics applied to Livestock Production*, 21, 509–516.

Orr, J. B. & Fraser, A. H. H. (1932) Restoring the fertility of Scottish sheep grazings. *Transactions of the Highland and Agricultural Society of Scotland* 5th Series, 44, 64–85.

Payer, D. C. & Coblentz, B. E. (1997) Seasonal variation in California bighorn ram (Ovis canadensis californiana) habitat use and group size. *Northwest Science*, 71, 281–288.

Paterson, I. W. & Coleman, C. D. (1982) Activity patterns of seaweed-eating sheep on North Ronaldsay, Orkney. *Applied animal Ethology*, 8, 137–146.

Penning, P. D., Parsons, A. J., Newman, J. A., Orr, R. J. & Harvey, A. (1993) The effects of group size on grazing time in sheep. *Applied Animal Behaviour Science*, 37, 101–109.

Picard, K., Thomas, D. W., Festa-Bianchet, M., Belleville, F. & Laneville, A. (1999) Differences in the thermal conductance of tropical and temperate bovid horns. *Ecoscience*, 6, 148–158.

Pickup, H. E. & Dwyer, C. M. (2002) Breed differences in the expression of maternal care at parturition persist throughout the lactation period in sheep. *Proceedings of the British Society of Animal Science*, p. 70.

Plant, J. W. (1978) The importance of predation as a cause of perinatal lamb loss in New South Wales. *New South Wales Veterinary Proceedings*, 14, 31–33.

Pollard, J. C., Shaw, K. J. & Littlejohn, R. P. (1999) A note on sheltering behaviour by ewes before and after lambing. *Applied Animal Behaviour Science*, 61, 313–318.

Portier, C., Festa-Bianchet, M., Gaillard, J. M., Jorgenson, J. T. & Yoccoz, N. G. (1998) Effects of density and weather on survival of bighorn sheep lambs (Ovis canadensis). *Journal of Zoology*, 245, 271–278.

Price, E. O. (1984) Behavioural aspects of animal domestication. *The Quarterly Review of Biology*, 59, 1–32.

Price, E. O. (1998) Behavioural genetics and the process of domestication. In: *Genetics and the Behaviour of Domestic Animals* (Ed. T Grandin), pp. 31–65. Academic Press, New York and London.

Price, E. O. (2002) Animal domestication and behavior. CAB International, Cambridge, MA.

Quirke, J. F., Hanrahan, J. P., Loughnane, W. & Triggs, R. (1986) Components of the breeding and non-breeding seasons in sheep: breed effects and repeatability. *Irish Journal of Agricultural Science*, 25, 167–172.

Rachlow, J. L. & Bowyer, R. T. (1994) Variability in maternial behaviour by Dall's sheep: environmental tracking or adaptive strategy? *Journal of Mammalogy* 75, 328–337.

Rachlow, J. L. & Bowyer, R. T. (1998) Habitat selection by Dall's sheep (Ovis dalli): maternal trade-offs. *Journal of Zoology*, 245, 457–465.

Rai, A. K., Singh, M. & More, T. (1979) Cutaneous water loss and respiration rates of various breeds of sheep at high ambient temperatures. *Tropical Animal Health and Production*, 11, 51–56.

Ramos, A. & Tennessen, T. (1992) Effect of previous grazing experience on the grazing behaviour of lambs. *Applied Animal Behaviour Science*, 33, 43–52.

Ratner, S. C. & Boice, R. (1975) Effects of domestication on behaviour. In: *The Behaviour of Domestic Animals* (Ed. E. S. E. Hafez), pp. 3–19. Williams and Wilkins, Baltimore.

Reale, D. & Bousses, P. (1995) Effect of ewe age and high population density on the early nursing behaviour of mouflon. *Ethology, Ecology & Evolution*, 7, 323–334.

Reale, D & Bousses, P. (1999) Effects of summer and winter birth on growth of lambs in a population of feral sheep. *Journal of Mammalogy*, 80, 1028–1037.

Reale, D., Bousses, P. & Chapuis, J. L. (1999) Nursing behaviour and mother-lamb relationships in mouflon under fluctuating population densities. *Behavioural Processes*, 47, 81–94.

Rhind, S. M., McMillen, S. R., Duff, E., Kyle, C. E. & Wright, S. (2001) Effect of long-term feed restriction on seasonal endocrine changes in Soay sheep. *Physiology & Behavior*, 71, 343–351.

Risenhoover, K. L. & Bailey, J. A. (1985) Foraging ecology of mountain sheep: implications for habitat management. *Journal of Wildlife Management*, 49, 797–804.

Robertson, A., Hiraiwa-Hasegawa, M., Albon, S. D. & Clutton-Brock, T. H. (1992) Early growth and sucking behaviour of Soay sheep in a fluctuating population. *Journal of Zoology*, 227, 661–671.

Romeyer, A. & Bouissou, M.-F. (1992) Assessment of fear reactions in domestic sheep, and influence of breed and rearing conditions. *Applied Animal Behaviour Science*, 34, 93–119.

Ross, T. T., Goode, L. & Linnerud, A. C. (1985) Effects of high ambient temperature on respiration rate, rectal temperature, fetal development and thyroid gland activity in tropical and temperate breeds of sheep. *Theriogenology*, 24, 259–269.

Russel, A. J. F., Gunn, R. G. & Doney, J. M. (1968) Components of weight loss in pregnant hill ewes during winter. *Animal Production*, 10, 43–51.

Rutberg, A. T. (1987) Adaptive hypotheses of birth synchrony in ruminants: an interspecific test. *American Naturalist*, 130, 692–710.

Ryder, M. L. (1973) The structure of, and growth cycles in, the coat of wild Mouflon sheep (Ovis musimon) and their crosses. *Research in Veterinary Science*, 15, 186–196.
Ryder, M. L. (1983) *Sheep and Man*. Duckworth, London.
Ryder, M. L. (1984) Sheep. In: *Evolution of Domesticated Animals* (Ed. I. L. Mason), Longman, London and New York.
Ryder, M. L. (1991) Domestication, history and breed evolution. In: *Genetic Resources of Pig, Sheep and Goat* (Ed. K. Majala), pp. 157–177. World Animal Series B8, Elsevier Science Publishers, Amsterdam.
Samson, D. E. & Slee, J. (1981) Factors affecting resistance to induced cooling in newborn lambs of 10 breeds. *Animal Production*, 33, 59–65.
Schaller, G. B. (1977) *Mountain Monarchs: Wild Sheep and Goats of the Himalaya*. University of Chicago Press, Chicago and London.
Scott, D. & Sutherland, B. L. (1981) Grazing behaviour of merinos on an undeveloped semi-arid tussock grassland block. *New Zealand Journal of Experimental Agriculture*, 9, 1–9.
Seip, D. R. & Bunnell, F. L. (1985) Foraging behaviour and food habits of Stone's sheep. *Canadian Journal of Zoology*, 63, 1638–1646.
Sevi, A., Taibi, L., Albenzio, M., Muscio, A., Dell'Aquila, S. & Napolitano, F. (2001) Behavioural, adrenal, immune and productive responses of lactating ewes to regrouping and relocation. *Journal of Animal Science*, 79, 1457–1465.
Shafie, M. M. & Abdelghany, F. M. (1978) Structure of the respiratory system of sheep as related to heat tolerance. *Acta Anatomica*, 100, 441–460.
Shannon, N. H., Hudson, R. J., Brink, V. C. & Kitts, W. D. (1975) Determinants of spatial distribution of Rocky Mountain Bighorn sheep. *Journal of Wildlife Management*, 39, 387–401.
Sharafeldin, M. A. & Shafie, S. S. (1965) Animal behaviour in the sub-tropics. II. Grazing behaviour of sheep. *Netherlands Journal of Agricultural Science*, 13, 239–247.
Sherwin, C. M. & Johnson, K. G. (1987) The influence of social factors on the use of shade by sheep. *Applied Animal Behaviour Science*, 18, 143–155.
Shillito-Walser, E. & Hague, P. (1981) Field observations on a flock of ewes and lambs made up of Clun Forest, Dalesbred and Jacob sheep. *Applied Animal Ethology*, 7, 175–178.
Sibbald, A. R., Elston, D. A. & Iason, G. R. (1996) Spatial analysis of sheep distribution below trees at wide spacing: a reappraisal. *Agroforestry Forum*, 7, 26–28.
Sibbald, A. M. & Hooper, R. J. (2003) Trade-offs between social behaviour and foraging by sheep in heterogeneous pastures. *Behavioural Processes*, 61, 1–12.
Sibbald, A. M., Shellard, L. J. F. & Smart, T. S. (2000) Effects of space allowance on the grazing behaviour and spacing of sheep. *Applied Animal Behaviour Science*, 70, 49–62.
Silanikove, N. (1992) Effects of water scarcity and hot environment on appetite and digestion in ruminants – a review. *Livestock Production Science*, 30, 175–194.
Silanikove, N. (2000) Effects of heat stress on the welfare of extensively managed domestic ruminants. *Livestock Production Science*, 67, 1–18.
Slee, J. (1968) Body temperature and vasomotor responses in Scottish Blackface and Tasmanian Merino sheep subjected to slow cooling. *Animal Production*, 10, 265–282.
Slee, J., Alexander, G., Bradley, L. R., Jackson, N. & Stevens, D. (1991) Genetic aspects of cold resistance and related characters in newborn Merino lambs. *Australian Journal of Experimental Agriculture*, 31, 175–182.
Slee, J. & Stott, A. W. (1986) Genetic selection for cold resistance in Scottish Blackface lambs. *Animal Production*, 43, 397–404.
Squires, V. R. (1974) Grazing distribution and activity of Merino sheep on a saltbrush community in south-east Australia. *Applied Animal Ethology*, 1, 17–30.
Squires, V. R. (1975) Ecology and behaviour of domestic sheep (*Ovis aries*): a review. *Mammal Review*, 5, 35–57.
Squires, V. R. & Wilson, A. D. (1971) Distance between food and water supply and its effect on drinking frequency and food and water intake in Merino and Border Leicester sheep. *Australian Journal of Agricultural Research*, 22, 283–290.

Stahl, P., Vandel, J. M., Herrenschmidt, V. & Migot, P. (2001) Predation on livestock by an expanding reintroduced lynx population: long-term trend and spatial variability. *Journal of Applied Ecology,* 38, 674–687.
Stolba, A., Hinch, G. N., Lynch, J. J., Adams, D. B., Munro, R. K. & Davies, H. I. (1990) Social organisation of Merino sheep of different ages, sex and family structure. *Applied Animal Behaviour Science,* 27, 337–349.
Stott, A. W. & Slee, J. (1985) The effect of environmental temperature during pregnancy on thermoregulation in the newborn lamb. *Animal Production*, 41, 341–347.
Taylor, R. G., Workman, J. P. & Browns, J. E. (1979) The economics of sheep predation in southwest Utah. *Journal of Range Management*, 32, 317–321.
Terrill, C. E. & Slee, J. (1991) Breed differences in adaptation of sheep. In: *Genetic Resources of Pig, Sheep and Goat* (Ed. K. Majala), pp. 195–233. World Animal Series B8, Elsevier Science Publishers, Amsterdam.
Tilton, M. E. & Willard, E. E. (1982) Winter habitat selection by mountain sheep. *Journal of Wildlife Management,* 46, 359–366.
Thomson, A. M. & Thomson, W. (1949) Lambing in relation to the diet of the pregnant ewe. *British Journal of Nutrition,* 2, 290–305.
Tobler, I., Jaggi, K., Arendt, J. & Ravault, J. P. (1991) Long-term 24-hour rest-activity pattern of sheep in stalls and in the field. *Experientia,* 47, 744–749.
Torres-Hernandez, G. & Hohenboken, W. (1979) An attempt to assess traits of emotionality in crossbred ewes. *Applied Animal Ethology,* 5, 71–83.
Tribe, D. E. & Tribe, E. M. (1949) North Ronaldsay sheep. *Scottish Agriculture*, 24, 1–4.
Turner, J. C. (1979) Osmotic fragility of desert bighorn sheep red blood cells. *Comparative Biochemistry and Physiology,* 64, 167–175.
Walker, A. B. D., Parker, K. L. & Gillingham, M. P. (2006) Behaviour, habitat associations and intrasexual differences of female Stone's sheep. *Canadian Journal of Zoology*, 84, 1187–1201.
Warren, J. T. & Mysterud, I. (1995) Mortality of domestic sheep in free-ranging flocks in southeastern Norway. *Journal of Animal Science*, 73, 1012–1018.
Warren, J. T., Mysterud, I. & Lynnebakken, T. (2001) Mortality of lambs in free-ranging domestic sheep (Ovis aries) in northern Norway. *Journal of Zoology*, 254, 195–202.
Waterhouse, A. (1996) Animal welfare and sustainability of production under extensive conditions – A European perspective. *Applied Animal Behaviour Science*, 49, 29–40.
Webster, A. J. F. (1983) Environmental stress and the physiology, performance and health of ruminants. *Journal of Animal Science*, 57, 1584–1593.
Webster, A. J. F., Hicks, A. M. & Hays, F. L. (1969) Cold climate and cold temperature-induced changes in heat production and thermal insulation of sheep. *Canadian Journal of Physiology,* 47, 553–562.
White, P. C. L., Groves, H. L., Savery, J. R., Conington, J. & Hutchings, M. R. (2000) Fox predation as a cause of lamb mortality on hill farms. *Veterinary Record*, 147, 33–37.
Wiener, G., Field, A. C. & Smith, C. (1977) Deaths from copper toxicity of sheep at pasture and the use of fresh seaweed. *Veterinary Record*, 101, 424–425.
Winfield, C. G. & Mullaney, P. D. (1973) A note on the social behaviour of a flock of Merino and Wiltshire Horn sheep. *Animal Production,* 17, 93–95.
Wolff, J. O. & van Horn, T. (2003) Vigilance and foraging patterns of American elk during the rut in habitats with and without predators. *Canadian Journal of Zoology,* 81, 266–271.
Woolf, A., O'Shea, T. & Gilbert, D. L. (1970) Movements and behavior of Bighorn sheep on summer ranges in Yellowstone National Park. *Journal of Wildlife Management,* 34, 446–450.
Xu, Z. Z., McDonald, M. F., McCutcheon, S. N. & Blair, H. T. (1991) Seasonal variations in testis size, gonadotrophin secretion and pituitary responsiveness to GnRH in rams of two breeds differing in the time of onset of the breeding season. *Animal Reproduction Science,* 26, 281–292.
Zohary, D., Tchernov, E. & Horwitz, L. K. (1998) The role of unconscious selection in the domestication of sheep and goats. *Journal of Zoology,* 245, 129–135.

Chapter 3
Behaviour and the Welfare of the Sheep

R. Nowak, R.H. Porter, D. Blache, and C.M. Dwyer

Abstract The most important features of the behaviour of sheep are their marked sociality and the bond formation between mother and young. Sheep show a strong need to stay with their group (or subgroup for some breeds), and become very vocal and agitated when separated from their flock mates. Social life requires rules that maintain the stability of the group and increase the fitness of each individual. Under wild or feral conditions, sheep populations contain a wide range of individuals including sexually mature females, juvenile males and females, and lambs, and their composition fluctuates over time. Adult males usually join a flock of females during the mating season. Under domestic conditions humans mainly control the social environment, and animals are usually maintained in single-sex groups of similar age or size, the main exceptions being the mother-young dyad, and male-female groups at mating. Social dominance is not as obvious as in other ruminants, unless the animals are confined and have to compete for resources. Sheep can nonetheless recognise other individuals of the group using various sensory modalities and this necessarily plays a role in group cohesion and bonding. From an animal welfare point of view, the important aspects of sheep behaviour are those related to social stability, abnormal forms of behaviour, and survival of the young. In general, mixing groups of sheep does not lead to increased agonistic behaviour between individuals and therefore social instability has never been a major concern. Separating mother and young is common practice even at an early age, especially in dairy breeds. Despite the existence of a strong affectional bond between the ewe and her lamb, a clear demonstration of behavioural deficits in lambs reared with peers, but without their dam, has never been reported. Early experience can, however, affect later sexual behaviour. Stereotyped behaviours, which are commonly used as an index of poor welfare, are rare in sheep compared to other species. Behavioural research has enormously advanced our understanding of the requirements of ewes and lambs at parturition, and the nature of the social bond that forms at birth. This has provided advances towards practical solutions to reduce lamb mortality. The issues associated with intensive farming relate to confinement and social restriction in pens, but there

R. Nowak
Equipe Comportement, Neurobiologie, Adaptation, Unité de Physiologie de la Reproduction et des Comportements, INRA, Nouzilly, France

C.M. Dwyer (ed.), *The Welfare of Sheep*,

is a need for further investigation. Animal-human relationships may also be important, especially for intensively farmed sheep that have more contact with humans than do sheep under extensive conditions.

Keywords Social behaviour · Recognition · Sexual behaviour · Maternal · Neonate · Abnormal behaviour

3.1 Introduction

Archaeological evidence shows that sheep were domesticated some 8–10000 years ago, and are thus one of the first animal species to have been tamed and used by mankind. Since that time, selection has resulted in more than 2000 breeds, which display greater morphological diversity than their wild ancestors. What made them so attractive for domestication? Sheep are multi-purpose animals; they have been traditionally used for their wool, pelts, meat, and milk. If more specialized breeds have been selected over time, approximately 25% of the current breeds have kept their multi-purpose nature (Mason 1969). From a behavioural point of view, sheep live in large social groups, they are herbivorous and can adapt to a wide range of environments. Adult males and females do not associate except at breeding time when they display promiscuous mating behaviour, and the ewe gives birth to a well-developed neonate that bonds to its mother at an early age and can easily socialize with humans. And finally they are particularly docile. These behavioural traits, along with their multi-purpose nature are believed to be the major reasons for the success of sheep domestication.

Surprisingly, unlike that of other species, sheep husbandry has changed very little over time. Most animals are still reared under extensive conditions throughout the world, whether they live in large unattended flocks, such as the Merino sheep in Australia, or as family units like Djallonké sheep in Ghana. Intensive production has mainly affected European dairy breeds and, to a lesser extent, meat breeds with the rearing of fat lambs. Even under these conditions animals are not submitted to dramatic changes of their housing conditions: adults live in social groups, and when penned, are kept on straw bedded floors. When separated from their mother at birth and reared with artificial milk, lambs can feed in a semi-natural manner by sucking the milk from a rubber teat. Overall, sheep husbandry contrasts markedly with the treatment of calves, pigs or poultry, under intensive production systems, and may partly explain why sheep welfare has never been considered a major issue. Other factors that may contribute to the lack of concern for sheep welfare are discussed in Chapter 1. In the book entitled 'Farm Animal Behaviour and Welfare' (Fraser & Broom 1997), only cattle, pigs and poultry have specific chapters on welfare problems. Abnormal sheep behaviour is only reported in other more general sections of the book.

This chapter addresses the behavioural responses of sheep and the relationship between behaviour and welfare. As flocking and social interactions are the dominant behaviour patterns of sheep these are given particular attention. The behaviours associated with reproduction, parturition and rearing of the young are discussed, with particular reference to the potential disruption of these behaviours by modern husbandry practices. As stated above, welfare issues (abnormal behaviours and behavioural needs) are less frequently described in sheep than in many other species, therefore, the relevant available information is reviewed in specific sections.

3.2 Social Organisation and Behaviour

3.2.1 Ontogeny of Social Organisation

Young domestic lambs pass from a neonatal phase where they mainly interact with their dams, to a phase where they spend more time associating with peers. Mothers and lambs remain in close proximity during the first week following birth, after which lambs aggregate in peer groups. When ewes give birth to two or more young, family groups develop in which a social bond arises between siblings. Twins usually stay together when grazing or resting even if the dam remains their preferred social partner (Shillito-Walser et al. 1981b, 1983). Preferential associations between familiar unrelated young animals may also be established but this occurs later. Such associations develop more rapidly if lambs are separated from their dam at birth. The formation of subgroups of lambs, in conjunction with the gradual decline of milk yield as lactation progresses, breaks the dominant social bond between mother and young, and a new social organisation arises. Thus, a hierarchy in social bonding develops over time in sheep: the first and strongest bond is between the dam and her young, the second association is between siblings, and then between unrelated peers. Under free-ranging conditions, young males disperse to join bachelor groups while daughters remain in their natal home-range group. In the feral Soay sheep of St Kilda in Scotland, groups consist of either all males or females with their offspring in extended families. Female offspring cease to associate with their mother a few weeks before the mother gives birth to her next lamb. The presence of a newborn lamb is responsible for the decline of the bond between ewes and their yearling daughters. Groups of males and females have overlapping home ranges but it is only during rutting that the males move to the female home range (Grubb & Jewell 1966). This pattern is very similar to that of wild Bighorn sheep (Geist 1971).

3.2.2 Social Behaviour

3.2.2.1 Spatial Relationships

Flocks of sheep maintain characteristic spatial relationships and individuals tend to remain at fixed distances from others. Spacing can be characterised by two measures: individual distance, which is the minimum distance observed between two individuals, and social distance, which is the maximum distance of dispersal. The former measures the degree of inter-individual tolerance while the latter is an index of the cohesion of the flock. The balance between individual and social distances determines the structure of the group. Spacing between individuals varies considerably across breeds. Sheep of mountain breeds usually tolerate a greater individual distance than sheep of lowland breeds. Thus, a distance between nearest neighbours of 6.9 m was observed in Welsh Mountain sheep and 7.5 m in Blackface, compared to 3.4 m in Suffolk and usually less than 1.5 m in Merinos (Lynch et al. 1992). Furthermore, the maximum distance of dispersal is also shorter in Merinos, giving their flock a more compact appearance. Subgroups consisting of family units or peers are sometimes observed for example, in the Dorset Horn. Subgroups of animals reared together may retain their group identity when mixed into a larger group. Mixing eventually occurs when animals are from the same breed, but when flocks are made up of groups of different breeds of sheep, breed segregation remains for a long time. Arnold & Pahl (1974) report that even after two years of grazing together, breed discrimination persists while animals are feeding or resting, and this discrimination is passed on to the offspring born into a flock of different breeds of ewes. There is some evidence of a gradual breakdown of breed links over time but even if this occurs the animals tend to associate with breeds that have a similar social structure (Winfield & Mullaney 1973).

The segregated social groupings observed among sheep/lambs could arise from various underlying mechanisms. For example, the separation of ewes of different breeds into same-breed flocks might reflect adaptations to different environments and resources (ecological segregation; Bon & Campan 1996). Thus, in free-ranging conditions there is often little overlap between flocks of lowland and hill breeds that prefer and exploit different habitats (Dwyer & Lawrence 1999b, 2000a). Ecological variables may likewise be implicated in the segregation between adult male and female sheep (Bon & Campan 1996). The nutritional requirements of reproductively active ewes presumably differ to some extent from those of adult males, and mothers with sucking young may seek habitats where potential exposure to predators is reduced (see Chapter 2). Behavioural incompatibility between males and females (e.g. differences in activity patterns, heightened sexual and agonistic behaviour by males) could further contribute to sexual segregation (Bon & Campan 1996).

3.2.2.2 Dominance and Leadership

In ruminants, demonstration of social status relies mainly on visual displays. Threats and aggression are common forms of agonistic behaviour that play a role in the social dynamics of wild mountain sheep. Horn size is also a major determinant of social status in this species. Dominance in domestic sheep is not as obvious as in their wild counterparts. The lack of hierarchy may be explained by the fact that it is expressed in situations of conflict when there is competition between individuals, such as access to food or to females. Domestic sheep are typically reared in single-sex flocks, or in groups of animals of similar ages, therefore competition for resources is rare. Most studies reporting social dominance relationships involved high stocking rates. For instance, the expression of sexual behaviour is influenced by the hierarchical position of the ram, but only under high density and in confined areas (Lindsay et al. 1976). Under extensive conditions, there is no evidence that dominant rams suppress the mating performance of subordinate rams. Lynch et al. (1985) reported a high level of aggression in Scottish Blackface ewes, particularly when the animals were resting in shelter, which may reflect competition for limited access to these areas. Weight is positively correlated with social rank (Dove et al. 1974), and when competing for supplementary food, juvenile females (1 year old) and old (7 years) ewes are less competitive than ewes of intermediate ages (Arnold & Maller 1974). Outside of competitive situations, dominance is not important in the organization of similar-age ewe flocks. The fact that agonistic behaviours have never been reported in studies mixing flocks from different origin supports this view.

Leadership is an additional major component of the social behaviour of sheep. It is expressed by animals that initiate movements of the group. However, a relationship between leadership and dominance has never been demonstrated in sheep, it is generally a more independent invidual that initiates movement (Arnold 1985) and no consistent movement order in a flock of sheep have been seen. The fact that older animals, which may be less concerned with maintaining close contact with the flock, may initiate movement within a flock suggests that the function of this following behaviour is to maintain familiarity with the environment.

3.2.3 Signals Used in Social Communication

In many instances, preferential social interactions appear to be mediated by discrimination between the phenotypic traits of individuals or higher order social categories, such as breed, sub-group or kin (also see Chapter 4). Such selective behavioural responsiveness to particular individuals or social classes is commonly cited as an operational definition of social recognition. That is, 'recognition' per se refers to unobservable neural processes whose existence can be inferred from social interactions. Experimental studies of the mechanisms and sensory processes involved in the discrimination between individual conspecifics (lambs and ewes) will be discussed further in the following sections. Outside the mother-lamb sphere, recognition of individuals has rarely been investigated. The number of individuals that can be identified by an adult animal is still unknown (but see also Chapter 4) and it is argued that the main force for sub-group cohesion is group identity rather than individual recognition.

3.2.3.1 Visual Signals

Sheep are predominantly visual animals; the span of their visual field is 270–280°, which allows individuals to maintain spatial relationships with animals not only in front but also behind them. Cohesion of the group, or subgroups, is therefore maintained by each animal adjusting its position and its behaviour relative to other members. Visual signals include movements such as pawing, stamping, or fleeing, as well as static body postures, and may involve only part of the body, usually the head. For instance, individuals that have sudden visual contact with a source of potential danger (a stockman, dog or predator) may signal alarm by adopting a rigid posture and remaining silent. The state of alertness spreads rapidly over the entire flock. Sheep maintain visual contact with the source of danger unless they turn to flee. The tendency of sheep to follow a leader is also controlled by visual signals; they are especially likely to follow individuals that move away from them. The view of the rump or a film showing sheep moving across the screen encourage sheep to move in the same direction (Franklin & Hutson 1982). In many societies, an animal trained to follow a stockman is used as a means of controlling the direction and movement of sheep, and even as a means of leading flocks to slaughter. Sheep also recognise individuals on the basis of visual cues from the face, and such recognition may last for at least 2 years (Kendrick et al. 2001a). Ewes respond differentially to projected images of conspecifics of their own versus a different breed, suggesting that they recognize visual characteristics of their breed (Bouissou et al. 1996).

3.2.3.2 Auditory Signals

Social interactions by mammalian and avian species commonly involve vocal signals, however, in sheep they are almost entirely confined to mother-young interactions and, to a lesser extent, the behaviour of rutting rams. Vocal signals include low-pitched bleats (rumbling sounds) made by the ewe and her newborn when at close range, but also by rams when courting a ewe. High-pitched bleats (loud calls) are considered to be contact or distress calls. They are emitted when mother and young are separated or when an animal is isolated from its social group. Under normal circumstances, flocks of rams, non-maternal ewes and weaned lambs are silent. This explains why vocal recognition of individuals has been uniquely investigated in the context of mother-young bonding, while there is still no evidence of vocal recognition between adult sheep. Breeds of sheep, like individuals, also differ in the characteristics of their bleats. Sonographic analyses show that several parameters of high-pitched bleats differ between Clun Forest, Jacob, Dalesbred and Border Leicester sheep (Shillito-Walser & Hague 1980), but again breed recognition by auditory signals has only been demonstrated between mothers and lambs. Ewes are more responsive to the bleats of lambs of their own breed than to those of a different breed (Shillito-Walser et al. 1982).

3.2.3.3 Olfactory Signals

The role of olfactory signals in individual identification of the lamb by its mother, and in the recognition of ewes' sexual state by rams, has been clearly demonstrated. Sheep can also discriminate between odours of conspecifics in an operant conditioning task (Baldwin & Meese 1977). Chemical signals can be transmitted via the secretion of various glands, urine, faeces and wool, yet their involvement in social structure and organisation is still unclear. There is evidence of fence-post marking by matriarchal Merino ewes near feeding areas (Stolba et al. 1990), a behaviour that may be dependent on the heterogeneity of the environment. According to Arnold (1985) group recognition may also be based on odours. He showed that a group of anosmic sheep, unlike intact animals, did not develop group affinity when mixed with other groups of sheep. The odour of the flock is a combination of the individual odours of each sheep and those of the environment (soil and vegetation) on the fleece and skin. A flock of sheep living in a specific environment will have its own olfactory signature and this may contribute to group cohesion.

3.2.4 Welfare Issues Associated with Social Behaviour

Little attention is generally given by farmers to the social needs of sheep, for example the optimal size and composition of the group, or the development of social relationships between individuals. Under farming conditions, the normal development of social organization is disrupted by management practices involving removal of lambs from the ewes before the time of natural weaning, and keeping sheep in groups of uniform age and sex. These flocks are often shifted from yard to yard when reared under intensive conditions, or from pasture to pasture under extensive conditions, so that the establishment of normal social structures may not occur.

3.2.4.1 Social Isolation

Sheep are highly social animals and separation from the flock is known to be particularly stressful (Kilgour & de Langen 1970). Rushen (1986) demonstrated that sheep find social isolation to be more aversive than capture and restraint within a group of sheep. Social isolation causes sustained elevations in plasma cortisol and heart rate (Cockram et al. 1994) and reductions in circulating lymphocytes (Minton et al. 1992). Further, isolation has a greater stimulatory effect on cortisol release than handling or restraint (Parrot 1990). Sheep show an initial increase in behavioural and vocal activity when socially isolated, followed by an increase in lying and behavioural withdrawal, which may be accompanied by reduction in food and water intake, if isolation is prolonged.

Domestic sheep are raised in groups of various sizes but there is no evidence that social stability and the welfare of individuals depends on an optimal flock size. Some authors recommend that the minimal group size is four or five animals (Lynch et al. 1992), but this cannot be generalized to all breeds of sheep nor to all types of environments (see Chapter 2). A state of "frustration" can be experienced when the need for social companions is not met. As a consequence, sheep should not be isolated unless it is an absolute necessity, and for only brief periods of time. In such cases, allowing flock mates to see one another may be beneficial. However, sheep may be kept in isolation when in quarantine or in hospital pens, males may be kept isolated outside the breeding season, and hobby farmers may keep only one or two sheep. In addition, sheep may be kept in metabolism crates or individual pens during experimental procedures. Thus, there are occasions when sheep are likely to experience poor welfare due to a lack of social companions.

3.2.4.2 Stocking Density and Crowding

At the other extreme is crowding which implies an excessive density of animals in a restricted space. Crowding can be found in intensive as well as extensive systems when there is competition for limited resources (water, food, shade), but it is difficult to evaluate. Increased competition may result in some animals being denied access to the resource(s) but crowding does not necessarily lead to heightened agonistic behaviour. In one trial, the percentage of sheep that failed to eat increased from 0% to 31% when feeding-trough space was reduced from 24 to 4 cm per animal (Lynch et al. 1992). Under intensive management a maximum of three adult ewes per metre of feeding-trough is usually recommended to ensure that all the animals will have access to food. This should be adjusted according to the physiological state of the sheep (pregnant or not) and to the size of the breed. However, when sheep are transported by ship this requirement may be considerably less at 12 sheep per metre of trough space. Various space allowances have also been prescribed as meeting the need of sheep in pens, which are usually designed to allow animals to lie down, stand up and circle without restriction. In France and the UK minimum recommended standards for housed sheep are that ewes are allowed 1–1.4 m^2 per head according to their size and physiological status, and lambs between 0.25 and 0.90 m^2 according to their size. It is recommended that rams are given more space (0.5–2 m^2). However, when sheep are transported on sea voyages, where they may spend 3 weeks on board ship, only 0.25–0.3 m^2 are legally required per animal. It is not clear whether these figures were chosen on objective grounds and truly satisfy the needs of various types of sheep, particularly considering the relatively few studies that have addressed this issue. Apparently this space allowance does not raise major practical problems for farmers in terms of production, although behavioural deficits have been reported when animals are kept indoors (Fraser 1983). When lying space is reduced from 1.0 to 0.5 m^2, total lying time and lying synchrony of sheep are decreased, and the frequency of displacement or disturbance of resting ewes increases (Bøe et al. 2006). This suggests that the ability of sheep on board ship to rest properly may well be compromised. In addition, low ranking animals are more frequently displaced and spend considerably less time lying, suggesting that the welfare of these animals may be compromised at higher stocking densities.

Considering the wide variation in inter-individual distances and flocking behaviour amongst breeds, standardized recommendations may not be possible or appropriate. More space should be provided if the behaviour of a group of sheep indicates that it is warranted, and mixing animals that differ markedly in size, or breeds that have marked differences in behaviour, should be avoided. As sheep cannot escape from their pen to seek shelter or avoid unpleasant environmental conditions, care must be taken not to expose them to excessive heat, cold, noise or dampness. Good ventilation is crucial as confinement may result in excess concentrations of ammonia and microbe populations, and high humidity rates. Deficient ventilation will increase the risk of spreading diseases.

3.2.4.3 Social Dominance and Aggression

Although social contacts are extremely important to sheep the presence of other sheep, or of particular individuals, can also act as a source of stress. Agonistic encounters and the increased importance of a social hierarchy may appear when sheep are crowded together and resources are limited (McBride et al. 1967). An increase in aggression is associated with sudden environmental change, lack of space, a large social group size and when food or feeder space may be restricted (Arnold & Maller 1974; Kiley-Worthington 1977; Done-Currie et al. 1984). However, social mixing and stocking density did not appear to affect aggressive interactions in prepubertal lambs (Ruiz-de-la-Torre & Manteca 1999a). Aggressive behaviours are more common in older sheep (Stolba et al. 1990), although younger sheep receive more aggression (Guilhem et al. 2000) and the frequency of agonistic interactions is greater in single age and sex groups than in mixed age groups (Stolba et al. 1990). Aggressive behaviour in male lambs after social mixing also increases with testosterone concentration (Ruiz-de-la-Torre & Manteca 1999b). Thus, aggressive behaviour may indicate welfare problems in certain groups of sheep under conditions of limited resources or high stocking density.

Many expressions of dominance in sheep are not necessarily associated with overtly aggressive behaviour. Sheep maintain social hierarchy through more subtle behaviours associated with head movement and eye contact. Subordinate sheep move away and do not retaliate when attacked, suggesting that, in stable groups, hierarchies are well defined and maintained without the need for agonistic encounters (Guilhem et al. 2000). Dominant sheep may displace subordinates from the feed troughs and from preferred lying positions by resting their chins on the backs of subordinate sheep, or by pawing (Done-Currie et al. 1984). There are a number of consequences for the subordinate sheep. When feeder space is limited the number of displacements or disturbances from the trough increases (Arnold & Maller 1974) and a progressively greater proportion of sheep cease to compete for food becoming non-feeders. These subordinate animals, which are usually the very young or older sheep (McBride et al. 1967), are likely to have a lower feed intake, are often at the tail of movement order and may eat the poorer quality or contaminated forage leading to higher worm burdens (Lynch & Alexander 1973). Subordinate sheep may also be displaced from shelter and shade during conditions of thermal extremes if space is limited (Sherwin & Johnson 1987; Deag 1996). They may, therefore, be chronically stressed, particularly when resources are limited and competition is great. Subordinates also display heightened behavioural and plasma cortisol responses to additional acute stressors (Kilgour & de Langen 1970) which can lead to total behavioural inhibition and learned helplessness.

Subordination may also have an impact on other behaviours. Subordinate rams mate the less preferred ewes, and in large groups they mate with fewer partners compared to dominant rams (Tilbrook et al. 1987). Although there are no differences in conception and lambing rates between dominant and subordinate Bighorn sheep (Hass 1991) the dominant animals spend more time than subordinates suckling their lambs (which may reflect better nutrition) and are more likely to adopt alien lambs. In domestic sheep lamb stealing and bullying around lambing time seems to be related to social dominance, with dominant animals stealing lambs from subordinates.

3.3 Reproductive Behaviour

The patterns of both ram and ewe sexual behaviour are similar in all breeds of sheep and have been extensively described for both wild and domesticated animals. In wild or feral sheep, the sexes mix only during the breeding season, do not form pairs and both are promiscuous (Main et al. 1996). Because rams are sexually active throughout the breeding season and in some breeds, such as the Merino, throughout the year, the sexual activity of ewes is the limiting factor and leads to the classification of the sheep as a seasonally polyoestrous species.

3.3.1 Hormonal Control of Gonadal Activity and Sexual Behaviour in Rams and Ewes

In sheep, gonadal activity in both sexes is controlled by a complex series of reciprocal hormonal messages between the central nervous system, the pituitary gland and the gonads (Fig. 3.1). Gonadotrophin-Releasing Hormone (GnRH), a neurohormone produced in the brain, controls the synthesis and release of the gonadotrophins, luteinising hormone (LH) and follicle-stimulating hormone (FSH), by the anterior pituitary gland. The gonadotrophins act on the gonads (ovary or testis) to control the maturation and release of the gametes (oocytes or spermatozoa). In turn, the gonadal steroids (progesterone plus oestradiol or testosterone) control gonadotrophin secretion by stimulating or inhibiting the activity of the hypothalamic-pituitary axis. The sex steroids are also responsible for the production of secondary sex characters and pheromones.

Activity of the hypothalamic-pituitary-gonadal axis and sensitivity of the brain to gonadal steroids are influenced by external factors such as photoperiod, nutrition, stress and social contact, and interactions between these factors lead to periods of sexual activity (breeding season) and quiescence (non-breeding season) (Fig. 3.1 and see below).

In male sheep, gamete production varies only slightly during the breeding season. By contrast, gamete production in ewes varies temporally, not only with the season, but also with the time of the cycle. The ewe is only fertile for a brief period around the time of ovulation, once each oestrous cycle – perhaps for only 2 days, at 17-day intervals. The normal oestrous cycle of the ewe is remarkably constant in length (17 $\pm$ 1 days) and is divided into two major phases:

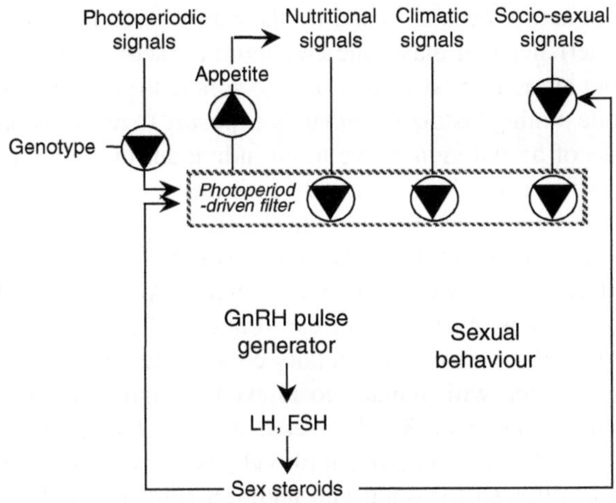

Fig. 3.1 Schematic representation of the relationships between photoperiodic, climatic, nutritional and social cues and their interaction with genotype and steroid feedback in the control of the hypothalamo-pituitary-gonadal axis and sexual behaviour in sheep. This model is adapted from the working model used to study the relationship between nutrition and reproduction in male sheep (Blache et al. 2003)

(i) The progestational or luteal phase is the longest phase of the cycle, occupying a period of ~14 days from ovulation until luteolysis. Secretions of the corpus luteum, particularly progesterone, dominate the endocrine and behavioural events during this part of the cycle.

(ii) The follicular phase, which can be conveniently sub-divided into early and late phases. In the early follicular phase, ovarian follicles enter their final stages of maturation; the late follicular phase begins with the onset of oestrous behaviour and ends with ovulation and the termination of oestrus. Selected follicles (Graafian follicles) ovulate and produce the large amounts of oestradiol that play a critical role in the coordinated and synchronised induction of ovulation and oestrus (Scaramuzzi et al. 1993).

3.3.2 Mating Behaviour

Sexual behaviour of sheep is best described by referring to the behaviour of ewes because they must be sexually active for the full spectrum of mating behaviour to be expressed by both partners. Ewes periodically exhibit specific sexual behaviour that Beach (1976) has dissected into three components: (i) physical changes to attract the attention of the ram (attractiveness); (ii) active search for, and attraction towards the ram (proceptivity); and (iii) acceptance of mating attempts (receptivity).

Attractiveness, as a change in physical appearance, is not very explicit in the ewe, at least to a human observer. However, rams find ewes more attractive when they are receptive and when they are not shorn (Tilbrook 1989). This suggests that both visual and olfactory cues are used by the ram to select ewes in a flock. The individual characteristics that make one ewe more attractive than another have not been identified but attractiveness is unrelated to solicitation by the ram, level of oestrogen, immediate mating history or previous exposure to rams (Tilbrook 1987a,b). The attractiveness of an individual ewe to an individual ram is relatively constant between sexual cycles and within a period of receptivity, suggesting that rams use specific criteria to choose a partner. Are these criteria genetically programmed or acquired during the animal's lifetime? Genetics seem to be important because rams show sexual preference for ewes of their own breed (Arnold & Dudzinski 1978). However, early contact with the mother can dramatically influence sexual preferences. In fact, a ram that was experimentally cross-fostered at birth onto a female goat preferentially mated with female goats even though it had been raised in a mixed group of sheep and goats (Kendrick et al. 2001b). The senses involved in this learning process are not known but, in adulthood, vision is the most important sense used by the rams, followed by smell and audition (Fletcher & Lindsay 1968). It seems likely that those same senses would also be critical during the early learning phase.

In ewes, vision is also the main sense used in the active search for the ram (proceptivity). Outside the time of ovulation ewes avoid rams but, beginning a few hours before ovulation, they prefer to spend time near rams rather than familiar ewes (Fabre-Nys & Venier 1989). Proceptivity may be so strong in some ewes that they initiate sexual encounters (Banks 1964). Inexperienced ewes can be more fearful of rams than are experienced ewes (Gelez et al. 2003). In large flocks, or when ewes are artificially synchronised, proceptivity is expressed by many ewes at the same time and this leads to the formation of harems in which females usually share the dominant ram (Mattner et al. 1971). Ewes may also compete for access to rams (Tomkins & Bryant 1972). Senses other than vision might be involved in the search for the ram and selection of a mating partner by the ewe. Olfaction is an obvious candidate because of its importance in maternal behaviour and stimulation of the reproductive axis of ewes by the smell of rams (see below).

For a limited period of time around ovulation, the ewe is receptive (oestrous) and accepts mounting. Oestrus lasts ~1.8 days, although the actual duration depends on the breed: 47 h in Prealpes du Sud; 38 h in Ile de France, 29.2 h in Merino; 35.5 h in Awassi (Joubert 1962; Banks 1964; Schindler & Amir 1972; Fabre-Nys & Venier 1989). However, the accuracy with which the duration of sexual activity can be measured is critically dependent on the type of behavioural test that is used. Quantitative behavioural tests that allow measurement of both proceptivity and receptivity in ewes (Fabre-Nys & Venier 1989) have shown that oestrous behaviour begins abruptly (the ewe becomes fully proceptive and receptive within eight hours), remains intense for a variable period and then terminates slowly. Around the time of ovulation, ewes display a well-defined response to the courtship of rams (Fabre-Nys et al. 1993). There is no ritualised sequence of events leading to mating, but the repertoire of behaviours expressed by both rams and ewes is quite standardised and similar between different breeds. The following example is typical of the sequence of behaviours leading to mating, although the chain of events may differ slightly between rams and mating attempts: the ram approaches the proceptive ewe who remains stationary. Alternatively, the ram and the ewe walk together in circles, the ewe following the ram and the ram trying to place himself behind her. Before any close contact with the ram, the ewe raises her tail and fans it a few times. For clarity, we describe three different behavioural patterns used by rams to approach a ewe – the ram can start with any of them and may express only one, or any combination of these behaviours:

1. The ram noses or sniffs the perineal region (tail and vulva) of the ewe, the ewe stands still, sometimes turning her head towards the ram. She may also crouch and urinate. The ram sniffs or licks the urine then arches his neck, lifts his nose and curls his upper lips showing his front teeth – this "flehmen" behaviour is expressed by males of many species of ungulate. Flehmen may also be displayed after sniffing a patch of urine on the ground.
2. The ram approaches the side of the ewe, his neck stretched horizontally and the muzzle raised in a straight line to his neck. In this position the ram's head is lower than, or at the same level, as the ewe's head. The ram then licks the flank of the ewe between her shoulder and hind legs (in woolly breeds, the ram sometimes pulls at the ewe's wool).
3. Standing near the hind legs of the ewe the ram kicks, in a paddling motion, the ewe's hind leg with his foreleg, vocalising and rubbing his head along or under the ewe's flank.

If the ewe remains immobile in response to any or all of the ram's approaches, he will abort the courtship and mount the ewe. The ram keeps his brisket in close contact with the ewe's rump, achieves intromission by a pelvic swing, then ejaculates after a few pelvic thrusts; at ejaculation, the ram lifts his head backward and sometime vocalises. Afterwards, the ram will dismount and both partners stand still for few seconds. During the behavioural exchange leading to mating, olfaction also plays an important role for the ram as illustrated by behaviours such as flehmen, nuzzling and sniffing. Rams can differentiate between urine from ewes in oestrus and those not in oestrus (Blissitt et al. 1990), and anosmic rams lose the capacity to detect receptive ewes (Fletcher & Lindsay 1968). However, flehmen is not necessary for the detection of oestrous ewes by their urinary odour (Blissitt et al. 1990). Olfaction does not seem to be involved in any of the sexual behaviours expressed by ewes. In contrast, vision may play a role in both sexes since 'head turning' (expressed by 81% of receptive ewes) and 'tail fanning' (91%) are the most frequent behaviours of the ewe during courtship (Lynch et al. 1992). Tactile exchanges such as nudging are also very frequent (96%) during sexual encounters. Audition seems to be of little importance given that very few vocalisations are produced by either sex during sexual encounters.

3.3.3 Hormonal Control of Sexual Behaviour

Sex steroids, progesterone and oestrogen, are implicated in the expression of ewes' sexual behaviour. Oestrogen triggers the period of receptivity and controls its duration, whereas progesterone is responsible for the synchronisation of ovulation with sexual receptivity (for review see Blache & Martin 1995).

Androgens are responsible for the expression of male sexual behaviour, but their action is slower than that of sex steroids on female sexual behaviour. Castration after puberty leads to a gradual decrease in sexual drive over a few weeks, or months in experienced rams, and androgen replacement restores sexual activity only after a delay of several days (D'Occhio & Brooks 1980). Additionally, natural (during the breeding season) or artificial (castration) withdrawal of androgen induces "irritable male syndrome" in Soay rams, a highly seasonal breed (Lincoln 2001).

3.3.4 External Factors Affecting Sexual Behaviour

The reproductive activity of sheep is influenced by a large number of external factors such as photoperiodic, social and nutritional cues, as well as stress. The effects of these factors in the control of gonadal activity of rams and ewes has been reviewed elsewhere (Goodman 1994; Blache et al. 2000) and the degree of control that any of them exerts on behaviour or gonadal activity depends on the breed (Blache et al. 2003) and, consequently, its geographical origin. Here, we address the effects of external factors on sexual behaviour, which in turn reflects their influences on the endocrine reproductive axis.

Most breeds of sheep have a limited breeding season and only become sexually active when the day length is shortening (late summer). Sheep from tropical regions, where the variation in day length is minimal, are sexually active throughout the year. The neuroendocrine mechanisms underlying these seasonal changes have been discussed elsewhere (Lincoln & Richardson 1998). Under temperate photoperiods, sexual behaviour rapidly increases in the late summer, remains high over autumn and then decreases in winter and spring. In Merino sheep, the effect of photoperiod on gonadal activity can be altered by the level of nutrition (Martin et al. 2002), however the sexual behaviour of males and females is only affected by severe decreases in intake. The effect of severe under-nutrition in the ram has been attributed to general weakness (Parker & Thwaites 1972). Similarly, over-nutrition decreases mating behaviour simply because the increase in weight leads to clumsiness (Okolski 1975). Among adult ewes, severe undernutrition may alter sexual behaviour because poor body condition leads to irregular oestrous or acyclicity (Hafez 1952; Allen & Lamming 1961). The mechanisms involved are not clear because undernutrition does not affect the preovulatory surge of LH and the release of oestrogen. Over-nutrition appears to have little influence on the sexual behaviour of the ewe because females treated in this manner mainly stand still during sexual encounters. However, proceptivity might be impaired since overweight ewes may not be able to actively search for a male. Exposure to high temperatures (42°C) for 5–6 days inhibits oestrous in 35% of Merino ewes, even though ovulation is not affected (Sawyer 1979a,b). Exposure to high temperatures reduces ram mating activity only in breeds adapted to cold climates, but sperm production is affected in most breeds (Lindsay 1969). Moreover, in the field during heat waves, interactions between the sexes decrease as rams and ewes actively seek shade (Fowler 1984). The effects of humidity on sexual behaviour have not received much attention. Heavy rain might induce mating inactivity because of the general unwillingness of sheep to walk in the rain (Fowler 1984).

Early postnatal experience, as mentioned above, has a dramatic effect on the sexual preferences of both sexes. However, the preference for a sexual partner from the fostered species is stronger in rams than in ewes (Kendrick et al. 2001b). Also, after being reared in single sex groups, rams present low levels of sexual activity, or homosexual behaviour, but these effects can be overcome by short exposure to ewes before puberty (Tilbrook & Cameron 1990).

The social environment can inhibit or stimulate the sexual activity of rams. The presence of an attractive ewe, as described above, is an example of a stimulatory effect. In contrast with bulls and bucks, watching sexually active males seems to have no effect on the level of ram sexual behaviour (Tilbrook & Cameron 1990). The ram's social status has been reported to influence the expression of sexual behaviour, with dominant rams suppressing the mating performance of subordinates (Lindsay et al. 1976). However, because of the confounding effect of the number of ewes, early experience and size of the paddock, reports are conflicting and competition between rams may also enhance mating performance (Lindsay & Ellsmore 1968). To our knowledge, there is no evidence of effects of rearing experience or dominance in ewes, but very little research has been done in the area. Interactions between rams and ewes can have dramatic effects on the expression of sexual behaviour, especially in females. After adults have been housed in single-sex flocks for several months, the introduction of individuals of the opposite sex activates the hypothalamic-pituitary axis, resulting in increased secretions of sex steroids and sexual behaviour (Walkden-Brown et al. 1999).

The effects of stress on sheep sexual behaviour have not been extensively studied. However, stress may stimulate or inhibit LH secretion in ewes, and probably affects their sexual behaviour, depending on the intensity (or nature) of the stressor and the duration of exposure (Dobson & Smith 2000). In males of other mammals, stress suppresses sexual behaviour, and it is therefore likely to have similar effects in rams (Moberg & Mench 2000). Moreover, fear of humans, which is known to profoundly affect sheep maternal behaviour, may influence the expression of sexual behaviour in both sexes (Murphy et al. 1994).

3.3.5 Welfare and Sexual Behaviour

In well-managed sheep farms, the duration of the mating season is controlled by humans and might not follow the natural breeding season of the animals. Inability to satisfy the sexual drive is considered to be a potential source of frustration in some domestic species and consequently a welfare problem (Webster 1995). The sex drive of female sheep appears to be quite strong since ewes in heat actively search for rams (proceptivity) and overcome their natural fear of males. Therefore, it is questionable whether the frustration of sexual restriction is higher in male sheep than in females. Moreover, since males are aroused in response to the sight or smell of females in heat, segregation of the sexes during the breeding season should not be a source of frustration for the ram. On the other hand, the confinement of males in a single sex group during the breeding season could result in increased aggressive interactions and therefore injuries. During the non breeding season, the separation of sexes should not be a problem for either rams or ewes because sheep, like most ungulates, naturally segregate outside of the mating period (Main et al. 1996). However, in equatorial latitudes, the artificial separation of sexes could lead to welfare issues because breeding is not seasonal.

3.4 Parturition and the Development of Maternal Responsiveness

3.4.1 Parturient Behaviour

3.4.1.1 Isolation and Shelter Seeking Behaviour

Being highly gregarious, sheep do not remain apart from conspecifics except for a very brief period of their life: at lambing. As parturition approaches, ewes tend to separate themselves from the flock. This trait is particularly evident in wild Rocky Mountain Bighorn ewes that move away from the flock for up to two weeks prior to parturition (Geist 1971), but is also observed in domestic sheep. Even under intensive rearing conditions prelambing ewes choose to isolate themselves if they are given the opportunity by providing cubicles in the shed (Gonyou & Stookey 1983, 1985). In paddocks, however, it is not always obvious whether parturient ewes actively seek isolation or are left behind by the flock. Soay, Suffolk, Cheviots, Scottish Blackface, Welsh Mountain, and Lacaune ewes all isolate themselves actively from the flock. Lécrivain and Janeau (1987) report that 48% of Lacaune ewes actively seek isolation and may stay away from the main flock for up to 48 h following lambing; this behaviour is particularly common in multiparous females. In contrast, only 2% of Merino ewes show a preference for lambing in isolation (Stevens et al. 1981). Most Merino ewes become isolated because of their inability to follow the flock as parturition approaches. Nonetheless, there is a true change in the social tendency of ewes around lambing. This reduction in social interest is not due to the bonding process with the neonate, but already exists prior to lambing (Poindron et al. 1997). It is likely to be related to the neurophysiological changes associated with parturition since this tendency is rapidly reversed after lambing (Poindron et al. 1994). This shift in social responsiveness is adaptive in highly gregarious animals, such as the sheep, since it favours interactions with the neonate and thereby maximises the establishment of satisfactory mother-offspring relationships, while minimising the risk of interference by other pre-parturient ewes.

Seeking shelter may contribute further to isolation from the flock. Rocky Mountain Bighorn ewes select broken rugged cliffs as lambing sites where they will be better protected from predators than in the high plains (Geist 1971), and feral Soay ewes choose to lamb beside sheltering walls (Lynch et al. 1992). However domestic preparturient ewes usually only seek shelter in cold, windy and wet weather. Again, this appears to vary according to the breed (Alexander et al. 1990a). In general, adult sheep with a full fleece make little use of shelter unless weather conditions are bad. Thus, Lacaune ewes, which do not have a thick fleece, choose to give birth in shrubby sites when the wind speed is over 10 km/h, even if there is no rain (Lécrivain and Janeau 1987). By contrast, there is little evidence of Merino ewes sheltering voluntarily when wind speed is less than 40 km/h unless they are shorn before lambing (Lynch & Alexander 1977; Alexander et al. 1979). Nevertheless, most studies indicate that the birth-site distribution is not random, and even Merino ewes show preferred lambing areas such as the high end of a paddock.

3.4.1.2 Onset of Maternal Behaviour and the Sensitive Period

The first sign of the imminence of birth is increasing restlessness. The ewe repeatedly lies down and stands up, paws the ground and walks in small circles. She also starts licking her lips with rapid movements of the tongue, and continues until the birth of the lamb. Pre-parturient ewes often display interest in amniotic fluids or in alien newborn lambs within two to eight hours prior to lambing. Pre-parturition attraction to newborn lambs varies from brief inspection, to active cleaning, and even nursing, and fades progressively as labour begins.

The ewe usually lies down on her side during labour, stretching her legs and neck, but she may stand up during the last stage of expulsion. The head and forelegs of the lamb normally emerge first at the vulva; the straining of the ewe increases in strength to force out the head and shoulders. Length of labour varies from a few minutes to more than three hours depending on parity, breed, litter size, and the lamb's birth weight and sex. However, most lambs are born within one hour. Once the lamb is expelled, the ewe stands up and begins to lick it (Fig. 3.2). Most ewes clean the head first and then move down the body. Grooming of the lamb may be delayed in primiparous ewes or those that have experienced difficult births (Arnold & Morgan 1975; Poindron & Le Neindre 1980; Poindron et al. 1984). The total duration of licking increases with litter size (Owens et al. 1985; Alexander et al. 1990b; O'Connor et al. 1992, Dwyer et al. 1998). With multiple births, the ewe seems to lose interest in the first-born lamb when the second lamb is born, therefore the former suffers a decrease in licking activity as the ewe focuses her attention on the newborn. Despite this shift of attention, second-born lambs do not receive as much grooming as first-born twins. Although there is an overall increase in licking activity by ewes with multiple offspring, each individual lamb receives less tactile stimulation than singletons. The intensive grooming period, when ewes may spend more than 80% of their time licking the lamb, declines over the first hour after birth, although ewes still spend up to 20% of their time licking the lamb four hours postpartum. Grooming behaviour cleans, dries and stimulates the lamb (see

Fig. 3.2 Scottish Blackface ewe licking her newborn lamb (photo: Cathy Dwyer)

Section 3.5.1), but also entrains additional maternal care and facilitates learning of the lamb's olfactory 'signature' (see below). While grooming, the dam emits frequent low-pitched bleats or rumbling noises with the mouth closed, and occasional high-pitched open-mouth bleats (Shillito & Hoyland 1971; Dwyer et al. 1998). Low-pitched maternal bleats are emitted almost exclusively in the presence of the lamb and like licking, decline over time. These are thought to reassure the newborn and to provide cues for later recognition of the dam. In contrast, the high-pitched bleats are rare immediately after parturition, but their rate increases thereafter. The initial rate of low-pitched bleats is an intrinsic process largely under hormonal control as it is affected by the breed and parity of the ewe, but not by litter size nor the lamb's own vocal activity (Dwyer et al. 1998). This is not the case once the mother has bonded to her litter (6–12 h after parturition) since bleating then increases with litter size (Pollard 1992).

If the mother is left undisturbed with her offspring, maternal behaviour develops normally and lasts for several months, until weaning. Attraction of ewes to neonates fades quickly, however, if they are separated from their young soon after birth. After four hours of separation starting at parturition, 50% of ewes still displayed maternal behaviour when the lamb was returned; this declined to 25% after 12 h of separation (Poindron & Le Neindre 1980). The establishment of maternal interest appears limited to a period of a few hours starting just before parturition. This is further demonstrated by the fact that 24 h of separation beginning one day following birth does not affect maternal care. The long lasting effect of experience during a brief perinatal interval implies that this may be a sensitive period for maternal receptivity. Recent evidence suggests that a sensitive period for maternal selectivity (that is the ability to recognise her own lamb) also exists as ewes separated from their lambs 4 hours after parturition lose their ability to be selective over time, whereas selectivity is maintained if ewes and lambs are separated after 7 days of contact (Keller et al. 2005). While internal factors mediate the mother's attraction to specific sensory stimuli from the neonate, continuing perception of these stimuli is necessary to maintain maternal responsiveness once the endocrine influence has disappeared.

3.4.2 Control of Immediate Maternal Responsiveness

3.4.2.1 Hormonal Factors

Prior to parturition ewes are indifferent, or sometimes aggressive, towards newborn lambs. However, at parturition, even naïve (primiparous) mothers immediately express maternal behaviour towards their newborn. The onset of maternal behaviour at birth is brought about by a combination of hormonal factors, peripheral stimulation associated with parturition and cues from the lamb/amniotic fluids. In naïve ewes the co-ordination and temporal sequence of these factors are crucial for the expression of normal maternal behaviour. Experience acquired through previous pregnancies and rearing of young has a positive effect on maternal care. To some extent, these effects may reflect maturation of the sensory and neuroendocrine mechanisms underlying maternal behaviour (Kendrick et al. 1992; Keverne et al. 1993), perhaps leading to enhanced sensitivity and responsiveness to lambs. Even in maternally experienced ewes, the expression of the full complement of maternal behaviours still relies on the presence of hormonal factors.

Exposure to the ovarian steroid hormones (oestradiol and progesterone) is essential for the induction of maternal care at parturition in the ewe; these hormones act as 'primers' for such behaviour (Kendrick & Keverne, 1991; Kendrick et al. 1997). Ovarian steroids appear to exert their effect by regulating the production of a number of important peptides and their receptors (particularly oxytocin and its receptor, the opioids and corticotrophin-releasing hormone). However, treating non-pregnant ewes with steroid hormones is ineffective in triggering maternal behaviour. Although non-parturient sheep do show a reduction in aggressive behaviours towards lambs following this treatment (Kendrick & Keverne, 1991), grooming or proactive maternal behaviour is not seen. Prolonged exposure to the lamb in oestradiol and progesterone-treated ewes sometimes elicits maternal behaviour. However, the immediate, short-latency expression of maternal care seen in ewes at birth requires additional sensory cues.

3.4.2.2 Sensory Mediation

Vaginocervical stimulation associated with labour and parturition triggers neurochemical processes that alter the significance of olfactory signals, which are vital for ewes' recognition of their offspring and the development of selective bonding. This occurs within the olfactory bulbs, a specialised area in the front of the brain that deals with the perception of odours. Before parturition, cells in the olfactory bulbs of pregnant ewes respond preferentially to odours associated with food (Kendrick et al. 1992). However, after ewes give birth and form a selective bond with their offspring, there is a large increase in the number of cells that respond to lambs' odours. A proportion of these cells fires discriminatively only in response to the odour of the ewe's own lamb. Thus, birth is associated with a general increase in responsiveness to lamb cues, as well as specific cellular responses related to the ewes' selective bond with their own young. The changes in firing patterns in olfactory bulb cells following birth are accompanied by changes in the release of specific neurotransmitters at synapses between different cell types (Kendrick et al. 1992).

In addition to the role played by the birth process, the ewe's maternal behaviour also requires relevant cues from the lamb itself. In particular, the smell and taste of amniotic fluids are important stimulants, especially for naïve, primiparous ewes. The presence of amniotic fluids on the coat of the lamb increases maternal licking, the number of low pitched bleats and udder acceptance by ewes (Lévy & Poindron 1984). For experienced (multiparous) ewes, washing the lamb to remove traces of amniotic fluids reduces maternal licking but otherwise does not affect maternal behaviour (Lévy & Poindron 1987). However, washing the lamb disrupts maternal behaviour of primiparous ewes and reduces the number that allows their newborn to suck. Other cues from the lamb such as its behaviour can enhance maternal responsiveness. Parturient ewes are very attracted to newborn lambs, but are less interested in lambs only a few hours older. Their maternal behaviour is disturbed if their own neonate is exchanged for a 12- to 24-hour-old lamb that is already dry and more active. In contrast, exchanging the ewe's own lamb with a newborn alien lamb does not result in noticeable perturbations (Poindron & Le Neindre 1980; Poindron et al. 1980). Primiparous ewes display normal behaviour as long as their newborn lamb is lying still, but they may become aggressive (butting) towards their own offspring when it tries to stand up or to suck (Poindron et al. 1984). Amniotic fluids and the lamb's immobility thus appear important in facilitating the initial contact between ewes and their neonate.

3.4.3 Maternal Selectivity and Recognition of the Lamb

Under natural conditions, ewes rapidly develop a selective bond with their newborn lamb. This maternal bond is characterised by the acceptance of sucking attempts by the ewe's own lamb along with rejection of alien young that approach her udder. Since ewes congregate in large flocks and births tend to be temporally synchronised, such early offspring recognition enhances the mother's reproductive fitness by enabling her to invest her limited resources in her own offspring alone rather than wasting energy on caring for the young of other ewes.

The primary developmental mechanism implicated in the establishment of the mother-young bond is rapid learning of the lamb's distinctive phenotypic "signature" (Poindron et al. 1993; Lévy et al. 1996). Immediately after parturition ewes are highly responsive to cues provided by new-born lambs and will respond maternally to alien young as well as their own. Recent studies demonstrate that 30–60 minutes of immediate post-partum contact with a neonate may be sufficient for the ewe to become familiar with that lamb's unique signature and discriminate subsequently between it and unfamiliar lambs (Keller et al. 2003). Olfaction plays a critical role in the selective acceptance of lambs at suckling, as described above. During sucking bouts, the two partners typically adopt a parallel-inverse orientation (Fig. 3.3) that enables the ewe to sniff the hind region of the lamb (Poindron 1974). Indiscriminate acceptance of alien lambs at the udder is observed among females that suffer

Fig. 3.3 Préalpes du Sud lamb sucking in the parallel-inverse position which allows the mother to smell it and accept it at the udder (photo: Alain Béguey)

olfactory deficits prior to parturition. On the other hand, elimination or alteration of the chemical signatures of familiar lambs has a negative effect on maternal acceptance (Alexander and Shillito 1981; Alexander et al. 1983a). The underlying basis of lambs' recognisable signatures was tested using dizygotic (fraternal) or monozygotic (identical) twins (Romeyer et al. 1993). Immediately after the first twin was born, it was placed into a wire-mesh cage that prevented the ewe from licking or nursing it, but still allowed access to vocal, visual and olfactory stimuli. The second-born twin was removed from the mother's pen at birth and isolated. During tests conducted 4–5 hours later, ewes more readily accepted their isolated monozygotic twin than an alien lamb, but this was not the case for the isolated dizygotic twins. Moreover, the isolated and familiar monozygotic twins did not elicit differential maternal responses, while the same categories of dizygotic twins were discriminated. It therefore appears that the recognizable signatures of monozygotic twins may be more similar than those of dizygotic twins, enabling the mothers to discriminate more effectively between fraternal twins. Once the mothers became familiar with the lamb that remained with them following parturition, they presumably discerned a resemblance in its monozygotic twin with which they had no prior (postnatal) contact, and therefore treated it more positively than an alien lamb. Thus, ewes were indirectly familiar with phenotypic traits of the isolated lamb that were shared with its familiar twin. The positive correlation that appears to exist between the resemblance of lambs' signatures and their degree of genetic relatedness supports the hypothesis that those recognisable phenotypes are (at least partially) genetically influenced.

Although initial attraction and maternal recognition are based on olfaction, mothers rapidly learn a more composite image of their lamb, including its appearance, behaviour and voice. Olfaction alone as a means of recognition declines in importance with lambs' age, and is probably only used for close contact discrimination after several days. Visual and/or auditory distance recognition of lambs appears to develop (somewhat) less rapidly than proximal olfactory discrimination. Intact multiparous ewes display a preference for their own offspring in two-choice recognition tests conducted 6 hours postpartum (Keller et al. 2003). In contrast, such offspring discrimination was not observed until 24 hours postpartum in a group of primiparous ewes, suggesting that prior maternal experience has a facilitating effect on the establishment of lamb recognition at a distance. In tests involving manipulation of various lamb cues, ewes appear to rely mainly on visual information, particularly from the head, as well as the lamb's behaviour and scent to identify their offspring (Alexander & Shillito 1977). Maternal ewes can also learn to discriminate between slide images of the faces of their own and alien lambs, but only after a lengthy training period (Kendrick et al. 1996). Vocal recognition has been assessed by observing the responses of ewes to recordings of lambs' bleats (Poindron & Carrick 1976). Overall, ewes bleat more during playback of their own lamb's voice and answer more frequently the bleats of their offspring rather than those of alien young (Shillito-Walser et al. 1981a). They can discriminate their young on the basis of individual vocal signatures as early as 24 h post-partum (Sebe et al. 2007).

3.4.4 Factors Influencing Maternal Behaviour

3.4.4.1 Effect of Parity

Primiparous ewes often respond differently than ewes that have previously delivered and reared young when some of the signals required for maternal behaviour are manipulated (see Section 3.4.2). Primiparous ewes, unlike multiparae, do not behave maternally when rendered anosmic (Lévy et al. 1995), following vaginocervical stimulation (Kendrick & Keverne, 1991; Dwyer & Lawrence 1997) or when amniotic fluids are removed from the lamb's coat (Lévy & Poindron 1987). They also show greater aversive responses to a newborn lamb when treated with oestradiol and progesterone (Dwyer & Lawrence 1997). Inexperienced mothers obviously require all the salient pre- and post-partum stimuli (hormonal factors, peripheral stimuli and lamb sensory cues) to develop appropriate maternal behaviour, whereas for multiparous ewes this same complete complement of factors is not necessary. Pregnant multiparous ewes are attracted to neonatal lambs several days before they give birth themselves and in the absence of the peripheral sensory information associated with parturition.

Even when the various factors and cues that normally contribute to the onset of maternal behaviour have not been manipulated, primiparous ewes often are less competent as mothers than are experienced ewes, and the mortality of their lambs is higher. Primiparous ewes tend to have a longer labour than experienced ewes and are slower to begin grooming their lambs after birth (Dwyer & Lawrence 1998). They are also more likely to show fearful behaviour towards the lamb, such as retreating from it, they may be more aggressive (i.e. butting or threatening the lamb) and in some cases they may fail to show maternal behaviour and abandon their lamb. In modern agriculture, sheep are often managed in peer groups until they join the breeding flock. First time mothers are therefore naïve about lambs, having never encountered them before, and the lamb may serve as a novel and potentially fearful stimulus when the ewe first gives birth. Poindron et al. (1984) suggested that as many as 50% of inexperienced ewes displayed various degrees of disturbance: delay in the onset of licking, aggressive behaviour, or circling when the lamb approached the udder, and 23% of primiparous ewes had not yet nursed their young after 3 hours. Experience of being a mother, gained during the initial contacts with her lamb, allows the ewe to learn to respond appropriately and she becomes less likely to prevent its subsequent sucking attempts (Dwyer & Lawrence 1998). After their first experience of maternal care ewes show consistent responses in their subsequent pregnancies: grooming behaviour is maintained and negative behaviours (rejections, retreats, lack of co-operation with suck attempts) decline or disappear (Dwyer & Lawrence 2000b).

3.4.4.2 Effect of Maternal Nutrition

Ewes that are undernourished during pregnancy give birth to light weight lambs with heightened mortality rates. Additional adverse effects of maternal undernutrition include reduced udder development, colostrum composition and milk production. Undernourished ewes also take longer to interact with their lambs (Thomson & Thomson 1949), display more aggression, spend less time grooming and more time eating after birth (Dwyer et al. 2003), and are more likely to desert their lambs (Putu et al. 1988). Moreover, ewes that are underfed during pregnancy have differing physiological profiles during gestation compared to well-fed ewes. Specifically, undernutrition is associated with higher plasma progesterone in late gestation (O'Doherty & Crosby 1996), and a lower ratio of oestradiol to progesterone at birth (Dwyer et al. 2003). Elevated plasma progesterone is negatively related to colostrum and milk yield, which may threaten the survival of newborn lambs. In addition, as described above, progesterone and oestradiol are involved in the onset of maternal behaviour, and high ratios of oestradiol to progesterone are correlated with maternal grooming behaviour (Shipka & Ford 1991; Dwyer et al. 1999). Thus, elevated progesterone in underfed ewes might contribute to poor maternal behaviour and the high level of maternal desertion seen in these animals.

3.4.4.3 Individual Differences: Effects of Breed and Temperament

Ewes may differ in the quantity and quality of their maternal care, including the amount of grooming behaviour, responses to the lamb's sucking attempts, and the likelihood of desertion. These differences are usually maintained over successive births (Dwyer & Lawrence 2000b), suggesting that they are intrinsic to the individual. One of the most frequently explored sources of individual variation is breed difference. Many breeds have been compared, and different behavioural measures recorded. The chosen breeds generally reflect the prevalent or commercially important lines in the countries where they were studied: for example, behavioural comparisons between the Merino, Perendale, Romney, Border Leicester, Cheviot, Dorset Horn and crossbreds in Australia and New Zealand (e.g. Whateley et al. 1974; Alexander et al. 1983b) between Romanov, Lacaune, Prealpes de Sud, and Ile-de-France in France (e.g. Poindron et al. 1984; Le Neindre et al. 1998) and between Dalesbred, Scottish Blackface, Suffolk and Soay breeds in the UK (Shillito-Walser 1980; Dwyer & Lawrence 1998, 2000b). Because these studies compared different breeds using a variety of behavioural measures, it is hard to draw any clear conclusions, except when comparing breeds within the same study. However, observations in Australia and New Zealand suggest that Merinos are generally poorer mothers than other breeds: as mentioned above they spend less time on the birth site, and have a much higher incidence of both permanent and temporary desertions of their lambs than other breeds (Alexander et al. 1983b, 1990b). When ewes' responses to handling of their lamb were scored, Merino ewes also rated lower than other breeds (Whateley et al. 1974). In French and British breeds, Romanov and Scottish Blackface ewes are considered to show better maternal care (more licking, grooming and lamb acceptance; less aggression) than the other breeds (Poindron et al. 1984; Le Neindre et al. 1998; Dwyer & Lawrence 1998). Thus, it is clear that considerable breed differences exist in the quality of expressed maternal behaviour. In general, hill, upland and more primitive breeds, which have been subjected to less human intervention, show the best quality of maternal care, whereas more intensively selected and reared animals display greater variability in maternal behaviour and make the poorest mothers.

Maternal behaviour has also been assessed using a composite measure of ewes' reactions when their lambs are handled by a shepherd – the Maternal Behaviour Score (O'Connor et al. 1985). This score shows variation within and between breeds (Whateley et al. 1974; O'Connor et al. 1985) and is related to both lamb survival and weaning weight. Heritability estimates of this measure for Scottish Blackface ewes (Lambe et al. 2001) are relatively low ($h^2 = 0.13$), but there is good repeatability (0.32). The consistency displayed by individuals may reflect their underlying emotivity or "temperament", as much as maternal behaviour per se. Romanov ewes, for example, are considered to be better mothers (in terms of licking, grooming and attachment to the lamb) in comparison to the Lacaune breed (Le Neindre et al. 1998). However, Romanov ewes displayed greater flight from humans and stood further from their handled lambs than Lacaunes (behaviour that would have earned them a lower Maternal Behaviour Score). These responses are believed to be a result of greater emotivity of the Romanov breed rather than a poorer quality of maternal care. In studies where Merino ewes were selected for temperament by measuring their responses to a variety of tests, the 'calm' ewes spent more time grooming their lambs than did 'nervous' ewes, and bleated more frequently to their lambs (Murphy et al. 1998). Lamb mortality in these lines was also lower in the 'calm' ewes compared to the 'nervous' animals. Ewes previously selected for their ability to rear lambs also show behavioural differences in an approach avoidance test, indicative of increased 'calmness' (Kilgour & Szantar-Coddington 1995). Thus, temperament may also contribute to individual differences in the quality of maternal care expressed by ewes. Differences in temperament between breeds, or between individuals, may also heighten the effects of incomplete neuroendocrine responses in post-parturient primiparous ewes (described in Section 3.4.2.1), and help explain some of the parity effects outlined in Section 3.4.4.1.

3.4.5 Welfare and Mother-Young Interactions

Neonate mortality, in both intensive and extensive systems, remains a welfare concern for sheep agriculture. Mortality of 15–25% is common in farming systems world-wide. Most preweaning lamb deaths occur within the first week of life (Nowak et al. 2000), emphasising the importance of the immediate post-partum period for lamb survival. In extensive systems the majority of lamb deaths are attributed to the starvation-mismothering-exposure complex, where the failure of ewes and lambs to form a strong attachment leads directly or indirectly to lamb death. Starvation and mismothering also contribute to lamb deaths in intensive systems although transmission of infectious disease, particularly in crowded lambing sheds, is also important (Binns et al. 2002). A poor ewe-lamb relationship may indirectly contribute to lamb infections since the immunologically-incompetent neonate lamb needs to suck colostrum from its dam both to close the gut to bacteria, and to obtain immunoglobulins. Figure 3.4 summarises the factors influencing lamb survival. From this it can be seen that both maternal and offspring factors need to interact to contribute to lamb survival. The role of lamb factors in survival will be considered in Section 3.5.4, the present section will deal with interactions between maternal behaviours and welfare.

3.4.5.1 Maternal Behaviour and Welfare

A number of ewe behaviours may contribute both directly and indirectly to the incidence of lamb mortality. Before birth, the shelter- and isolation-seeking behaviours of the ewe play a role in lamb survival. In studies with Merino ewes provision of shelter reduced lamb mortality in poor weather by up to 50% (Alexander & Lynch 1976; Lynch et al. 1980; Alexander et al. 1980). In wild sheep, the ewe and her lamb may remain segregated from the flock for several days (Geist 1971). Even in intensive systems, where parturient ewes were offered access to cubicles in lambing sheds, ewes showed a marked preference to lamb in the cubicles rather than in the open pen (Gonyou & Stookey 1983), and ewe-lamb separations and lamb stealing by preparturient ewes were reduced compared to control pens without cubicles (Gonyou & Stookey 1985). Time spent at the birth site has also been shown to correlate with lamb survival in Merino ewes (Stevens et al. 1982; Alexander et al. 1983b, 1984; Putu et al. 1988; Cloete 1992). Characteristics of the birth site per se appear to be less important than the ewe remaining undisturbed with her lambs for at least 6 hours (Murphy et al. 1994). Thus, isolation from the flock at parturition serves to facilitate the formation of the ewe-lamb bond without interference from other ewes, and appears to be an important ewe trait for the survival of the

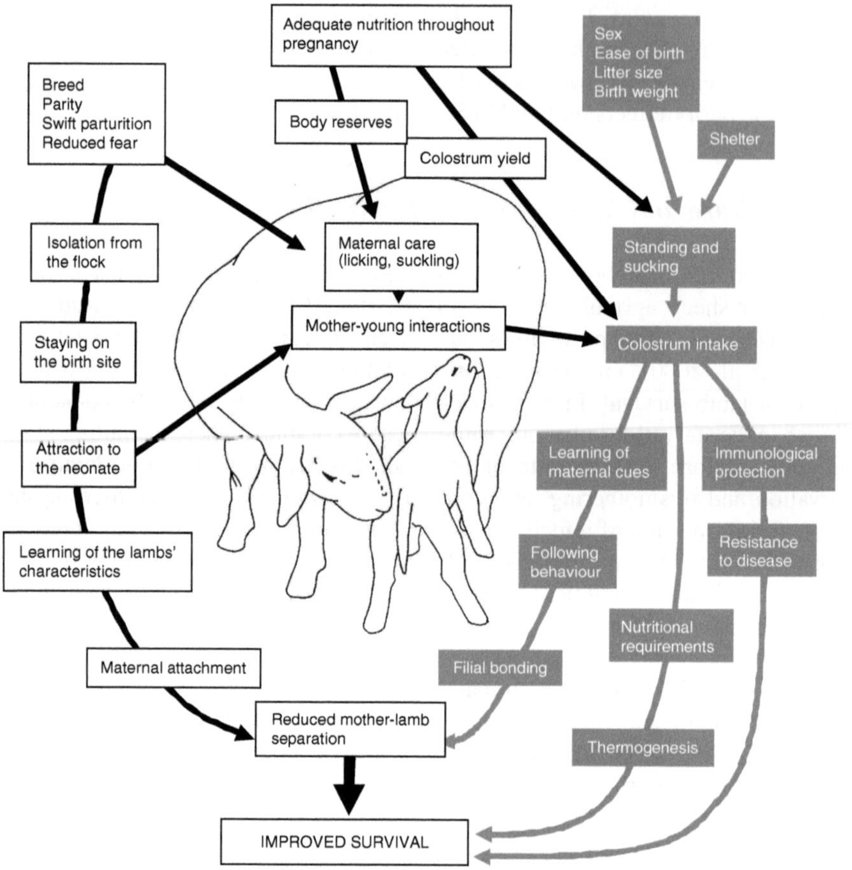

Fig. 3.4 Behavioural and environmental factors influencing lamb survival. White boxes: maternal factors, Grey boxes: offspring factors

lamb. Management practices where ewes and lambs are moved from the birth site to small pens may disrupt the normal transfer of maternal attention from the birth site to the lamb and hinder the formation of the ewe-lamb bond. This is likely to be particularly detrimental to primiparous mothers where these processes take longer to mature than in experienced ewes (see Section 3.4.2). Additionally, in ewes highly motivated to isolate themselves at parturition, the lack of an opportunity to seek shelter or isolation may act as a source of social stress.

A prolonged labour can increase the possibility of brain trauma and hypoxia in the neonate (Haughey 1993) and impairs sucking, locomotor activity and thermoregulation in lambs (Haughey 1980; Eales & Small, 1981; Dwyer 2003). A swift parturition is clearly important for lamb survival, and in flocks selected to improve lamb survival to weaning, the main outcome of the selection process is an increase in the speed and ease of parturition (Cloete & Scholtz 1998). Similar responses are seen in 'easy-care' Romney ewes (Knight et al. 1988; Kilgour & de Langen 1980).

Fear, stress or disturbance is known to cause involuntary suppression of uterine contractions in mammals during labour. For ewes unaccustomed to human presence, close supervision may act as a source of stress and unnecessarily delay or prolong parturition. A low stress environment for lambing ewes is likely to be associated with better welfare for the ewe, and improved lamb survival.

As most ewes are selective for their own offspring, a lamb that fails to form an attachment with its dam will not be cared for by any other ewe and will not survive. Likewise, the offspring of a non-selective ewe will not thrive, as it is unlikely that the ewe will produce sufficient milk to feed several lambs. The maternal behavioural traits expressed at birth associated with selectivity and hence lamb survival include: maternal licking and grooming, low-pitched bleating, absence of aggression and lamb desertion, co-operation with lamb sucking attempts, ewe selectivity and lamb recognition, and maintenance of close contact between ewe and lamb. The expression of these behaviours are affected by maternal nutrition in pregnancy, ewe parity, breed and temperament (see Section 3.4.4). Thus, lamb welfare will be compromised by under and over feeding of the ewe in pregnancy, and by ewe management and handling.

Within-breed studies indicate that ewe maternal behaviour is affected by genotype even when that characteristic was not included in the selection criteria (e.g. Cloete & Scholtz 1998; Kuchel & Lindsay 2000; Dwyer et al. 2001). For example, Merino ewes selected for superfine wool were less maternally responsive and had higher lamb mortality than broader wool Merinos (Kuchel & Lindsay 2000). Blackface ewes selected for low carcass fat were quicker to groom their newborn lambs and stayed closer to them immediately after delivery than ewes selected for more carcass fat (Dwyer et al. 2001). In other studies where improved maternal ability was the aim, lamb survival was greater for Merino ewes selected for fertility and success in rearing multiple offspring, than for unselected or divergently selected lines (Atkins 1980; Cloete & Scholtz 1998). This selection criterion decreased desertion of lambs (Cloete et al. 2005), although the main effect appeared to be an improvement in parturition, such as ease and speed of delivery (Cloete & Scholtz 1998). The 'easy-care' Romney sheep produced in New Zealand show a similar ability to give birth unaided (Kilgour & de Langen 1980). These data suggest that aspects of maternal behaviour are under genetic control, although few estimates of genetic parameters (heritability, phenotypic and genetic correlations) exist. Similarly, in studies where Merino ewes were selected for temperament, the 'calm' ewes spent longer grooming their lambs than did 'nervous' ewes, and bleated more frequently to their lambs (Murphy et al. 1998). Lamb mortality in these lines was also lower in the 'calm' ewes in comparison to the 'nervous' animals. Potentially, therefore, it may be possible to improve maternal care through genetic means.

3.4.5.2 Fostering

Lambs that have been rejected by their own mothers, or triplet lambs that are unlikely to be reared successfully by their mothers, may be either reared artificially or fostered onto a suitable surrogate (generally a ewe whose own lambs have died, or a ewe with a singleton lamb capable of rearing twins). Arguably, a foster mother is preferable for lamb welfare, since this provides the lamb with a social environment similar to remaining with its own mother, and the benefits thereof, rather than isolation or peer-rearing. Similarly, the welfare of the parturient ewe that has lost her litter, but remains highly maternally motivated, might be improved by provision of a foster lamb. However, the welfare, particularly of the surrogate ewe, may also be impaired by the methods (often restraint and confinement) used to affect a ewe-lamb bond after the normal sensitive period described above has waned. Unsuccessful fostering attempts may also expose the lamb to injury from the rejection behaviours of the surrogate.

Fostering in sheep usually involves one of the three methods: (i) bonding the alien lamb to the dam during the period of maternal responsiveness (known as 'wet-foster'). Variations on this method include artificially stimulating the birth canal of a recently parturient ewe to mimic the birth process (Kendrick et al. 1992; Dwyer & Lawrence 1997) which can cause the ewe to become responsive to newborn lambs in a similar manner to having given birth naturally. This method can be facilitated by applying birth fluids to the foster lamb and by presenting them to ewes shortly after they have given birth. (ii) In ewes which have already bonded to their lamb fostering can be achieved by giving the alien characteristics of the dam's own offspring. Because olfaction is the main sensory modality used in proximal recognition and acceptance at udder, matching the odour of the alien lamb to that of the ewe's offspring is often a successful means of fostering. This principle has been used for centuries by shepherds who placed the skin of dead lambs on the body of orphan lambs. Attempts involving methods that change the odour of the ewe's own lamb by smearing it with foreign odorous substances have been less successful. Fostering mothers still discriminate between their own and alien lambs with similar applied odours. (iii) Finally the initial rejection of the alien young by the dam can be reduced or prevented to allow a bond to form gradually by cohabitation. This frequently involves restraint of the ewe (e.g. see Fig. 3.5) to prevent her butting the lamb

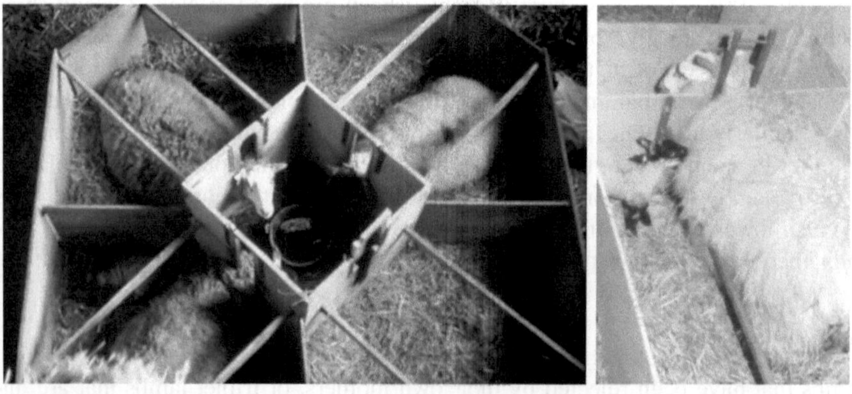

Fig. 3.5 Ewes confined in foster crates demonstrating their restricted ability to move and lack of olfactory interactions with their foster lambs (Photos: David Henderson, John Vipond)

whilst allowing the lamb free access to the udder. There are a number of welfare concerns with this method: the ewe is often restricted in her access to feed and water, the pen may become contaminated with faeces thus increasing the incidence of disease to both ewe (e.g. mastitis) and lamb, and the ewe is unable to turn round or lie comfortably. In addition, this method is often poorly effective since the ewe is prevented from sniffing the lamb, the main sensory route for bond formation. Poorly attached lambs are particularly vulnerable when removed from the fostering device and allowed out to pasture.

3.5 Behaviour of Lambs

3.5.1 Early Post-Natal Behaviour Leading to Sucking

Within minutes after birth the lamb raises and shakes its head, moves its legs, turns its body onto its sternum and bleats. It then kneels on its forelegs, tries to push up onto its hind legs, and eventually gets up by extending its forelegs and rapidly learns to stand steadily (Fig. 3.6). The lamb finds the udder by exploring the underneath of the ewe's body from the chest to the udder. In particular it spends time nosing the axillary and inguinal areas of the udder until it finds the teat. Most lambs stand up within the first 30-min of delivery and begin to suck 1–2 hours post-partum. However, significant differences between breeds in the early postnatal behaviour of

Fig. 3.6 Early postnatal behaviour of a Merino lamb: (**1**) lying head raised up, (**2**) pushing up onto its hind legs, (**3**) exploring the mother's body, (**4**) sucking successfully (photos: Raymond Nowak)

lambs have been reported (Slee & Springbett 1986; Alexander et al. 1990b; Dwyer et al. 1996; Dwyer & Lawrence 1998, 1999a). Birth weight may also influence the time to stand as do gender and litter size. Male lambs are usually slower than females in the expression of early behaviours, and twins slower than singletons, although some authors suggest that this twin effect is a function of reduced birth weight.

Visual and acoustic cues are dominant in guiding the neonate into close contact with the dam. The first directional response is oriented towards the nearest large object, especially if it moves and bleats (Vince et al. 1985), and lack of vision markedly reduces the lamb's motor activity and ability to locate its mother (Vince et al. 1987). High-pitched bleats stimulate general activity of lambs: they stand sooner and display more movements when exposed to such maternal stimulation (Vince et al. 1985). On the other hand, the low "rumble"-type vocalisations of the ewe have a quietening effect on the lamb (Vince 1986). The importance of vocal stimulation is further supported by the fact that the first hours are a period of intense maternal vocal activity (Dwyer et al. 1998) that may play a role in ewe-lamb bonding (Nowak 1990a). Investigations of the influence of maternal grooming on early behavioural development of offspring show conflicting results; both facilitatory and inhibitory effects have been reported. Lambs' responses to experimental massage that mimics licking depend on the part of the body that is stimulated. Stroking the head is associated with forward and downward head movements while massage of the back elicits leg movements (Vince 1993). Therefore, if the lamb is licked more frequently on the back, it may stand sooner. It is well known that standing attempts increase when mothers lick the anogenital region of their newborn lambs (and sometimes push them from behind). During the initial stage of the lamb's exploratory activity the ewe tends to move in order to keep the lamb in front of her, and continues to clean it, focusing on the anal region. Then, the ewe allows the lamb to move towards the udder and experienced ewes arch their backs and spread their hind legs, or lift a hind leg as their lamb approaches the inguinal area, to help the lamb locate the udder more easily (Vince 1993).

Tactile cues play a crucial role in the search for the teat as lambs make munching movement once they come into contact with the mother's body. Vince (1993) showed that tactile stimulation applied to the face, forehead and eyes elicits vigorous head and neck movement as well as oral activity which resembles that of newborn lambs when nuzzling the body of their dam. This 'teat-seeking' activity is modulated by the characteristics of the explored surface. Lambs maintain longer contact with a warm, smooth surface than with a cold one, and they also nuzzle and make more oral movements against the warm surface. When presented with surfaces differing in their degree of yield, lambs nose the intermediate (most udder-like) and high-yielding surfaces more than low-yielding ones. Lambs also respond to the smell of amniotic fluids and inguinal wax with head movements, oral activity, exploration, and increased breathing and heart rates (Vince & Ward 1984; Schaal et al. 1995). Thus, stimuli in various modalities emanating from the mother clearly direct the behaviour of the newborn over the surface of her body until the teat is found.

3.5.2 Behaviour Patterns at Sucking

As the lamb grows older there is a clear change in the frequency and duration of sucking. During the first week post-partum the lamb is allowed to suck as often and for as long as it wishes. Later, the ewe begins to control the suckling pattern by moving away while the lamb is feeding or attempting to do so. Daytime sucking frequencies are greater than night-time frequencies, reflecting the general activity pattern of ewes (Gordon & Siegmann 1991). With time there is an increased tendency for ewes with twins to nurse only if both lambs are present (Ewbank 1967; Hinch 1989). Studies on the effect of litter size on sucking activity are inconsistent but this may reflect differences in methods (number and age of lambs, rearing conditions, differences in the definition of 'suckling') between the various studies. Under unconfined conditions, there is often a higher frequency of sucking by multiple born lambs than singletons because of competition for the limited milk/teat resource. Even though patterns of suckling behaviour are not clearly related to milk transfer to the young (Cameron 1998), lambs suck more frequently when their mothers have a low milk yield (Robertson et al. 1992). Ewes rearing twins produce only 30–50% more milk than those with singles (Treacher 1983). Thus, a lower milk intake would logically lead to increased sucking attempts, at least during the early postnatal period when maternal milk is the sole source of nutrients. The increased sucking activity by lambs of primiparous ewes (Dwyer 2003) probably reflects similar causes. Three positions of the lamb have been observed at suckling: parallel-inverse, perpendicular and between the hind legs of the ewe (Poindron & Signoret 1977; Poindron & Le Neindre 1980). Lambs usually suck in the parallel-inverse position after passing near the front of the ewe. This allows identification of the lamb by its mother before she allows sucking: the ewe smells the lamb and rejects it if it is not her own. Sucking in a perpendicular position or between the hind legs of the ewe is observed when lambs, especially twins, attempt to feed from alien mothers. This type of udder approach usually occurs when the ewe is nursing her own offspring, and is thus unable to smell the alien lamb and reject it. Cross sucking is very common when flocks are kept indoors, at high stocking density (Hess et al. 1974; Poindron & Le Neindre 1980), but may also be observed in the paddock, although lambs are less successful at gaining access to the udder of a ewe that is not their mother in this environment (Hinch 1989).

3.5.3 Recognition of the Ewe by Her Lamb

3.5.3.1 Mechanisms

Survival of the lamb depends upon the maintenance of contact with its dam since 'maternal' care is not available from other ewes that have selectively bonded with their own young. This is probably the basis of the early development of a preference for the mother by the lamb. Most lambs can discriminate between their mother and an alien maternal ewe by 24 hours after birth. Preferential orientation to their own mother improves markedly during the first few days postpartum (Shillito and Alexander 1975; Nowak et al. 1989). While recognition of the mother at 24 hours is based primarily on cues that lambs can perceive at close quarters (<50 cm), they can clearly discriminate their mother from a distance of several metres when 3 days old (Nowak 1990b). The development of early recognition of the dam varies according to breed, sex, and litter size. Border-Leicester x Merino lambs recognise their mother at 12 hours regardless of litter size, and a proportion of them can even identify their dam from a distance. In Merinos, good recognition performance is not achieved before lambs are 24 hours old, and at that stage, singles are much better than twins. Likewise, in the British breeds, Scottish Blackface lambs show better recognition of their mothers at 24 hours after birth than Suffolk lambs (Pickup 2003). Twins commonly take more than two days before they begin to respond discriminatively to their dam. Even then, behavioural differences still persist between singles and twins (Nowak 1990b).

Although lambs respond specifically to some maternal cues immediately after birth, the development of mother preference is based, to a great extent, on postnatal learning during the first mother-young interactions. Rewards associated with parental care (food intake, licking, warmth) provide an early basis for learning in mammalian neonates. The preference of 24-hour old lambs for their own mother over an alien ewe depends mainly on the first suckling interactions. When sucking is prevented during the first 6 hours after birth, the lambs ability to discriminate its mother is impaired at 24 hours, even if they had access to the maternal udder from 6 hours onwards and gained weight (Nowak et al. 1997). This effect is unique to this neonatal period as it is not observed if sucking is temporarily prevented later in life. In a related study (Napolitano et al. 2003), lambs that were allowed to suck only during the first day after birth, remaining with their dam but prevented from sucking thereafter, still displayed a strong preference for their mother at one month of age. Within the first hours following birth, newborn lambs associate sucking reinforcement with the mother and consequently develop a preferential relationship with her. As with maternal care and selectivity, the first sucking bouts and their physiological consequence are important in the early development of the mother-young relationship. Thereafter the influence of sensory factors provided by the mother becomes important in the maintenance of the lamb's relationship with her.

3.5.3.2 Sensory Mediation

Newborn lambs can recognise maternal odours shortly after birth (Vince & Ward 1984; Schaal et al. 1995), and they soon develop the ability to discriminate their mother at a distance via auditory and visual cues (Arnold et al. 1975; Alexander 1977; Nowak 1991). However, selective orientation towards the mother may initially be based on more general signals, since 24-hour-old lambs seem to respond to ewes' expressions of acceptance (low-pitched bleats and allowing access to the udder) and rejection behaviours (high-pitched bleats, aggression; Terrazas et al. 2001). The behaviour and posture of the ewe when bleating may further help the lamb identify its mother (Shillito-Walser et al. 1985). As mother-offspring interactions continue, the lamb learns various individual features of its mother that enable her to be recognized from a distance of several metres (beginning about 3 days postpartum).Vocal recognition of the mother plays an important role in the maintenance of the ewe-lamb contact. Within days after birth most lambs can find their mother when hidden behind a canvas (Shillito 1975; Nowak 1991), or during playback experiments (Sebe et al. 2007). Discrimination of the mother on the sole basis of visual cues develops between the second and third weeks of life (Alexander & Shillito-Walser 1978). The importance of visual cues relative to auditory cues increases between the second and fourth week postpartum (Alexander 1977). The performance of lambs in tests of mother recognition varies across breeds. Clun and Jacob lambs, unlike Dalesbred lambs, seem to use visual cues more than auditory cues to find their own dams (Shillito-Walser 1980). Shillito and Alexander (1975) suggest that Jacob lambs rely more on sight than do Dalesbred lambs because of greater differences in the visual appearance between Jacob individuals (brown and white wool, two or four horns). However, as Dalesbred lambs born to Jacob ewes, after embryo transfer, are also less capable of recognising their mother's voice than are Dalesbred lambs born to Dalesbred ewes, this suggests that characteristics of the maternal voice may be important, or that lamb learn to rely on the most readily identifiable cue.

3.5.4 Lamb Behaviour and Welfare

To suck successfully, the lamb must be able to stand and move to the ewe's abdomen, while the mother stimulates and helps it orient to the udder (Alexander & Williams 1964). Several studies have shown that survival is enhanced in lambs that stand and suck quickly (Alexander 1958; Owens et al. 1985; Cloete 1993; Dwyer et al. 2001). Lambs that are slow to stand and suck have greater difficulty in maintaining body temperature after birth than lambs that suck quickly (Dwyer & Morgan 2006), and are thus more vulnerable to hypothermia. The ability of the lamb to stand and suck effectively is affected by a difficult or protracted delivery, breed, sex, litter size, birth weight (both very heavy and very light lambs are slow to stand and suck), ewe parity, nutrition in pregnancy, and sire (Dwyer 2003). Management practices that are designed to increase productivity through an increase in lambs born per ewe are likely, therefore, to carry a welfare cost in terms of lamb deaths (e.g. through poor lamb sucking or ewe-lamb separation; Fig. 3.7) unless specific steps are taken to prevent this.

As described above for maternal behaviours, appropriate maternal nutrition throughout pregnancy will also enhance lamb survival through its effect on lamb neonatal behaviour, as well as maximising maternal colostrum production. A number of studies have considered whether deficiencies or supplementation of specific trace elements can improve lamb survival and some of these studies have also considered lambs' behaviour. Significant cobalt deficiency is reported to reduce lamb vigour at birth and lead to reductions in lamb survival (Fisher & MacPherson 1991).

Fig. 3.7 Separation of Merino ewes from one of their twins in Western Australia. Top: winter lambing. Bottom: autumn lambing (photos: Raymond Nowak)

However, marginal cobalt deficiency has no impact on lamb neonatal behaviour or survival, although periconception cobalt deficiency reduces postnatal lamb activity (Mitchell et al. 2007). Supplementation of pregnant ewes with selenium may reduce lamb mortality (Kott et al. 1983; Langlands et al. 1991; Munoz et al. 2006), although other studies suggest no effect on survival. However, no data on lamb behaviour are given in these studies. Supplementation with vitamin E and/or fatty acids are also considered to improve lamb survival, and reduce the time taken by lambs to reach the udder and suckle (Williamson et al. 1996; Merrell 1999; Capper et al. 2005, 2006). Although equivocal, these studies do suggest that trace elements, and deficiencies therefore, may play an important role in neonatal lamb behaviour and survival.

The effects of breed on lamb behaviours suggest a breeding route to improve lamb survival. Within-breed strain differences in neonatal lamb behaviours have also been reported (Cloete & Scholtz 1998; Kuchel & Lindsay 2000; Dwyer et al. 2001). In Merino sheep divergently selected for multiple rearing ability, the lambs born to ewes selected for this trait were quicker to progress from standing to sucking than the unselected line (Cloete & Scholtz 1998), and had improved survival. In another study, selection for other characteristics, e.g. lean tissue content (Dwyer et al. 2001), had an unexpectedly positive effect in improving lamb activity at birth. These studies suggest that there is a genetic component to neonatal lamb behaviour, which appears to be independent of breed effects on other traits influencing lamb behaviour. In addition, the strong effect of sire on lamb behaviour suggests that appropriate selection of sire breeds or individual rams is likely to improve lamb survival. This may be particularly effective if applied to vulnerable groups, such as first parity ewes, where breeding management for vigorous lambs is likely to have the most significant welfare benefits.

3.5.5 Twin/Agemate Recognition by Lambs

In the field twin lambs often remain in close physical proximity, even when they are not near their mother (Shillito-Walser et al. 1981b). The closeness of sibling bonds varies across breeds (Arnold & Pahl 1974) and close associations by same-sex twins tend to be more common than for mixed-sex twins. Young lambs respond preferentially to their twin rather than alien lambs during simultaneous choice tests. The presence of their sibling also appears important in stressful situations as lambs that are paired with their familiar twin after being removed from their mother emit fewer distress bleats than do lambs in unfamiliar-agemate pairs (Nowak 1990b; Porter et al. 1995). These data suggest that siblings may provide a special form of social companionship, and that welfare may be impaired if twins are separated.

3.5.5.1 Mechanisms

Learned familiarity plays an important role in twin and agemate recognition. Whereas 3 day old lambs responded randomly to their familiar twin versus an alien agemate, lambs tested at 7 days of age displayed a significant twin preference (Nowak 1990b). Preferential association was also observed between 'twin' lambs of two different breeds that were born of the same ewe following embryo transplantation, which clearly demonstrates that genetic relatedness per se is not necessary for the development of such social discrimination (Shillito-Walser et al. 1981b). This conclusion is supported by accounts of the behaviour of unrelated lambs (born of their respective biological mothers) that have been housed together in small groups for varying periods of time prior to testing. Five days after the sub-groups were established, the presence of a familiar penmate was more effective than the presence of an unfamiliar lamb in reducing the rate of distress bleating associated with maternal separation (Porter et al. 2001). Discrimination of individuals of one's own group is still evident in similar tests conducted five days after the group members were separated and housed in different pens (Ligout et al. 2002). The development of peer interactions and recognition may be further influenced by the lambs' mother. In a recent series of experiments (Ligout & Porter 2003), lambs that were raised with their mother responded more positively to their twin than to other agemates from the same pen.

In contrast, such preferential interactions between twins as compared to non-twin penmates are not observed amongst lambs that had been separated from their mother and artificially reared. Moreover, recognition of unrelated agemates appears to develop more rapidly when lambs are raised in the absence of their mother.

3.5.5.2 Sensory Mediation

Descriptive accounts of the behaviour of animals in free-ranging conditions provide only limited insights concerning the sensory basis of agemate recognition by lambs. Shillito-Walser et al. (1981b) report that twin lambs "often looked for each other and responded to each other's bleats" while they were grazing, and therefore suggested that vision and hearing play a role in sibling recognition. Preferential responses to twins in choice-test situations support this hypothesis; when released at a distance from two caged stimulus lambs, a majority of subject lambs initially approached their sibling rather than an alien (Nowak 1990b; Shillito-Walser et al. 1983). However, the age at which twin discrimination first became evident in this context varied across breeds, ranging from 7 days for Merino lambs to 6 weeks for Dalesbred lambs. Since individual odours are unlikely to be involved in recognition from a distance, it is reasonable to assume that the initial choice was based upon visual and/or auditory cues from the stimulus lambs. Nonetheless, subsequent interactions between subject and stimulus lambs suggest that olfactory cues may be involved in more proximal discrimination of twins. After approaching a stimulus lamb, the subject lambs engage in mutual nosing with that lamb across the bars of its pen. The test lambs remain close to their twin after it was nosed, whereas they move away from an alien following nose-to-nose contact.

Analysis of bleating frequencies by pairs of 3-week-old lambs confined together in a small pen indicates that the presence of a familiar twin more effectively reduces distress in this novel situation than does the presence of an unfamiliar agemate (Porter et al. 1997). This calming effect was obtained (i) when lambs were separated by a fence that prevented physical contact, but allowed visual, auditory and olfactory communication, (ii) in the presence of a tranquillized twin displaying marked behavioural disturbances including suppressed bleating, or (iii) in complete darkness. These results suggest that the involvement of no single sensory modality is necessary for individual recognition by lambs. That is, evidence of twin or familiar agemate discrimination is found in testing situations that prevent physical contact, as well as those in which visual, auditory or olfactory cues are not available. Social discrimination in lambs therefore appears to be influenced by functionally overlapping multi-sensory cues.

3.5.6 Social Relationship Between the Ewe and Her Lamb Beyond the Neonatal Period

Lambs remain in close physical proximity to their mother during the period immediately following birth. The distance between ewes and their lambs increases beginning 3–9 days postpartum and reaches a pleateau by 28–36 days after birth (Morgan & Arnold 1974). As the lambs get older they suck less and spend more time grazing. Playing increases from day 3 to 10, then gradually declines, however, lambs associate increasingly in peer groups until they are approximately 9 weeks old. At that age, peer group activity lessens and close associations with the mother resume (Arnold & Grassia 1985; Hinch et al. 1987). This suggests the existence of a cyclical pattern in the spatial relationship between ewes and lambs. The mother's interest in her young is initially very strong, but it decreases as the lamb begins to assume a greater role in maintaining the bond. Nonetheless, the ewe-lamb relationship and their interactions with other conspecifics appear to be influenced strongly by the behavioural characteristics of the mother. In an experiment using embryo transfer between two breeds of sheep, Dwyer & Lawrence (1999b, 2000a) found that Blackface ewes maintained closer spatial proximity to their newborn lamb than did Suffolk ewes, regardless of the lambs' breed. Also, lambs with Blackface mothers were more active than lambs with Suffolk ewes, and this difference persisted after weaning. It appears that while the neonatal activity is mainly controlled by infantile factors, the maternal environment plays a major role in shaping the behaviour of the lamb beyond the third postnatal day. Recent investigation suggests that the ewe uses postural communication, particularly the performance of the 'head-up' posture (the alert or attention posture), to maintain proximity to her lamb. Breed variation in this posture (which may also reflect maternal vigilance and attentiveness to her lamb) probably contributes to the differences in ewe-lamb distance described above (Pickup & Dwyer 2002). The close relationship that exists during the first week after birth maximises the chance of survival of young lambs that are completely dependent on milk provided by their mother. Subsequently, association with the ewe maintains the existing social structure of the flock, stimulates lambs' exploration of their physical environment and facilitates transmission of acquired knowledge across generations. Although most young ungulates preferably follow and imitate their mothers, generalised following responses (young following any adult) are observed in natural conditions even after the ability to recognise the dam has developed (Lent 1974). Such behaviour has obvious advantages for the survival of individuals of species living in flocks. For example, the flight response to potential predators could be learned from conspecifics other than the mother alone.

Many reports indicate that an infant is more likely to explore a novel environment or a stranger in the presence of its caregiver. Thus, under natural conditions, the mother remains the primary model that influences exploration of the environment, including the choice of particular solid food items. When observing two breeds of sheep with different feeding habits, Key and MacIver (1980) concluded that the grazing pattern of 6–7 month-old lambs was determined by the breed of ewe that reared them. Welsh Mountain lambs cross-fostered onto Clun Forest ewes grazed like Clun Forest sheep, likewise the grazing habits of Clun Forest lambs reared by Welsh Mountain ewes reflected the pattern of the latter breed. Early exposure of lambs to novel food in the presence of their mothers results in a marked increase in the subsequent intake of that same food (Lobato et al. 1980; Lynch et al. 1983).

Fig. 3.8 A three-day-old Merino lamb investigating lupin grain as its mother ingests it (photo Raymond Nowak)

Weaned lambs readily consume wheat even though their experience with that grain in the mother's presence was confined to the first four days of life, and it was unlikely that they had ingested any of it. Beginning at a very young age, lambs are extremely sensitive to factors that influence subsequent feeding behaviour. Overall, weaned lambs eat more solid food if they have experienced it in the presence of their mother than in the presence of a dry ewe (Thorhallsdottir et al. 1990) despite reports that 7-week-old lambs are able to learn from any experienced adult ewe (Chapple et al. 1987). It is not surprising that the mother is the best model given the affective bond that the young develop with her; i.e. lambs would presumably be more attentive to her than to other adults. This learning does not simply entail passive observation of the model, rather lambs also need to explore the food in the presence of their dam, although actual ingestion is not necessary (Lynch et al. 1983; Fig. 3.8). Such acquisition of learned 'traditions' by imitating other animals ensures rapid adjustment to a changing environment, while at the same time conserving adaptive behavioural responses. Thus, learning specific dietary choices by imitation avoids the potential risks associated with both hard-wired (inborn) preferences (i.e. inability to adapt to unpredictable environments), and reliance on trial and error learning (i.e. sampling inappropriate items/substances could result in illness or even be fatal).

3.5.6.1 Social Disruption During Ontogeny

Lambs are always weaned abruptly by the intervention of the shepherd; mothers and young are physically separated and reallocated into flocks of peers. Weaning usually occurs when lambs are 3 months of age, although there may be considerable variation since an entire flock is usually weaned at the same time so later-born lambs may well be younger. The exception is dairy breeds in which lambs are separated from their dam 4 weeks after birth, although immediate separation is practiced in some systems (see Chapter 6). Mothers and lambs emit frequent vocalisations but this response decreases rapidly during the two days following weaning. Paradoxically, studies by Orgeur et al. (1998, 1999) revealed no increases in plasma cortisol after weaning in either ewes or lambs. Weaning at 3 months may not be considered a major stressor since by this age the ewe-lamb bond is naturally weakening, although natural weaning would not take place for another month or more. However, there is evidence that weaning at 3 months can disrupt the social structure in which the lambs lived prior to being separated from their dam. Lambs re-associate in temporary groups, and an increase in pair association between twins is observed during the first few days after separation from the mother. It may take months before an adult form of flocking behaviour is achieved. Arnold and Pahl (1967) have shown that Merino lambs aggregate in sub-groups of 8 animals at four months of age, and in sub-groups of 35 when 11 months old, but it takes another year before the flock can be considered as a cohesive unit. Weaning may be more traumatic when performed at an earlier age, however data on this matter are lacking. Early weaning is routinely adopted by the dairy industry and lambs can be successfully switched to solid food at 28 days of age. However, they are likely to suffer a delay in growth resulting from early mother separation (Moberg et al. 1980), and because the rumen has not necessarily reached mature proportions.

Separation from the dam and post-weaning practices can have other adverse effects on the behaviour of lambs since social contact during early development influences adult behaviour. In sheep, male-like sexual behaviour is frequently observed during infancy and defined as sexual play. During adolescence (before puberty), males' patterns of sexual behaviour are progressively organised into the copulatory sequence and are selectively oriented towards females (Orgeur & Signoret 1984). If females are absent during this period, the rams will direct their sexual behaviour exclusively toward males (Katz et al. 1988; Price et al. 1988). Frequent passive reactions or even immobility of target males to nudging and mount attempts, associated with no achievement of mating, could explain the emergence of homosexual preferences or sexual inhibition. Surprisingly, there is no information on the impact of mother-deprivation on the subsequent development of maternal behaviour by female lambs. Lambs, like other young mammals, are influenced by social deprivation and when reared in isolation they do not respond in a normal manner to agemates or ewes (Price et al. 1988). Lambs reared in isolation withdraw from novel environments, are slow to initiate movements and investigate new objects, and vocalize less than group-reared animals (Moberg & Wood 1982). However, the behaviour of lambs reared in peer groups without a mother is not very different from that of lambs housed with their dams, thus the presence of peers can compensate for the absence of mothers. On commercial farms, artificially-reared lambs are usually separated from their dam 24 h after birth, once they have ingested a sufficient amount of colostrum. The high rate of artificial group-rearing of lambs in parts of Europe with a substantial dairying industry suggests that major behavioural consequences of this practice may not be observed, although some effects (e.g. subsequent reactivity or maternal behaviour) may not yet have been investigated in detail.

3.6 Behaviour and the Assessment of Welfare

As described in Chapter 1, behavioural measures are important tools in the assessment of animal welfare, even though interpretation of behaviours may be as fraught with difficulty as interpretation of physiological measures. A major difficulty is to define 'normal behaviour' and consequently 'abnormal behaviour' or behavioural disturbances in farm animals that have adapted to the constraints of captivity through domestication (Price 2002). One option that may appear appealing is to compare the behaviour of domesticated animals to their wild counterparts (as in Chapter 2). In this way the behaviour of wild sheep, or that of domesticated sheep that have returned to the wild (feral), could be a good source of information to define normal behaviour (Keeling & Jensen 2002). Fraser (1985) claims that: "Restrictive environment reduces social activity and, as a result of this, highly organized behaviour cannot find expression in confined animals". Substantial ethological deficits due to intensive production of sheep in comparison to extensive systems have been reported (Fraser 1983). However, this does not necessarily imply that the behaviour of the animals is anomalous nor that their welfare is affected. Domestic sheep have been selected for various traits that facilitate adaptation to their confined environment. Behaviours displayed in their rearing environment can be normal and therefore, the norm defined from the behaviour of wild sheep may no longer be applicable to domesticated animals. Nonetheless, behaviours expressed by domestic sheep might not be normal even though they occur frequently.

The welfare issues are of a different nature in extensive farming systems, such as Australian sheep stations, and in intensive farming systems, such as medium sized European farms. Under intensive husbandry problems arise from confinement, management and social instability, or social restriction in pens. By contrast, the welfare issues associated with extensive farming relate to lack of shelter, undernutrition, predation, or lamb mortality. Animal-human relationships may also be important, especially for intensively farmed sheep that have more contact with humans than do sheep under extensive conditions (Lynch et al. 1992).

3.6.1 Stereotypies and Abnormal Behaviours

In general sheep appear to have the lowest rate of stereotyped behaviour of the various species of farm animal. This may be due to two factors. Firstly, although they were amongst the first species to be domesticated, the farming methods do not involve the high levels of intensity found for pigs, poultry and dairy calves. Secondly, sheep spend a large portion of their time ruminating and it is noteworthy that ungulates that do not ruminate (horses, pigs, veal calves) are more likely to display stereotypies. Nonetheless, it may be that their stereotypies are expressed in a more discrete manner compared to those of other domestic animals, and may therefore not be noticed. Individually-housed sheep have, however, been shown to demonstrate stereotypical oral behaviours, such as mouthing bars, chewing slats or chains, rattling or chewing buckets, biting and chewing pen fixtures, mandibulation (licking lips and mouthing air), and repetitive licking. Locomotor stereotypies have also been reported including rearing against the pen, repetitive butting, star-gazing (arching the head and neck over the back), leaping vertically up and down, weaving and route-tracing (Done-Currie et al. 1984; Marsden & Wood-Gush 1986). These studies suggest that sheep do perform stereotypies, although they may not be as frequent as in other species. Feed restriction, particularly of energy, or a diet lacking in fibre, increases the frequency of abnormal oral behaviours (Done-Currie et al. 1984; Marsden & Wood-Gush 1986; Cooper et al. 1994; Yurtman et al. 2002). Providing hay or increased fibre in the diet reduce oral stereotypy (Done-Currie et al. 1984; Cooper et al. 1995), and increase lying and rumination (Cooper & Jackson 1996) although not in all studies (Yurtman et al. 2002).

Whether stereotypies exist under farming conditions or are a feature of animals reared under extremely impoverished conditions is worth further investigation. The only aberrant behaviour that is described in farmed sheep is wool pulling or wool eating (Sambraus 1985; Lynch et al. 1992; Fraser & Broom 1997) and occurs both in adults and lambs. Wool-pulling in adult sheep is only reported in intensive rearing systems. Typically, an individual pulls with its mouth on the strands of wool of others, usually on the back, and may ingest it. As a consequence, the afflicted animal becomes progressively denuded on the back area. Eating wool by itself does not lead to any serious impairment of health; it may, however, increase the risk of parasitic infestation as wool-pulling animals ingest soiled wool. The causes of this behavioural disorder have not been satisfactorily established but it is related to overcrowding and social dominance since low-ranking animals are usually the targets. Nutritional deficiencies have also been suspected but the involvement of specific nutrients has not been proven. However, providing hay, but not straw, is believed to reduce the occurrence of wool pulling. A recent study (Vasseur et al. 2006) suggests that wool-biting can be decreased by provision of fibre in the diet, and that this behaviour is primarily a redirected oral response of housed sheep deprived of activity or oral stimulation. Young lambs may eat the wool from their mother and this has been observed as early as one week of age. This may lead to digestive problems as strands of wool accumulate in the lambs' stomach and form compact fibrous balls called bezoars. Affected animals become anaemic, suffer from colic and lose weight. In some cases bezoars cause complete intestinal obstruction which is fatal to the lamb.

3.6.2 Welfare and the Human-Animal Relationship

Do sheep habituate to contact with humans in normal intensive farming systems? Is it better overall for the sheep to have contact with humans only twice in a life-time as typically happens in very extensive Australian systems? These questions could be answered if the relative importance of repetition and intensity of a particular stress could be measured in welfare terms. However, it has been demonstrated that early contact with humans may be beneficial in reducing the fear they induce in sheep. Lambs that have become accustomed to frequent handling and close contact with people are less stressed by subsequent human handling than animals that have less experience of human contact (Boivin et al. 2000). Lambs are also able to distinguish between familiar and unfamiliar shepherds after 3 weeks of positive treatment, making fewer vocalisations in the presence of the familiar shepherd (Boivin et al. 1997). Lambs were still able to discriminate between shepherds after 6 weeks, although by 3 months of age the lambs were less likely to approach and contact any shepherd and were more active and vocal in the presence of a human. It seems that regular human contact is required to maintain approach behaviour towards humans. In contrast, Markowitz et al. (1998) demonstrated that only 2 days of handling, applied when the lamb was between 1–3 days old, increased lamb affinity for humans that appeared to persist for up to 3 months. These lambs were separated from their mothers for only 48 hours, during the treatment period, and results suggest a sensitive period exists in sheep where only 40 minutes of positive human contact is sufficient to reduce lamb timidity. The experiments of Boivin et al. (2000) also demonstrated that lambs reacted to both the appearance and the disappearance of the human, suggesting that the artificially reared lambs had formed a social bond with their handlers. The separation of the young animal from its mother that accompanies these treatments, even if this is only temporary, may interact with the responses of the lamb to the treatment. However, if these treatments are carried out in the presence of the dam they are much less effective (Boivin et al. 2002); it seems that the lamb can form only one primary social attachment, to either its dam or a human substitute. The importance of separation of the animal from its mother to enhance docility in the presence of humans suggests these methods are not particularly useful from a practical viewpoint of reducing stress to handling procedures in farmed sheep.

Handling of older animals demonstrates that sheep that were stroked and fed over a number of days approached a human more readily, had shorter flight distances and lower heart rates than unhandled sheep (Hargreaves & Hutson 1990; Mateo et al. 1991). Gentling, therefore, appears to have a taming effect and apparently reduces the fear that the sheep had towards humans. However, gentled sheep have similar struggling responses to restraint, or aversion to sham-shearing, as sheep that have not been gentled. Thus, although gentling reduced fear of human contact, this did not generalise to other procedures. The importance of the human-animal relationship to sheep welfare will be discussed further in Chapter 8.

3.7 Conclusions

Despite its long history of domestication and association with man, the domestic sheep has retained well-developed and complex social behaviours, perhaps because these very behaviours made it an attractive candidate for domestication in the first place. The relatively fixed nature of sheep farming over the centuries, and the generally extensive nature of sheep farms has meant that consideration of the behavioural requirements of sheep has been largely ignored. In addition, the low levels of intra-specific agonistic interactions in sheep flocks, their relative docility and ease of handling, and the low level of stereotypy in ruminants in general has meant that their psychological welfare, apart from disease and certain management practices, has not been considered in detail. In this chapter we have highlighted both the complexity of sheep social relationships and pointed to some of the possible causes for welfare concern. Perhaps the most studied area of welfare compromise is that of neonatal lamb mortality. Here behavioural research has enormously advanced our understanding of the requirements of ewes and lambs at parturition, and the nature of the social bond that forms at birth. This has provided considerable information and advances towards practical solutions to reduce lamb mortality.

References

Alexander, G. (1958) Behaviour of newly born lambs. *Proceedings of the Australian Society for Animal Production*, 2, 123–125.

Alexander, G. (1977) Role of auditory and visual cues in mutual recognition between ewes and lambs in Merino sheep. *Applied Animal Ethology*, 3, 65–81.

Alexander, G., Kilgour, R., Stevens, D. & Bradley, L. R. (1984) The effect of experience on twin care in New Zealand Romney sheep. *Applied Animal Behaviour Science*, 12, 363–372.

Alexander, G. & Lynch, J. J. (1976) Phalaris windbreaks for shorn and fleeced lambing ewes. *Proceedings of the Australian Society for Animal Production*, 11, 161–164.

Alexander, G., Lynch, J. J. & Mottershead, B. (1979) Use of shelter and selection of lambing sites by shorn and unshorn ewes in paddocks with closely or widely spaced shelters. *Applied Animal Ethology*, 5, 51–69.

Alexander, G., Lynch, J. J., Mottershead, B. E. & Donnelly, J. B. (1980) Reduction in lamb mortality by means of grass wind-breaks: Results of a five-year study. *Proceedings of the Australian Society for Animal Production*, 13, 329–332.

Alexander, G. & Shillito, E. E. (1977) Importance of visual cues from various body regions in maternal recognition of the young in Merino sheep (Ovis aries). *Applied Animal Ethology*, 3, 137–143.

Alexander, G. & Shillito, E. E. (1981) Recognition of washed lambs by Merino ewes. *Applied Animal Ethology*, 7, 77–86.

Alexander, G. & Shillito-Walser, E. E. (1978) Visual discrimination between ewes by lambs. *Applied Animal Ethology*, 4, 81–85.

Alexander, G., Stevens, D. & Bradley, L. R. (1983a) Washing lambs and confinement as aids to fostering. *Applied Animal Ethology*, 10, 251–261.

Alexander, G., Stevens, D. & Bradley, L. R. (1990a) Distribution of field birth-sites of lambing ewes. *Australian Journal of Experimental Agriculture*, 30, 759–767.

Alexander, G., Stevens, D., Bradley, L. R. & Barwick, S. A. (1990b) Maternal behaviour in Border-Leicester, Glen Vale (Border Leicester derived) and Merino sheep. *Australian Journal of Experimental Agriculture*, 30, 27–38.

Alexander, G., Stevens, D., Kilgour, R., de Langen, H., Mottershead, B. E. & Lynch, J. J. (1983b) Separation of ewes from twin lambs: Incidence in several sheep breeds. *Applied Animal Ethology*, 10, 301–317.

Alexander, G. & Williams, D. (1964) Maternal facilitation of sucking drive in newborn lambs. *Science*, 146, 665–666.

Allen, D. M. & Lamming, G. E. (1961) Nutrition and reproduction in the ewe. *Journal of Agricultural Science (Cambridge)*, 56, 69–79.

Arnold, G. W. (1985) Associations and social behaviour. In: *Ethology in Farm Animals. A Comprehensive Study of the Behavioural Features of the Common Farm Animals* (Ed. A. F. Fraser), pp. 233–248. Elsevier Science, Amsterdam.

Arnold, G. W., Boundy, C. A. P., Morgan, P. D. & Bartle, G. (1975) The roles of sight and hearing in the lamb in the location and discrimination between ewes. *Applied Animal Ethology*, 1, 167–176.

Arnold, G. W. & Dudzinski, M. L. (1978) *Ethology of Free-Ranging Domestic Animals*. Elsevier Scientific Publishing, Amsterdam.

Arnold, G. W. & Grassia, A. (1985) Spatial relationships between ewes and lambs. *Applied Animal Behaviour Science*, 14, 253–261.

Arnold, G. W. & Maller, R. A. (1974) Some aspects of competition between sheep for supplementary feed. *Animal Production*, 19, 309–319.

Arnold, G. W. & Morgan, P. D. (1975) Behaviour of the ewe and lamb at lambing and its relationship to lamb mortality. *Applied Animal Ethology*, 2, 25–46.

Arnold, G. W. & Pahl, P. J. (1967) Sub-grouping in sheep flocks. *Proceedings of the Ecological Society of Australia*, 2, 183–189.

Arnold, G. W. & Pahl, P. J. (1974) Some aspects of social behaviour in domestic sheep. *Animal Behaviour*, 22, 592–600.

Atkins, K. D. (1980) The comparative productivity of five ewe breeds. 1. Lamb growth and survival. *Australian Journal of Experimental Agriculture and Animal Husbandry*, 20, 272–279.

Baldwin, B. A. & Meese, G. B. (1977) The ability of sheep to distinguish between conspecifics by means of olfaction. *Physiology and Behavior*, 18, 803–808.

Banks, E. M. (1964) Some aspects of sexual behavior in domestic sheep, Ovis aries. *Behaviour*, 23, 249–279.

Beach, F. (1976) Sexual attractivity, proceptivity, and receptivity in female mammals. *Hormones and Behavior*, 7, 105–138.

Binns, S. H., Cox, I. J., Rizvi, S. & Green, L. E. (2002) Risk factors for lamb mortality on UK sheep farms. *Preventive Veterinary Medicine*, 52, 287–303.

Blache, D., Chagas, L. M., Blackberry, M. A., Vercoe, P. E. & Martin, G. B. (2000) Metabolic factors affecting the reproductive axis in male sheep. *Journal of Reproduction and Fertility*, 120, 1–11.

Blache, D. & Martin, G. B. (1995) Neural and endocrine mechanisms underlying the synchronicity of sexual behaviour and ovulation in the sheep. *Oxford Review of Reproductive Biology*, 17, 205–254.

Blache, D., Zhang, S. & Martin, G. B. (2003) Fertility in males: Modulators of the acute effects of nutrition on the reproductive axis of male sheep. In: *Reproduction in Domestic Ruminants V* (Ed. B. K. Campbell, R. Webb, H. Dobson & C. Doberska), pp. 387–402. Society for Reproduction and Fertility, Cambridge.

Blissitt, M. J., Bland, K. P. & Cottrel, D. F. (1990) Discrimination between odours of fresh oestrous and non-oestrous ewe urine by rams. *Applied Animal Behaviour Science*, 25, 51–59.

Bøe, K. E., Berg, S. & Andersen, I.L. (2006) Resting behaviour and displacements in ewes – effects of reduced lying space and pen shape. *Applied Animal Behaviour Science*, 98, 249–259.

Bon, R. & Campan, C. (1996) Unexplained sexual segregation in polygamous ungulates: A defense of an ontogenetic approach. *Behavioural Processes*, 38, 131–154.

Boivin, X., Boissy, A., Nowak, R., Henry, C., Tournadre, H. & Le Neindre, P. (2002) Maternal presence limits the effects of early bottle feeding and petting on lambs' socialisation to the stockperson. *Applied Animal Behaviour Science*, 77, 311–328.

Boivin, X., Nowak, R., Desprès, G., Tournadre, H. & Le Neindre, P. (1997) Discrimination between shepherds by lambs reared under artificial conditions. *Journal of Animal Science*, 75, 2892–2898.

Boivin, B., Tournadre, H. & Le Neindre, P. (2000) Hand-feeding and gentling influence early-weaned lambs' attachment responses to their stockperson. *Journal of Animal Science*, 78, 879–884.

Bouissou, M. F., Porter, R. H., Boyle, L. & Ferreira, G. (1996) Influence of a conspecific image of own vs different breed on fear reactions of ewes. *Behavioural Processes*, 38, 37–44.

Cameron, E. Z. (1998) Is suckling behaviour a useful predictor of milk intake? A review. *Animal Behaviour*, 56, 521–532.

Capper, J. L., Wilkinson, R. G., Kasapidou, E., Pattinson, S. E., Mackenzie, A. M. & Sinclair, L. A. (2005) The effect of dietary vitamin E and fatty acid supplementation of pregnant and lactating ewes on placental and mammary transfer of vitamin E to the lamb. *British Journal of Nutrition*, 93, 549–557.

Capper, J. L., Wilkinson, R. G., Mackenzie, A. M. & Sinclair, L. A. (2006) Polyunsaturated fatty acid supplementation during pregnancy alters neonatal behavior in sheep. *Journal of Nutrition*, 136, 397–403.

Chapple, R. S., Wodzicka-Tomaszewska, M. & Lynch, J. J. (1987) The learning behaviour of sheep when introduced to wheat. II. Social transmission of wheat feeding and the role of the senses. *Applied Animal Behaviour Science*, 18, 163–172.

Cloete, S. W. P. (1992) Observations on litter size, parturition and maternal behaviour in relation to lamb mortality in fecund Dormer and South African Mutton Merino ewes. *South African Journal of Animal Science*, 22, 214–221.

Cloete, S. W. P. (1993) Observations on neonatal progress of Dormer and South African Mutton Merino lambs. *South African Journal of Animal Science*, 23, 38–42.

Cloete, S. W. P. & Scholtz, A. J. (1998) Lamb survival in relation to lambing and neonatal behaviour in medium wool Merino lines divergently selected for multiple rearing ability. *Australian Journal of Experimental Agriculture*, 38, 801–811.

Cloete, S. W. P., Scholtz, A. J., Cloete, J. J. E. & van Wyk, J. B. (2005) The ability of merino ewes to reunite after separation, as affected by divergent selection for ewe multiple rearing capacity. *Australian Journal of Experimental Agriculture*, 45, 1131–1137.

Cockram, M. S., Ranson, M., Imlah, P., Goddard, P. J., Burrells, C. & Harkiss, G. D. (1994) The behavioural, endocrine and immune response of sheep to isolation. *Animal Production*, 58, 389–399.

Cooper, J. J., Emmans, G. E. & Friggens, N. C. (1994) Effect of diet on behaviour of individually penned sheep. *Animal Science*, 58, 44.

Cooper, J. & Jackson, R. (1996) A comparison of the feeding behaviour of sheep in straw yards and on slats *Applied Animal Behaviour Science*, 49, 99.

Cooper, J., McCullam, J. & Shanks, M. (1995) The effect of fibrous diet on abnomral oral behaviour in stall housed lambs. *Proceedings of the British Society for Animal Science*, p. 207.

Deag, J. M. (1996) Behavioural ecology and the welfare of extensively farmed animals. *Applied Animal Behaviour Science*, 49, 9–22.

Dobson, H. & Smith, R. F. (2000) What is stress, and how does it affect reproduction? *Animal Reproduction Science*, 60, 743–752.

D'Occhio, M. J. & Brooks, D. E. (1980) Effects of androgenic and oestrogenic hormones on mating activity in rams castrated before or after puberty. *Journal of Endocrinology*, 86, 403–411.

Done-Currie, J. R., Hecker, J. F. & Wodzicka-Tomaszewska, M. (1984) Behaviour of sheep transferred from pasture to an animal house. *Applied Animal behaviour Science*, 12, 121–130.

Dove, E. H., Beilharz, R. G. & Black, J. L. (1974) Dominance pattern and positional behaviour of sheep in yards. *Animal Production*, 19, 157–168.

Dwyer, C. M. (2003) Behavioural development in the neonatal lamb: Effect of maternal and birth-related factors. *Theriogenology*, 59, 1027–1050.

Dwyer, C. M., Dingwall, W. S. & Lawrence, A. B. (1999) Physiological correlates of maternal and offspring behaviour in sheep. *Physiology and Behavior*, 67, 443–454.

Dwyer, C. M. & Lawrence, A. B. (1997) Induction of maternal behaviour in non-pregnant hormone-primed ewes. *Animal Science*, 65, 403–408.

Dwyer, C. M. & Lawrence, A. B. (1998) Variability in the expression of maternal behaviour in primiparous sheep: Effects of genotype and litter size. *Applied Animal Behaviour Science*, 58, 311–330.

Dwyer, C. M. & Lawrence, A. B. (1999a) Does the behaviour of the neonate influence the expression of maternal behaviour in sheep? *Behaviour*, 136, 367–389.

Dwyer, C. M. & Lawrence, A. B. (1999b) Ewe-ewe and ewe-lamb behaviour in a hill and a lowland breed of sheep: A study using embryo transfer. *Applied Animal Behaviour Science*, 61, 319–334.
Dwyer, C. M. & Lawrence, A. B. (2000a) Effects of maternal genotype and behaviour on the behavioural development of their offspring in sheep. *Behaviour*, 137, 1629–1654.
Dwyer, C. M. & Lawrence, A. B. (2000b) Maternal behaviour in domestic sheep (Ovis aries): Constancy and change with maternal experience. *Behaviour*, 137, 1391–1413.
Dwyer, C. M., Lawrence, A. B. & Bishop, S. C. (2001) Effects of selection for lean tissue content on maternal and neonatal lamb behaviours in Scottish Blackface sheep. *Animal Science*, 72, 555–571.
Dwyer, C. M., Lawrence, A. B., Bishop, S. C. & Lewis, M. (2003) Ewe-lamb bonding behaviours at birth are affected by maternal undernutrition in pregnancy. *British Journal of Nutrition*, 89, 123–136.
Dwyer, C. M., Lawrence, A. B., Brown, H. E. & Simm, G. (1996) Effect of ewe and lamb genotype on gestation length, lambing ease and neonatal behaviour of lambs. *Reproduction, Fertility, and Development*, 8, 1123–1129.
Dwyer, C. M., McLean, K. A., Deans, L. A., Chirnside, J., Calvert, S. K. & Lawrence, A. B. (1998) Vocalisations between mother and young in sheep: Effects of breed and maternal experience. *Applied Animal Behaviour Science*, 58, 105–119.
Dwyer, C. M. & Morgan, C. A. (2006) Maintenance of body temperature in the neonatal lamb: Effects of breed, birth weight and litter size. *Journal of Animal Science*, 84, 1093–1101.
Eales, F. A. & Small, J. (1981) Effects of colostrum on summit metabolic rate in Scottish Blackface lambs at five hours old. *Research in Veterinary Science*, 30, 266–269.
Ewbank, R. (1967) Nursing and suckling behaviour amongst Clun Forest ewes and lambs. *Animal Behaviour*, 15, 251–258.
Fabre-Nys, C., Poindron, P. & Signoret, J.-P. (1993) Reproductive behaviour. In: *Reproduction in Domesticated Animals* (Ed. G. J. King), pp. 147–194. Elsevier Science, Amsterdam.
Fabre-Nys, C. J. & Venier, G. (1989) Quantitative analysis of oestrous behaviour through the breeding season in two breeds of sheep. *Animal Reproduction Science*, 21, 37–51.
Fisher, G. J. & MacPherson, A. (1991). Effect of cobalt deficiency in the pregnant ewe on reproductive performance and lamb viability. *Research in Veterinary Science*, 50, 319–327.
Fletcher, I. C. & Lindsay, D. R. (1968) Sensory involvement in the mating behaviour of domestic sheep. *Animal Behaviour*, 16, 410–414.
Fowler, D. G. (1984) Reproductive behaviour of rams. In: *Reproduction in Sheep* (Ed. D. R. Lindsay & D. T. Pearce), pp. 39–46. Australian Academy of Science, Canberra.
Franklin, J. R., & Hutson, G. D. (1982) Experiments in attracting sheep to move along a laneway. III. Visual stimuli. *Applied Animal Ethology* 8, 457–478.
Fraser, A. F. (1983) The behaviour of maintenance and the intensive husbandry of cattle, sheep and pigs. *Agriculture, Ecosystems and Environment*, 9, 1–23.
Fraser, A. F. (1985) Deprivation of maintenance behaviour in modern farm animal hysbandry. In: *Ethology of Farm Animals. A Comprehensive Study of the Behavioural Features of the Common Farm Animals* (Ed. A. F. Fraser), pp. 377–389. Elsevier Science, Amsterdam.
Fraser, A. F. & Broom, D. M. (1997) *Farm Animal Behaviour and Welfare*. CAB International, Oxon, UK.
Geist, V. (1971) Behavior of ewes and lambs. In: *Mountain Sheep* (Ed. G. B. Schaller), pp. 239–255. University of Chicago Press, Chicago.
Gelez, H., Lindsay, D. R., Blache, D., Martin, G. B. & Fabre-Nys, C. (2003) Temperament and sexual experience affect female sexual behaviour in sheep. *Applied Animal Behaviour Science*, 84, 81–87.
Gonyou, H. W. & Stookey, J. M. (1983). Use of lambing cubicles and the behavior of ewes at parturition. *Journal of Animal Science*, 56, 787–791.
Gonyou, H. W. & Stookey, J. M. (1985) Behavior of parturient ewes in group-lambing pens with and without cubicles. *Applied Animal Behaviour Science*, 14, 163–171.

Goodman, R. L. (1994) Neuroendocrine control of the ovine estrous cycle. In: *The Physiology of Reproduction* (Eds. E. Knobil, & J. Neill), pp. 1929–1969. Raven Press, New York.

Gordon, K. & Siegmann, M. (1991) Suckling behavior of ewes in early lactation. *Physiology and Behavior*, 50, 1079–1081.

Grubb, P. & Jewell, P. A. (1966) Social grouping and home range in feral Soay sheep. *Symposium of the Zoological Society of London*, 18, 179–210.

Guilhem, C., Bideau, E., Gerard, J. F. and Maublanc, M. L. (2000) Agonistic and proximity patterns in enclosed mouflon (*Ovis gmelini*) ewes in relation to age, reproductive status and kinship. *Behavioural Processes*, 50, 101–112.

Hafez, E. S. E. (1952) Studies on the breeding season and reproduction of the ewe. *Journal of Agricultural Science, Cambridge*, 42, 189–265.

Hargreaves, A. L. & Hutson, G. D. (1990) The effect of gentling on heart rate, flight distance and aversion of sheep to a handling procedure. *Applied Animal Behaviour Science*, 26, 243–252.

Hass, C. C. (1991) Social status in female bighorn sheep (*Ovis canadensis*): Expression, development and reproductive correlates. *Journal of Zoology*, 225, 509–523.

Haughey, K. G. (1980) The effect of birth injury to the foetal nervous system on the survival and feeding behaviour of lambs. In *Reviews in Rural Science* No. 4. (Eds. M. Wodzicka-Tomasczewska, T. N. Edey & J. J. Lynch). pp. 109–111, University of New England, Armidale.

Haughey, K. G. (1993) Perinatal lamb mortality – Its investigation, causes and control. *Irish Veterinary Journal*, 46, 9–28.

Hess, C. E., Graves, H. B. & Wilson, L. L. (1974) Individual preweaning suckling behavior of single, twin and triplet lambs. *Journal of Animal Science*, 38, 1313–1318.

Hinch, G. N. (1989) The sucking behaviour of triplet, twin and single lambs at pasture. *Applied Animal Behaviour Science*, 22, 39–48.

Hinch, G. N., Lécrivain, E., Lynch, J. J. & Elwin, R. L. (1987) Changes in maternal-young associations with increasing age of lambs. *Applied Animal Behaviour Science*, 17, 305–318.

Joubert, D. M. (1962) Sex behaviour of purebred and crossbred Merino and Blackhead Persian ewes. *Journal of Reproduction & Fertility*, 32, 41–49.

Katz, L. S., Price, E. O., Wallach, S. J. R. & Zenchak, J. J. (1988) Sexual performance of rams reared with or without females after weaning. *Journal of Animal Science*, 66, 1166–1173.

Keeling, L. & Jensen, P. (2002) Behavioural disturbance, stress and welfare. In: *The Ethology of Domestic Animals: An Introductory Text* (Ed. P Jensen), pp. 79–98. CAB International, Wallingford UK.

Keller, M., Meurisse, M., Poindron, P., Nowak, R., Shayit, M., Ferreira, G. & Lévy, F. (2003) Maternal experience influences the establishment of visual/auditory, but not of olfactory recognition of the newborn lamb by ewes at parturition. *Developmental Psychobiology*, 43, 167–176.

Keller, M., Meurisse, M. & Lévy, F. (2005) Mapping of brain netowkrs involved in the consolidation of lamb recognition memory. *Neuroscience*, 133, 359–369.

Kendrick, K. M., Atkins, K., Hinton, M. R., Heavens, P. & Keverne, B. (1996) Are faces special for sheep? Evidence from facial and object discrimination learning tests showing effects of inversion and social familiarity. *Behavioural Processes*, 38, 19–35.

Kendrick, K. M., Da Costa, A. P., Broad, K. D., Ohkura, S., Guevara, R., Lévy, F. & Keverne, E. B. (1997) Neural control of maternal behaviour and olfactory recognition of offspring. *Brain Research Bulletin*, 44, 383–395.

Kendrick, K. M., da Costa, A. P., Leigh, A. E., Hinton, M. R. & Peirce, J. W. (2001a) Sheep don't forget a face. *Nature*, 414, 165–166.

Kendrick, K. M., Haupt, M. A., Hinton, M. R., Broad, K. D. & Skinner, J. D. (2001b) Sex differences in the influence of mothers on the sociosexual preferences of their offspring. *Hormones and Behavior*, 40, 322–338.

Kendrick, K. M. & Keverne, E. B. (1991) Importance of progesterone and estrogen priming for the induction of maternal behavior by vaginocervical stimulation in sheep: Effects of maternal experience. *Physiology and Behavior*, 49, 745–750.

Kendrick, K. M., Lévy, F. & Keverne, E. B. (1991) Importance of vaginocervical stimulation for the formation of maternal bonding in primiparous and multiparous parturient ewes. *Physiology and Behavior*, 50, 595–600.

Kendrick, K. M., Lévy, F. & Keverne, E. B. (1992) Changes in the sensory processing of olfactory signals induced by birth in sheep. *Science*, 256, 833–836.

Keverne, E. B., Lévy, F., Guevara-Guzman, R. & Kendrick, K. M. (1993) Influence of birth and maternal experience on olfactory bulb neurotransmitter release. *Neuroscience*, 56, 557–565.

Key, C. & MacIver, R. M. (1980) The effects of maternal influences on sheep: Breed differences in grazing, resting and courtship behaviour. *Applied Animal Ethology*, 6, 33–48.

Kiley-Worthington, M. (1977) *Behaviour Problems of Farm Animals*. Oriel Press Ltd, Northumberland, UK.

Kilgour, R. & de Langen, H. (1970) Stress in sheep resulting from management practices. *Proceedings of the New Zealand Society of Animal Production*, 30, 65–76.

Kilgour, R. & de Langen, H. (1980) Neonatal behaviour in 'easycare' sheep. In *Reviews in Rural Science: Behaviour in Relation to Reproduction, Management and Welfare of Farm Animals*. (Eds. M. Wodzicka-Tomasczewska, T. N. Edey & J. J. Lynch), pp. 109–111, University of New England, Armidale.

Kilgour, R. J. & Szantar-Coddington, M. R. (1995) Arena behaviour of ewes selected for superior mothering ability differs from that of unselected ewes. *Animal Reproduction Science*, 37, 133–142.

Knight, T. W., Lynch, P. R., Hall, D. R. H. & Hockey, H. U. P. (1988) Identification of factors contributing to the improved lamb survival in Marshall Romney sheep. *New Zealand Journal of Agricultural Research*, 31, 259–271.

Kott, R. W., Ruttle, J. L. & Southward, G. M. (1983). Effects of vitamin E and selenium injections on reproduction and preweaning lamb survival in ewes consuming diets marginally deficient in selenium. *Journal of Animal Science*, 57, 557–558.

Kuchel, R. C. & Lindsay, D. R. (2000) Maternal behaviour and the survival of lambs in superfine wool sheep. *Reproduction, Fertility and Development*, 11, 391–394.

Lambe, N. R., Conington, J., Bishop, S. C., Waterhouse, A. & Simm, G. (2001) A genetic analysis of maternal behaviour score in Scottish Blackface sheep. *Animal Science*, 72, 415–425.

Langlands, J. P., Donald, G. E., Bowles, J. E. & Smith, A. J. (1991) Subclinical selenium deficiency 2. The response in reproductive performance of grazing ewes supplemented with selenium. *Australian journal of Experimental Agriculture*, 31, 33–35.

Le Neindre, P., Murphy, T., Boissy, A., Purvis, I. W., Lindsay, D. R., Orgeur, P., Bouix, J. & Bibé, B. (1998) Genetics of maternal ability in cattle and sheep. *6th World Congress on Genetics Applied to Livestock Production*, 27, 23–30.

Lécrivain, E. & Janeau, G. (1987) Comportement d'isolement et de recherche d'abri de brebis agnelant en plein air dans un système d'élevage à caractère extensif. *Biology of Behaviour*, 12, 127–148.

Lent, P. C. (1974) Mother-infant relationship in ungulates. In: *The Behaviour of Ungulates and Its Relation to Management* (Ed. V. Geist & F. Walther), pp. 14–55. U.I.C.N, Morgues, Switzerland.

Lévy, F., Kendrick, K., Keverne, E. B., Porter, R. H. & Romeyer, A. (1996) Physiological, sensory and experiential factors of parental care in sheep. *Advances in the Study of Behavior*, 25, 385–473.

Lévy, F., Locatelli, A., Piketty, V., Tillet, Y. & Poindron, P. (1995) Involvement of the main but not the accessory olfactory system in maternal behavior of primiparous and multiparous ewes. *Physiology and Behavior*, 57, 97–104.

Lévy, F. & Poindron, P. (1984) The influence of amniotic fluid on maternal behaviour in parturient sheep. *Biology of Behaviour* 9, 65–88.

Lévy, F. & Poindron, P. (1987) The importance of amniotic fluids for the establishment of maternal behaviour in experienced and inexperienced ewes. *Animal Behaviour*, 35, 1188–1192.

Ligout, S. & Porter, R. H. (2003) Social discrimination in lambs: The role of indirect familiarization and methods of assessment. *Animal Behaviour*, 65, 1109–1115.

Ligout, S., Porter, R. H. & Bon, R. (2002) Social discrimination in lambs: Persistence and scope. *Applied Animal Behaviour Science*, 76, 239–248.

Lincoln, G. A. (2001) The irritable male syndrome. *Reproduction, Fertility and Development*, 13, 567–576.

Lincoln, G. A. & Richardson, M. (1998) Photo-neuroendocrine control of seasonal cycles in body weight, pelage growth and reproduction: Lessons from the HPD sheep model. *Comparative Biochemistry and Physiology. Part C Pharmacology, Toxicology, Endocrinology*, 119, 283–94.

Lindsay, D. R. (1969) Sexual activity and semen production of rams at high temperatures. *Journal of Reproduction and Fertility*, 18, 1–8.

Lindsay, D. R., Dunsmore, D. G., Williams, J. D. & Syme, G. J. (1976) Audience effects on mating behaviour of rams. *Animal Behaviour*, 24, 818–821.

Lindsay, D. R. & Ellsmore, J. (1968) The effect of breed, season and competition on mating behaviour of rams. *Australian Journal of Experimental Agriculture and Animal Husbandry*, 8, 649–652.

Lobato, J. F. P., Pearce, G. R. & Beilharz, R. G. (1980) Effects of early familiarization with dietary supplements on the subsequent ingestion of molasses-urea blocks by sheep. *Applied Animal Ethology*, 6, 149–161.

Lynch, J. J. & Alexander, G. (1973) Animal behaviour and the pastoral industries. In *Pastoral Industries of Australia* (Eds. G. Alexander and O. B. Williams), pp. 371–400. Sydney University Press, Sydney.

Lynch, J. J. & Alexander, G. (1977) Sheltering behaviour of lambing Merino sheep in relation to grass hedges and artificial windbreaks. *Australian Journal of Agricultural Research*, 28, 691–701.

Lynch, J. J., Hinch, G. N. & Adams, D. B. (1992) *The Behaviour of Sheep. Biological Principles and Implications for Production*. CAB International and CSIRO Australia, Melbourne.

Lynch, J. J., Keogh, R. G., Elwin, R. L., Green, G. C. & Mottershead, B. E. (1983) Effects of early experience on the post-weaning acceptance of whole grain wheat by fine wool Merino lambs. *Animal Production*, 36, 175–183.

Lynch, J. J., Mottershead, B. E. & Alexander, G. (1980) Sheltering behaviour and lamb mortality amongst shorn Merino ewes lambing in paddocks with a restricted area of shelter or no shelter. *Applied Animal Ethology*, 6, 163–174.

Lynch, J. J., Wood-Gush, D. G. M. & Davies, H. I. (1985) Aggression and nearest neighbours in a flock of Scottish Blackface. *Biology of Behaviour*, 10, 215–225.

McBride, G., Arnold, G. W., Alexander, G. & Lynch, J. J. (1967) Ecological aspects of the behaviour of domestic animals. *Proceedings of the Ecological Society of Australia*, 2, 133–165.

Main, M. B., Weckerly, F. W. & Bleich, V. C. (1996) Sexual segregation in ungulates: New directions for research. *Journal of Mammalogy*, 77, 449–461.

Markowitz, T. M., Dally, M. R., Gursky, K. & Price, E. O. (1998) Early handling increases lamb affinity for humans. *Animal Behaviour*, 55, 573–587.

Marsden, D. & Wood-Gush, D. G. M. (1986) A note on the behaviour of individually-penned sheep regarding their use for research purposes. *Animal Production*, 42, 157–159.

Martin, G. B., Hötzel, M. J., Blache, D., Walkden-Brown, S. W., Blackberry, M. A., Boukliq, R., Fisher, J. A. & Miller, D. W. (2002) Determinants of the annual pattern of reproduction in mature male Merino and Suffolk sheep: Modification of response to photoperiod by annual cycle of food supply. *Reproduction, Fertility and Development*, 14, 165–175.

Mason, I. L. (1969) *A World Dictionary of Livestock Breeds, Types and Varieties*. CAB International, Wallingford, UK.

Mateo, J. M., Estep, D. Q. & McCann, J. S. (1991) Effects of differential handling on the behaviour of domestic ewes (*Ovis aries*). *Applied Animal Behaviour Science*, 32, 45–54.

Mattner, P. E., Braden, A. W. H. & Turnbull, K. E. (1971) Studies of flock mating of sheep. *Australian Journal of Experimental Agriculture and Animal Husbandry*, 7, 102–109.

Merrell, B. G. (1999) Effects of supplementing ewes with vitamin E in late pregnancy on sheep performance. *Proceedings of workshop on vitamin E requirement in sheep*, Harper Adams University College, 20pp.

Minton, J. E., Coppinger, T. R., Reddy, P. G., Davis, W. C. & Blecha, F. (1992) Repeated restraint and isolation stress alters adrenal and lymphocyte functions and some leukocyte differentiation antigens in lambs. *Journal of Animal Science* 70, 1126–1132.

Mitchell, L. M., Robinson, J. J., Watt, R. G., McEvoy, T. G., Ashworth, C. J., Rooke, J. A. & Dwyer, C. M. (2007). Effects of cobalt/vitamin B12 status in ewes on ovum development and lamb viability at birth. *Reproduction, Fertility and Development*, 19, 553–562.

Moberg, G. P., Anderson, C. O. & Underwood, T. R. (1980) Ontogeny of the adrenal and behavioral responses of lambs to emotional stress. *Journal of Animal Science*, 51, 138–142.

Moberg, G. P. & Mench, J. A. (2000) *The Biology of Animal Stress: Basic Principles and Implications for Animal Welfare*. CAB International, Wallingford, UK.

Moberg, G. P. & Wood, V. A. (1982) Effect of differential rearing on the behavioral and adrenocortical response of lambs to a novel environment. *Applied Animal Ethology*, 8, 269–279.

Morgan, P. D. & Arnold, G. W. (1974) Behavioural relationships between Merino ewes and lambs during the four weeks after birth. *Animal Production*, 19, 169–176.

Munoz, C., Carson, A. F., McCoy, M., Dawson, L. E. R. & Gordon, A. (2006) Nutritional status of ewes in early and mid pregnancy 2: Effect of selenium supplementation on ewe reproduction and offspring performance. *Proceedings of the British Society of Animal Science*, p. 7 (abstract).

Murphy, P. M., Lindsay, D. R. & Le Neindre, P. (1998) Temperament of Merino ewes influences maternal behaviour and survival of lambs. *Proceedings of the 32nd Congress of the International Society for Applied Ethology*, 32, 131.

Murphy, P. M., Purvis, I. W., Lindsay, D. R., Le Neindre, P., Orgeur, P. & Poindron, P. (1994) Measures of temperament are highly repeatable in Merino sheep and some are related to maternal behaviour. *Proceedings of the Australian Society of Animal Production*, 20, 247–250.

Napolitano, F., Annicchiarico, G., Caroprese, M., De Rosa, G., Taibi, L. & Sevi, A. (2003) Lambs prevented from suckling their mothers display behavioral, immune and endocrine disturbances. *Physiology and Behavior*, 78, 81–89.

Nowak, R. (1990a) Lamb's bleats: Important for the establishment of the mother-young bond? *Behaviour*, 115, 14–29.

Nowak, R. (1990b) Mother and sibling discrimination at a distance by three- to seven-day-old lambs. *Developmental Psychobiology*, 23, 285–295.

Nowak, R. (1991) Senses involved in the discrimination of Merino ewes at close contact and from a distance by their newborn lambs. *Animal Behaviour*, 42, 357–366.

Nowak, R., Murphy, T. M., Lindsay, D. R., Alster, P., Andersson, R. & Uvnäs-Moberg, K. (1997) Development of a preferential relationship with the mother by the newborn lamb: Importance of the sucking activity. *Physiology and Behavior*, 62, 681–688.

Nowak, R., Poindron, P. & Putu, I. G. (1989) Development of mother discrimination by single and multiple newborn lambs. *Developmental Psychobiology*, 22, 833–845.

Nowak, R., Porter, R.H., Lévy, F., Orgeur, P. & Schaal B. (2000) Role of mother-young interactions in the survival of offspring in domestic mammals. *Reviews of Reproduction*, 5, 153–163.

O'Connor, C. E., Jay, N. P., Nicol, A. M. & Beatson, P. R. (1985) Ewe maternal behaviour score and lamb survival. *Proceedings of the New Zealand Society of Animal Production*, 45, 159–162.

O'Connor, C. E., Lawrence, A. B. & Wood-Gush, D. G. M. (1992) Influence of litter size and parity on maternal behaviour at parturition on Scottish Blackface sheep. *Applied Animal Behaviour Science*, 33, 345–355.

O'Doherty, J. V. & Crosby, T. F. (1996) The effect of diet in late pregnancy on progesterone concentration and colostrum yield in ewes. *Theriogenology*, 46, 233–241.

Okolski, A. (1975) Effect of different amounts of protein in the diet on sexual behaviour and properties of semen in rams. *Acta Agraria et Silvestria, Zootechnica*, 15, 101–121.

Orgeur, P., Bernard, S., Naciri, M., Nowak, R., Schaal, B. & Lévy, F. (1999) Psychobiological consequences of two different weaning methods in sheep. *Reproduction, Nutrition, Development*, 39, 231–244.

Orgeur, P., Mavric, N., Yvore, P., Bernard, S., Nowak, R., Schaal, B. & Levy, F. (1998) Artificial weaning in sheep: Consequences on behavioural, hormonal and immuno-pathological indicators of welfare. *Applied Animal Behaviour Science*, 58, 87–103.

Orgeur, P. & Signoret, J. P. (1984) Sexual play and its functional significance in the domestic sheep (Ovis aries L.). *Physiology and Behavior*, 33, 111–118.

Owens, J. L., Bindon, B. M., Edey, T. N. & Piper, L. R. (1985) Behaviour at parturition and lamb survival of Booroola Merino sheep. *Livestock Production Science*, 13, 359–372.

Parker, G. V. & Thwaites, C. J. (1972) The effects of undernutrition on libido and semen quality in adult Merino rams. *Australian Journal of Agricultural Research*, 23, 109–115.

Parrot, R. F. (1990) Physiological responses to isolation in sheep. In: *Social Stress in Domestic Animals* (Ed. R. Zayan & R. Dantzer), pp. 212–226. Kluwer Academic Publishers, Brussels.

Pickup, H. E. (2003) *Maternal behaviour in the ewe: Consistency in the expression of maternal behaviour during lactation and the effect of variation in maternal care on the development of offspring*. PhD Thesis, University of Edinburgh.

Pickup, H. E., Dwyer, C. M. (2002) Breed differences in the expression of maternal care at parturition persist throughout the lactation period in sheep. *Proceedings of the British Society of Animal Science*, p. 70.

Poindron, P., Caba, M., Gomora Arrati, P., Krehbiel, D. & Beyer, C. (1994) Responses of maternal and non-maternal ewes to social and mother-young separation. *Behavioural Processes*, 31, 97–110.

Poindron, P. (1974) Etude de la relation mere-jeune chez des brebis Ovis aries, lors de l'allaitement. C. R. Hebd. Seances Academie Science Paris Series D, Science Naturelles, 278, 2691–2694.

Poindron, P. & Carrick, M. J. (1976) Hearing recognition of the lamb by its mother. *Animal Behaviour*, 24, 600–602.

Poindron, P. & Signoret. J. P. (1977) Etude de comportement maternel de la brebis: Mechanismes de reconnaissance mutuelle de l'agneau et de sa mere. Annales de Medecine Veterinaire, 121, 27–33.

Poindron, P. & Le Neindre, P. (1980) Endocrine and sensory regulation of maternal behavior in the ewe. *Advances in the Study of Behavior*, 11, 75–119.

Poindron, P., Le Neindre, P., Raksanyi, I., Trillat, G. & Orgeur, P. (1980) Importance of the characteristics of the young in the manifestation and establishment of maternal behaviour in sheep. *Reproduction, Nutrition, Développement*, 20, 817–826.

Poindron, P., Nowak, R., Lévy, F., Porter, R. H. & Schaal, B. (1993) Development of exclusive mother-young bonding in sheep and goats. *Oxford Reviews of Reproductive Biology*, 15, 311–364.

Poindron, P., Raksanyi, I., Orgeur, P. & Le Neindre, P. (1984) Comparaison du comportement maternel en bergerie à la parturition chez des brebis primipares ou multipares de race Romanov, Préalpes de Sud et Ile-de-France. *Génétique, Sélection, Evolution*, 16, 503–522.

Poindron, P., Soto, R. & Romeyer, A. (1997) Decrease of response to social separation in preparturient ewes. *Behavioural Processes*, 40, 45–51.

Pollard, J. C. (1992) Effects of litter size on the vocal behaviour of ewes. *Applied Animal Behaviour Science*, 34, 75–84.

Porter, R. H., Désiré, L., Bon, R. & Orgeur, P. (2001) The role of familiarity in the development of social recognition by lambs. *Behaviour*, 138, 207–219.

Porter, R. H., Nowak, R. & Orgeur, P. (1995) Influence of a conspecific agemate on distress bleating by lambs. *Applied Animal Behaviour Science*, 45, 239–244.

Porter, R. H., Nowak, R, & Orgeur, P. (1997) Twin/non-twin discrimination by lambs: An investigation of salient stimulus characteristics. Behaviour, 134, 463–475.

Price, E. O. 2002. *Animal Domestication and Behavior*. CAB International, Wallingford, UK.

Price, E. O., Katz, L. S., Wallach, S. J. R. & Zenchak, J. J. (1988) The relationship of male-male mounting to the sexual preferences of young rams. *Applied Animal Behaviour Science*, 21, 347–355.

Putu, I. G., Poindron, P. & Lindsay, D. R. (1988) A high level of nutrition during late pregnancy improves subsequent maternal behaviour of Merino ewes. *Proceedings of the Australian Society for Animal Production*, 17, 294–297.

Robertson, A., Hiraiwa-Hasegawa, M., Albon, S. D. & Clutton-Brock, T. H. (1992) Early growth and sucking behaviour of Soay sheep in a fluctuating population. *Journal of Zoology*, 227, 661–671.

Romeyer, A., Porter, R. H., Lévy, F., Nowak, R., Orgeur, P. & Poindron, P. (1993) Maternal labelling is not necessary for the establishment of discrimination between kids by recently parturient goats. *Animal Behaviour*, 46, 705–712.

Ruiz-de-la-Torre, J. L. & Manteca, X. (1999a) Behavioural effects of social mixing at different stocking densities in prepubertal lambs. *Animal Welfare*, 8, 117–126.

Ruiz-de-la-Torre, J. L. & Manteca, X. (1999b) Effects of testosterone on aggressive behaviour after social mixing in male lambs. *Physiology and Behavior*, 68, 1098–1113.

Rushen, J. (1986) Aversion of sheep for handling treatments: paired-choice studies. *Applied Animal Behaviour Science*, 16, 360–370.

Sambraus, H. H. (1985) Mouth-based anomalous syndromes. In: *Ethology of Farm Animals: A Comprehensive Study of the Behavioural Features of the Common Farm Animals* (Ed. A. F. Fraser), pp. 394–422. Elsevier Science, Amsterdam.

Sawyer, G. J. (1979a) The influence of radiant heat load on reproduction in the Merino ewe. I. The effect of timing and duration of heating. *Australian Journal of Agricultural Research*, 30, 1133–1141.

Sawyer, G. J. (1979b) The influence of radiant heat load on reproduction in the Merino ewe. II. The relative effect of heating before and after insemination. *Australian Journal of Agricultural Research*, 30, 1143–1149.

Scaramuzzi, R. J., Adams, N. R., Baird, D. T., Campbell, B. K., Downing, J. A., Findlay, J. K., Henderson, K. M., Martin, G. B., McNatty, K. P., McNeilly, A. S. & Tsonis, C. G. (1993) A model for follicle selection and the determination of ovulation rate in the ewe. *Reproduction, Fertility and Development*, 5, 459–478.

Schaal, B., Orgeur, P. & C., Arnold. (1995) Olfactory preferences in newborn lambs : Possible influence of prenatal experience. *Behaviour*, 132, 351–365.

Schindler, H. & Amir, D. (1972) Length of oestrus, duration of phenomena related to oestrus, and ovulation time in the local fat-tailed Awassi ewe. *Journal of Agricultural Science (Cambridge)*, 78, 151–156.

Sebe, F. Nowak, R., Poindron, P., & Aubin, T. (2007) Establishment of vocal communication and discrimination between ewes and their lamb in the first two days after parturition. *Developmental Psychobiology*, 49, 375–386.

Sherwin, C. M. & Johnson, K. G. (1987) The influence of social factors on the use of shade by sheep. *Applied Animal Behaviour Science*, 18, 143–155.

Shillito, E. E. (1975) A comparison of the role of vision and hearing in lambs finding their own dams. *Applied Animal Ethology*, 1, 369–377.

Shillito, E. E. & Alexander, G. (1975) Mutual recognition amongst ewes and lambs of four breeds of sheep. *Applied Animal Ethology*, 1, 151–165.

Shillito, E. E. & Hoyland, V. J. (1971) Observation on parturition and maternal care in Soay sheep. *Journal of Zoology, London*, 165, 509–512.

Shillito-Walser, E. E. (1980) Maternal recognition and breed identity in lambs living in a mixed flock of Jacob, Clun Forest and Dalesbred sheep. *Applied Animal Ethology*, 6, 221–231.

Shillito-Walser, E. E. & Hague, P. (1980) Variation in the structure of bleats from sheep of four different breeds. *Behaviour*, 75, 21–35.

Shillito-Walser, E. E., Hague, P. & Walters, E. (1981a) Vocal recognition of recorded lambs voices by ewes of three breeds of sheep. *Behaviour*, 78, 260–272.

Shillito-Walser, E. E., Hague, P. & Yeomans, M. (1983) Preferences for sibling or mother in Dalesbred and Jacob twin lambs. *Applied Animal Ethology*, 9, 289–297.

Shillito-Walser, E. E., Walters, E. & Ellison, J. (1982) Observations on vocalisation of ewes and lambs in the field. *Behaviour*, 91, 190–203.

Shillito-Walser, E. E., Walters, E., Hague, P. & Williams, T. (1985) Responses of lambs to model ewes. *Behaviour*, 95, 110–120.

Shillito-Walser, E. E., Willadsen, S. & Hague, P. (1981b) Pair association between lambs of different breeds born to Jacob and Dalesbred ewes after embryo transplantation. *Applied Animal Ethology*, 7, 351–358.

Shipka, M. P. & Ford, S. P. (1991) Relationship of circulating oestrogen and progesterone concentrations during late pregnancy and the onset phase of maternal behaviour in the ewe. *Applied Animal Behaviour Science*, 31, 91–99.

Slee, J. & Springbett, A. (1986) Early post-natal behaviour in lambs of ten breeds. *Applied Animal Behaviour Science*, 15, 229–240.

Stevens, D., Alexander, G. & Lynch, J. J. (1981) Do Merino ewes seek isolation or shelter at lambing? *Applied Animal Ethology*, 7, 149–155.

Stevens, D., Alexander, G. & Lynch, J. J. (1982) Lamb mortality due to inadequate care of twins by Merino ewes. *Applied Animal Ethology*, 8, 243–252.

Stolba, A., Hinch, G. N., Lynch, J. J., Adams, D. B., Munro, R. K. & Davies, H. I. (1990) Social organisation of Merino sheep of different ages, sex and family structure. *Applied Animal Behaviour Science*, 27, 337–349.

Terrazas, A., Nowak, R., Serafin, N., Ferreira, G., Lévy, F. & Poindron, P. (2001) Twenty-four-hour-old lambs rely more on maternal behavior than on the learning of individual characteristics to discriminate between their own and an alien mother. *Developmental Psychobiology*, 40, 408–418.

Thomson, A. M. & Thomson, W. (1949) Lambing in relation to the diet of the pregnant ewe. *British Journal of Nutrition*, 2, 290–305.

Thorhallsdottir, A. G., Provenza, F. D. & Balph, D. F. (1990) Ability of lambs to learn about novel foods while observing or participating with social models. *Applied Animal Behaviour Science*, 25, 25–33.

Tilbrook, A. J. (1987a) Physical and behavioural factors affecting the sexual attractiveness of the ewe. *Applied Animal Behaviour Science*, 17, 109–115.

Tilbrook, A. J. (1987b) The influence of factors associated with oestrous on the sexual attractiveness of ewes to rams. *Applied Animal Behaviour Science*, 17, 117–128.

Tilbrook, A. J. (1989) Ram mating preferences for woolly rather than shorn ewes. *Applied Animal Behaviour Science*, 24, 301–312.

Tilbrook, A. J. & Cameron, A. W. N. (1990) The contribution of the sexual behaviour of rams to sucessful mating of ewes under field conditions. In: *Reproductive Physiology of Merino Sheep* (Ed. C. M. Oldham, G. B. Martin & I. W. Purvis), pp. 143–160. School of Agriculture, The University of Western Australia, Perth.

Tilbrook, A. J., Cameron, A. W. N. & Lindsay, D. R. (1987) The influence of ram mating preferences and social interaction between rams on the proportion of ewes mated at field joining. *Applied Animal Behaviour Science*, 18, 173–184.

Tomkins, T. & Bryant, M. J. (1972) Mating behaviour in a small flock of lowland sheep. *Animal Production*, 15, 203–210.

Treacher, T. T. (1983) Nutrient requirements for lactation in the ewe. In: *Sheep Production* (Ed. W. Haresign), pp. 133–153. Butterworths, London.

Vasseur, S., Paull, D. R., Atkinson, S. J., Colditz, I. G. & Fisher, A. D. (2006) Effects of dietary fibre and feeding frequency on wool biting and aggressive behaviours in housed Merino sheep. *Australian Journal of Experimental Agriculture*, 46, 777–782.

Vince, M. A. (1986) Response of the newly-born Clun Forest lamb to maternal vocalisation. *Behaviour*, 96, 164–170.

Vince, M. A. (1993) Newborn lambs and their dams: The interaction that leads to sucking. *Advances in the Study of Behavior*, 22, 239–268.

Vince, M. A., Lynch, J. J., Green, G. C. & Elwin, R. L. (1987) Interactions between normal ewes and newly born lambs deprived of visual, olfactory and tactile sensory information. *Applied Animal Ethology*, 19, 119–136.

Vince, M. A., Lynch, J. J., Mottershead, B., Green, G. & Elwin, R. (1985) Sensory factors involved in immediately postnatal ewe/lamb bonding. *Behaviour*, 94, 60–84.

Vince, M. A. & Ward, T. M. (1984) The responsiveness of newly-born Clun Forest lambs to odour sources in the ewe. *Behaviour*, 89, 117–127.

Walkden-Brown, S. W., Martin, G. B. & Restall, B. J. (1999) Role of male-female interaction in regulating reproduction in sheep and goats. *Journal of Reproduction and Fertility Supplement*, 54, 241–255.

Webster, J. (1995) *Animal Welfare. A Cool Eye Towards Eden*. Blackwell Scientific Publications, Oxford.

Whateley, J., Kilgour, R. & Dalton, D. C. (1974) Behaviour of hill country sheep breeds during farming routines. *Proceedings of the New Zealand Society of Animal Production*, 34, 28–36.

Williamson, J. K., Riley, M. L., Taylor, A. N. & Sanson, D. W. (1996) Performance of nursing lambs receiving vitamin E at birth or from dams that received vitamin E. *Sheep and Goat Research Journal*, 12, 69–73.

Winfield, C. G. & Mullaney, P. D. (1973) A note on the social behaviour of a flock of Merino and Wiltshire horn sheep. *Animal Production*, 17, 93–95.

Yurtman, I. Y., Savas, T., Karaagac, F. & Coskuntuna, L. (2002) Effects of daily protein intake levels on the oral stereotypic behaviours in energy restricted lambs. *Applied Animal Behaviour Science*, 77, 77–88.

Chapter 4
Sheep Senses, Social Cognition and Capacity for Consciousness

K.M. Kendrick

Abstract Sheep are generally held in low regard as far as cognition and social skills are concerned. However, there is now increasing evidence from studies of their behaviour and brain function that they have highly sophisticated social and emotional recognition skills using faces, voices and smells. They are able to recognize and remember many different sheep and humans for several years or more and appear to have some capacity for forming mental images of the faces of absent individuals. The presence of such social cognition abilities in this species means that we must pay careful attention to welfare factors such as the composition and stability of their social environment as well as the nature of our own interactions with them.

Keywords Cognition · Consciousness · Emotional cues · Faces · Hearing · Mental imagery · Olfaction · Social recognition · Vision

4.1 Introduction

To many humans, sheep are regarded as being as close to an automaton and mindless animal species as can be imagined and any serious consideration of their cognitive, social and general mental faculties deemed futile. As such, few give serious consideration to their welfare. However, such lowly opinions of the sheep 'mind' are influenced primarily by the overt behaviour patterns of a species that is highly fearful of predation, since it has few defences, and which seems content to be led rather than to lead and to adopt a safe group rather than an individual mentality. As such, what I will now outline about the cognitive, social and mental abilities of this species in this chapter will come as something of a shock to many although hopefully less so to those who have spent considerable time looking after and interacting with sheep.

Before embarking on demonstrating cognitive skills and capacity for conscious awareness the most natural starting point to consider is how sheep can actually use

K.M. Kendrick
Cognitive and Behavioural Neuroscience, The Babraham Institute, Babraham, Cambridge, CB22 3AT, UK

C.M. Dwyer (ed.), *The Welfare of Sheep*,

their different senses to perceive the world around them and the other individuals that populate it. As we will see, this may also serve to dispel certain widely held preconceptions about which senses are most important for this species.

4.2 Sensory Discrimination Abilities and Their Uses

Sheep, like other species, have adapted their senses to achieve remarkable discrimination skills. These allow them to identify important individuals and objects in their environment and communication of social signals. Popular belief tends to emphasise the absolute importance of smell for sheep in line with many other mammals, however as we shall see this is another misconception with all three major senses playing essential roles.

4.2.1 Sheep Olfaction

Like most mammals sheep have an impressive representation of odour detection hardware in their noses and brain. As with all sensory systems the range and acuity of the olfactory sense can easily be ascertained from the number and types of receptors present in the olfactory epithelium in the nose. While fully detailed studies have not been carried out in sheep, the large size of the epithelium and extensive projections into the brain probably put this species on a par with most mammals like rodents, cats and dogs. In these species around 1000 different types of receptors in the epithelium can in combination allow discrimination between literally hundreds of thousands of different airborne odours. So, on the face of it, one might be forgiven for thinking that the sense of smell is all they really need.

There has also been a large amount of work investigating olfactory recognition of both objects and individuals by sheep. Sheep can, for example, during operant choice tasks (where they have to press panels with their nose or feet to indicate which of two smells they have learned is associated with a food reward), distinguish readily between odours from samples of wool, faeces, saliva and secretions from the interdigital pouch, the inguinal pouch and the infraorbital pouch collected from different individuals (Baldwin & Meese 1977). In other contexts, rams are able to distinguish between oestrous and non-oestrous ewes using olfaction (Blissett et al. 1990) and the smell of a ram or their wool can induce oestrus in ewes (Knight 1983). Similarly ewes are actually attracted to the odours of rams when they are sexually receptive during oestrus.

The area smell plays arguably its most important role in sheep is where post-partum ewes learn to recognise their lambs by their individual odour characteristics within 1–2 h of giving birth. The behavioural aspects of this recognition are considered in more detail in chapter 3. We have also made detailed studies of the way the sheep brain processes lamb odours to allow this remarkably effective recognition memory to occur so fast and with such accuracy (Kendrick 2001). The source of the lamb's odour signature that the ewe learns to recognise is thought to be from the wool and skin rather than the amniotic fluid (Alexander & Stevens 1981; Poindron, & Levy 1990). The odour signature must be highly individual since a ewe will reject one of its own twin lambs if it is removed from her at birth and re-introduced to her at a later date. The individuality of this odour recognition can even be seen within the brain at the level of single cells in the primary brain region for processing smell, the olfactory bulb. Here cells can be found that change their electrical activity selectively to the smell of a mother's lamb or its wool (Kendrick et al. 1992).

So why is smell not important for every aspect of recognition? In the first place a number of studies that have deprived sheep of the ability to smell by a variety of different means have actually found that this has little impact on their ability to recognise important objects and individuals. Although this stops maternal ewes from rejecting suckling attempts from strange lambs they are still able to recognise their own lambs specifically using other senses. Indeed, it would appear that it is mainly the sense of smell that can elicit the strong aggressive responses to unrecognised lambs. Maternal ewes that have been rendered incapable of smelling their lambs fail to reject suckling attempts by strange lambs even though they can recognise their own lambs by sight and sound (Baldwin & Shillito 1974). Sheep without the ability to smell also appear to have normal social relationships with other adult animals and have no problems with diet selection (Baldwin et al. 1977). Thus sheep do not rely that much on being able to smell things and, although they are capable of very fine discrimination, it is over a very limited physical range. For example, current estimates suggest that maternal ewes are unable to recognize the odours of their lambs at distances of greater than half a metre (Ferreira et al. 2000).

4.2.2 Sheep Hearing

As with most animal species we know relatively little in detail about how well sheep hear and their ability to distinguish, for example, between individual voices and the different types of vocalisations used. As far as we can tell from observing sheep they are indeed very sensitive to sounds and will quickly orientate their ears towards any new sound source. Current estimates are that sheep have a similar auditory sensitivity to humans (around 10 dB) and that while their low frequency range is slightly higher (125 Hz) compared with us (20–40 Hz) their high frequency range extends well into the ultrasonic domain (42 KHz) which is somewhat above that of humans (20 KHz) and only marginally less than that of dogs (50 KHz). So sheep should be able to hear dog whistles! We should also be mindful of the potential stressful effects of exposing the animals to sources of ultrasonic noise that we may be oblivious to (notably defective machinery, firework noise etc). Surprisingly sensitivity of sound localisation is not that good in ungulates compared with other mammals. Goats, for example, which should presumably be similar to sheep, have an acuity of 19° compared with 5–7° in cats and dogs and 1.5° in humans (see Heffner & Heffner 1992 for comparative auditory ranges and acuities).

Work has established that maternal ewes learn to recognise the voices of their lambs and *vice versa* quite quickly after birth (Shillito 1975, 1978). However, very little is known about the abilities of sheep to distinguish between different adult animals, or human caretakers for example, from their voices although it would indeed be curious if this were only present in mothers and lambs and did not extend to adults. We have been able to demonstrate in behavioural choice maze experiments some ability of sheep to distinguish between sheep and human voices (Kendrick et al. 1995). More recently we have started to analyse sheep voices in more detail using sound spectrographic analysis approaches.

For most people the vocalisation we associate with sheep is the high pitched bleat. This vocalisation, as its description suggests, is at a higher frequency than the low pitched bleat and used in a wide variety of contexts ranging from excitement in anticipating or receiving food to warning the flock about the presence of intruders to signaling that they are experiencing stress, fear or pain. To a casual human ear the call appears similar in all of these contexts. So for sheep can one call really say different things in different circumstances, or is it some form of stereotyped response that does little other than identify who the caller is and that there may be something worth paying attention to?

If one looks at the sound spectrograms from different individuals producing high pitch bleats when they are excited by food as opposed to stressed or fearful during a short period of social isolation the picture is very clear. In the first place it may come as no surprise to see that there are significant differences in the fundamental spectrographic patterns between individual animals (Fig. 4.1). Thus the animals should be able to identify specific individuals from their high pitch bleats.

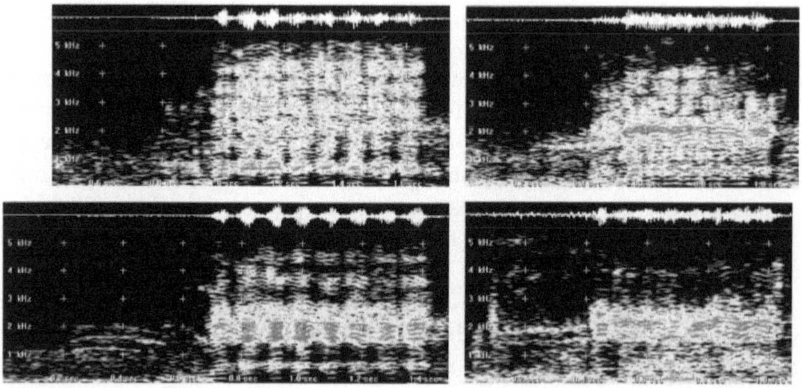

Fig. 4.1 Sound spectrograms of high pitch bleats from two different Clun Forest sheep (*left and right panels*) when they are excited by food (but with a low heart rate – <80 bpm; *top*) or stressed by isolation (heart rate >110 bpm; *bottom*). Note that there are clear individual differences between the animals and that in both cases the bleat under stressed conditions covers a broader frequency and loses the clear bands of modulation that characterise the bleat when the animal is unstressed but excited by food.

What is more interesting however is that the sound spectrograph of the same sheep producing a high pitch bleat when it is excited as opposed to stressed/fearful is also clearly different. When the vocalisation is produced in a positive state of excitement it uses a broader frequency range, especially at the higher frequencies and shows distinct regular bands of alternating high and low intensity. When it is produced during stressful circumstances it has less representation in the higher frequency ranges and almost completely loses the bands of intensity modulation. So this one call may indeed be able to communicate different things to other sheep in different circumstances. Indeed, preliminary behavioural observations in choice mazes have shown that sheep have a preference for choosing high pitch bleat sounds produced by animals in positive emotional states. From a sheep welfare point of view this raises the obvious point that they should be capable of communicating their current emotional state very accurately to others by the different sounds of their bleats. Thus hearing the voice of another individual in pain or distress may also be distressing for any animal that can hear it.

Other interesting observations from analysing the sound spectrograms of sheep vocalisations are that most of the primary components range between 0.5 and 5 KHz, which are similar to those for human speech. Thus the animals should also be reasonably good at distinguishing between human voices and vocal tones.

At this time we do not have evidence that cells in the auditory regions of the sheep brain can respond differentially to vocalisations made by different individuals during distinct emotional states. However, there is every likelihood that they could be found either within the auditory cortex or parts of association cortex dealing with the integration of sensory information and behavioural action.

4.2.3 Sheep Vision

With their laterally displaced eyes sheep have almost all round vision which in fact extends to around 290°. However, although this extensive field of vision is impressive, acuity in the periphery is relatively low and is primarily for detecting movement. Indeed, as soon as the animal detects some movement in this peripheral field it turns its head rapidly to bring the moving object into the field of view where both eyes can see it. When sheep view objects in this frontal binocular eye-field (which is around 40–60°of their visual field) then they are capable of considerable acuity. In psychophysical terms this acuity has been estimated to be in the region of 3–4′ (see Piggins 1992). This estimate places their visual acuity in between that of a cat and a monkey. Again however, as with the peripheral visual field, it is possible that for the sheep their visual acuity even in the frontal eye field may be better for moving rather than static objects (Backhaus 1959; Clarke & Whitteridge 1976). The large eye ball and well developed tapetum, which acts to increase retinal sensitivity by reflecting light back through the photoreceptor layer (Ollivier et al. 2004), also means that sheep are likely to have good vision even at low light intensities (Piggins and Phillips 1996).

With binocular vision (Clarke et al. 1976) sheep have no problems in determining depth and this not only aids discrimination of differences between complex objects but also important environmental features such as sudden drops and cliff-edges. Visual discrimination tests of visual acuity using operant choice tasks (where one of two different visual objects must be chosen to gain a food reward) have shown that sheep can learn to discriminate between different geometric shapes or between the same shape presented at different orientations (Baldwin 1981). They also appear to rely on their visual sense to distinguish between different kinds of grasses and clovers to allow them to exercise their individual grazing preferences (Kendrick 1992; Bazely 1988).

Fig. 4.2 Scenes and individuals viewed from a sheep's-eye point of view (*left panel*) as a dichromat with 20:60 vision and from a human point of view (*right panel*) as a trichromat with 20:20 vision. NB the slight blurring effect that the reduced acuity of the sheep has is most noticeable when viewing the human face, and that the green of the grass appears as yellow to the sheep.

Like horses and other ungulates, sheep would appear to be dichromats rather than trichromats like humans (Carroll et al. 2001). This gives them colour vision in the yellow-green-blue portion of the colour spectrum but not the red. Dichromatic vision is supposed to be better for movement detection compared with trichromatic, and the latter is better for acuity and for detecting objects in the orange/red areas of the spectrum including some fruits, flowers and of course many human skin tones. Thus, the world viewed through the eyes of an average dichromat sheep with a 3′ visual acuity (equivalent to human with 20:60 vision) does appear slightly different from the way it would appear to a human trichromat with 1′ acuity (i.e. 20:20 vision) (Fig. 4.2). However the colour representation would be similar to a human who is red-green colour blind. If this representation of the visual world of the sheep is accurate, it is interesting that the deep green hues that we experience when viewing fields of grass actually would appear more as a range of yellow hues and, as such, hay and grass do not look that dissimilar (Fig. 4.2). Different shades of red actually appear as grey or yellow hues depending on the intensity.

4.3 Recognising Others by Sight

Behavioural observations of wild American Bighorn sheep have revealed remarkable distance vision ability in this species with coyotes and humans being detected at ranges of up to one kilometre. One of the practical advantages that this gives these wild sheep is that they are notoriously difficult for human hunters to shoot (Geist 1968, 1971).

As with both olfaction and hearing, the ability of sheep to use vision to recognise each other was also first established for mother ewes recognising their lambs. Here it was shown that mothers avoided their normally white lambs if either their whole body was coloured black or just their heads were blackened (Alexander & Shillito Walser 1977, 1978). The implication from this is that visual cues from the head are important for recognition. The same researchers also used this strategy to show that the animals could recognise different colours on their lambs (Alexander & Stevens 1979). Following on from this is the question of whether sheep normally recognise each other from their faces. Their abilities to discriminate faces and the way their brains are organised to do this have now received extensive attention in my own laboratory at the Babraham Institute.

Humans (McCarthy et al. 1997; Sergent et al. 1992) and monkeys have evolved specialised parts of the brain for helping recognise faces. The main brain region involved is the temporal cortex where some cells respond selectively to faces. Damage to this region of the brain can lead to problems in recognising faces (prosopagnosia) while recognition of other visual objects is unaffected (Sergent & Signoret 1992). In 1987 we established that this region in the sheep brain also has cells specialised for responding to visual images of faces (Kendrick & Baldwin 1987). Not only did sheep have these specialised cells but they could be shown to classify faces into emotionally distinct categories (whether faces had horns and how big they were– an important index of dominance and gender; whether faces were of members of the same breed and how familiar they were – sheep prefer the company of their own breed and are known to strike up long-term individual friendships; whether faces were from species that could pose a threat – humans and dogs). The cells respond very quickly (usually in 100–200 ms) which is faster than it is possible to make any behavioural response to what is being seen. This may explain why we, and sheep, often move automatically towards an individual that looks familiar only to find that they are not after all. Alternatively, the sight of a face resembling that of your boss, or an enemy, may trigger immediate evasive tactics without real cause!

In terms of the way their brains are organised the face recognition system in sheep is mainly designed for identifying categories of individual that have a specific emotional significance. It implies close interactions between the brain systems dealing with detection of faces and those associated with making emotional responses. Interestingly, it is often just this link that appears to break down in humans with schizophrenia and autism. The system is also optimised for speed, resulting in a speed-accuracy trade-off. This makes sense from a survival point of view because if there is a chance that you might get eaten or beaten up by another individual, it is better to optimise the speed with which you escape from them even if you get it wrong from time to time (better embarrassed at being wrong and alive, than chuffed at being correct but injured or even killed!).

Over the next years we set about the task of showing systematically whether sheep could distinguish between categories of individual and then the extent to which they could actually identify specific individuals from their faces. The first step for doing this was to construct a choice maze apparatus that allowed sheep to choose between face images in order to gain access to the real individual whose face-picture had been seen (Fig. 4.3). To do this we gave the sheep pairs of faces that had different attractions to them (i.e. sheep vs. human; familiar vs. unfamiliar animal or breed; male vs. female). If the sheep normally used faces to distinguish between categories of individuals we argued that they would not have to learn to do this task and would always chose the face that was most attractive to them. This is exactly what happened (Kendrick et al. 1995). The sheep chose sheep faces over human ones and familiar sheep faces over unfamiliar ones. We mainly used female sheep for these studies and they showed a clear ability for distinguishing gender. When they were not sexually interested in males they chose female faces every time, but switched to choosing male faces for a couple of days during each cycle when sex was on the agenda.

4.3.1 Recognising and Remembering Other Sheep and Humans

To establish the full extent of individual face recognition powers of sheep we had to use faces of the same breed and sex and that had more equivalent levels of attraction. In this case it was necessary to reward the animals for making a correct

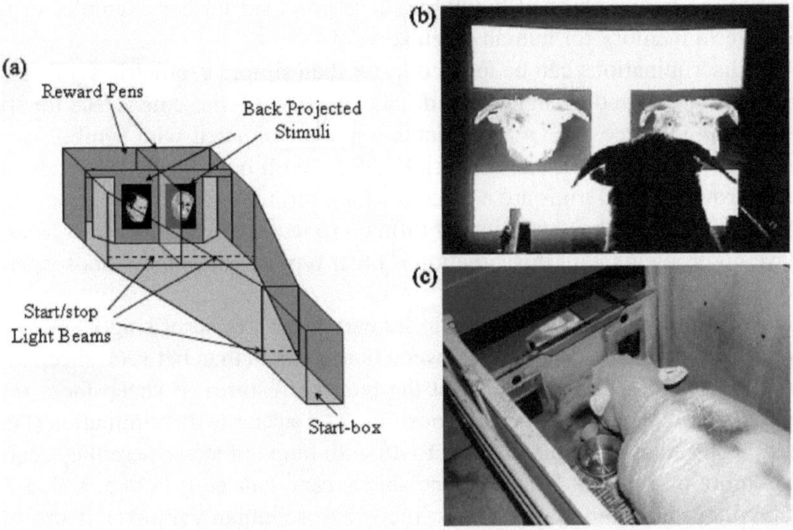

Fig. 4.3 Pictures of the different test apparatus used for sheep to display their face discrimination skills (**a**) a schematic representation of the Y-choice maze, (**b**) and (**c**) photographs of sheep discriminating between two face pictures by pressing one of two operant panels to gain a food reward

choice by giving them food and they obviously had to learn which of the two face pictures got them the reward. We used the same choice maze for this and also a more sophisticated apparatus where the animals indicated which of the two pictures they had chosen by pressing one of two different panels with their nose (see Fig. 4.3). This allowed us to use computer technology to systematically alter the appearance of faces either by showing missing, selective or rearranged features or by blending two faces progressively into one another using morphing programmes to assess how good they were at telling two faces apart. In all cases the face pictures were edited so that no other part of the body was shown.

Using these approaches we have now established a number of facts about sheep face recognition skills which underline what an impressive use of this facility they are capable of making:

(1) Both sheep and human faces can be discriminated although sheep are better at recognising sheep than human faces (the opposite is of course the case for humans!). Our current experiments have indicated that up to 50 different sheep and 10 human faces can be discriminated at any one time (probably many more – Kendrick et al. 2001a).
(2) After the animals have learned to recognise different faces they can remember whether or not they were associated with food for over two years (Kendrick et al. 2001a). Our own experience is that sheep show signs of remembering specific sheep or humans after absences of several years and anecdotal reports

sent to me by members of the public have provided further examples of such long-term memory for human caretakers.

(3) Face discriminations can be learned faster than simple geometric symbols like squares, circles and triangles (Kendrick et al. 1996). The same is true for speed of learning to recognise novel palatable foods compared with symbols or objects associated with them (Kendrick 1992). With both faces and foods often only around 10–20 trials are needed for long-term recognition to occur.

(4) As with humans, sheep find it very difficult to recognise upside-down faces but have no problems with this situation if other types of objects or landscapes are used (Kendrick et al. 1996).

(5) Sheep faces are learned faster than human ones. Not surprisingly sheep take longer to learn to discriminate between human faces than between sheep faces. Indeed, whereas they can use just the internal features of sheep faces (eyes, nose and mouth and their relative positions) for accurate discrimination (Peirce et al. 2000) they find this difficult to do with human faces where they seem to rely more on external features (face shape, ears, hair etc) (Peirce et al. 2001). Thus they may find it difficult to recognise a human caretaker if the latter changes their appearance (change in hair style, wearing a hat etc). The same is not true for familiar sheep within a stable flock where, for example, shearing has no significant impact on recognition since the appearance of internal face features is not affected. However, for recognition of unfamiliar sheep shearing would have an impact since in this case external cues are relied upon in the same way as for recognising humans.

(6) For recognition of familiar sheep the most important feature is the appearance of the eyes (as it is for humans), although the nose and mouth also contribute (Kendrick et al. 1995). Although for horned breeds discrimination of horns and their size is an important behavioural focus and involves specialised encoding in the areas of the brain processing faces, horns are not used by the animals to recognise each other. Thus, an animal trained to respond to the face picture of a particular horned sheep will happily continue to recognise it if the horns are digitally edited out using a computer.

(7) Like humans, the appearance of the half of the face that is seen in the left visual field (i.e. the left side of the face as seen by the observer or the right side of face of the observed) is used more for recognition than the right (Broad et al. 2000; Peirce et al. 2000). This implies that sheep have the same right brain hemisphere advantage for faces as in humans since visual information from this part of the visual field is preferentially routed to the right hemisphere.

(8) The visual acuity of sheep for faces is quite remarkable. In the first place they are still capable of discriminating between sheep face pictures that are 25% of normal size. However, even more impressive than this is the fact that they are still able to distinguish between two sheep faces that only differ from one another by 5–10% difference. This is done using systematic computer morphing programmes that will progressively merge two faces together. Figure 4.4 shows what these faces actually look like and compares sheep and human

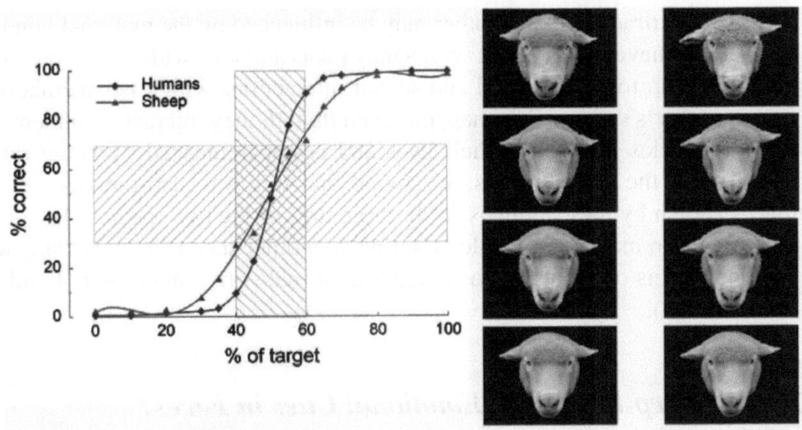

Fig. 4.4 Graph shows the average face discrimination acuity in sheep vs. humans for 10 different pairs of sheep faces. Acuity is measured by progressively merging the two face pictures using computer morphing programmes. To discriminate better than chance both sheep and humans have to achieve >70% or <30% choice. Examples of one of the face pairs showing 100, 70, 60 and 55% levels of the face that the sheep is normally rewarded for choosing are also given

performances on the same discrimination sets. Humans are only slightly better than sheep in this difficult task!

(9) Similar to human children, lambs are not born with an immediate ability to distinguish between faces, they require experience to do this. Thus, lambs cannot recognise the faces of their mothers accurately for nearly 4 weeks after they are born (Kendrick 1994). Maternal ewes are therefore more responsible for locating their young than the other way around. Interestingly, the mothers also have trouble in recognising the faces of their lambs for the first 2–3 weeks after birth, which is why odour and vocal cues are more important at this time (Kendrick et al. 1996). The slow speed for mothers recognising the faces of their lambs is probably due to their high level of homogeneity in appearance and small size, since learning to recognise new adult faces can be achieved in a few hours (Kendrick et al. 1996).

4.3.2 Do sheep Find Faces Attractive?

While we know that sheep are more attracted to familiar than unfamiliar faces the real test of whether the animals find one face more attractive than another is where the choice is made between faces of unfamiliar individuals. We have found that female sheep exhibit clear individual preferences for the faces of specific males and that when the female sees their face it more strongly activates areas of the brain controlling sexual and emotional responses (Fabre-Nys et al. 1997). Interestingly, more mature males seem to be preferred over younger ones!

Face cues are therefore clearly being used as a source of individual attraction as well as recognition in sheep and we have even shown that what determines whether one face is more attractive than another may be influenced by the maternal bond. In studies where we have raised lambs with nanny goats and kids with ewes from birth, the young grow up to prefer social and sexual interactions with other members of their foster mother's species. This happens even though they interact with members of their own species throughout their lives and is not influenced by being raised with a sibling of the same species. We found that this was also the case for the preferences shown by these animals for face pictures of the two species. The effect is much stronger in male than female offspring so mother sheep and goats may well have lifelong effects on what type of females their male offspring prefer (Kendrick et al. 1998, 2001b).

4.3.3 Can Sheep Recognise Emotional Cues in Faces?

Charles Darwin in "The Expression of the Emotions in Man and Animals" (1872) was one of the first biologists to assert that animals involuntarily communicate their emotional state to others by subtle muscular and nervous changes affecting both physical appearance and vocalisations. He ends by stating: "To understand, as far as is possible, the source or origin of the various expressions which may be hourly seen on the faces of the men around us, not to mention our domesticated animals, ought to possess much interest for us" and "the object...deserves still further attention, especially from any able physiologist". It has taken nearly 100 years for the physiological control of emotional behaviour and learning to receive this further attention (LeDoux 1996) but few studies have focussed on how communication of emotional states occurs using visual or vocal cues in mammals other than humans. From an animal welfare point of view it is now reasonable to accept that many species can experience both positive and negative emotional states. However, if they can also communicate these emotions readily to others, who can then empathise with them, this adds a further dimension to be considered when assessing both the complexity of their social environment and their subsequent welfare needs.

The links between perception and emotion processing are important for any social animal species to function normally. When they break down in humans it can result in the distress of being unable to communicate or respond appropriately to important social signals, leading to problems with integration into society (Blair 2003). While there has been a major focus on the neural substrates for fear conditioning and aggression within the amygdala in the limbic system (LeDoux 1996), little is known about how perception and emotion processing in social contexts is regulated at a behavioural or neural level.

The high level of acuity that sheep have for discriminating between faces should allow them both to distinguish between facial expressions on human faces and to detect small changes in the appearance of sheep faces (enlarged protruding eyes showing the whites, flared nostrils and skin wrinkling around the nose, flattened ears, open mouth, etc – see Fig. 4.5) denoting a range of different positive and negative emotional states. In preliminary operant choice experiments we have confirmed this in four sheep using human faces. Here the animals showed a

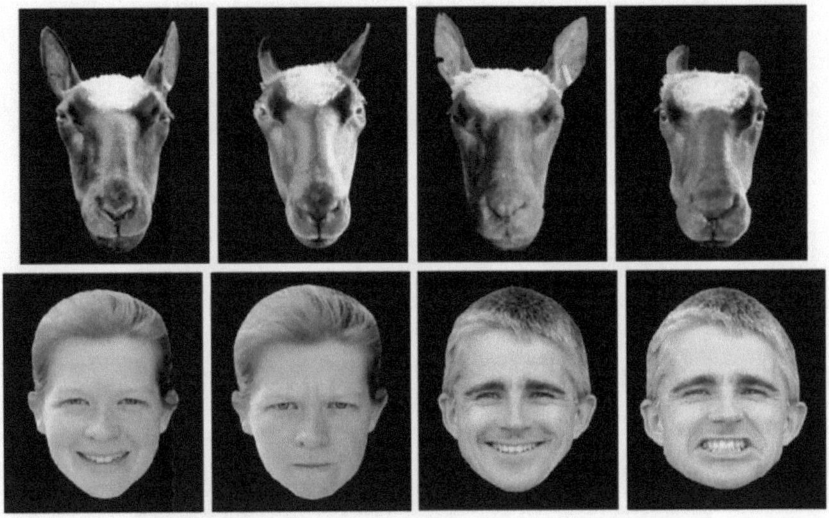

Fig. 4.5 Face pictures of the same two sheep in calm (*left*) vs. stressed (*right*: isolated with heart rate >110 bpm) conditions and smiling/angry humans. Sheep normally show a preference for choosing the calm and smiling versions respectively. NB the characteristic bulging eyes, flattened ears and slightly flared nostrils of the sheep when they are stressed

significant preference for pressing panels to gain a food reward that were associated with a smiling as opposed to angry or neutral versions of the same familiar human face (70–80% choice). This choice of smiling faces also occurred if the faces used were of unfamiliar individuals but the effect was less robust, suggesting a learning or motivation component. We have also accumulated evidence that when sheep are initially presented with new pairs of sheep faces to discriminate between, they persistently avoid choosing face pictures of individuals that are vocalising (showing open mouth) or have flattened ears or enlarged protruding eyes and pupils which also show the whites, indicative of being stressed. We have now confirmed that animals prefer to choose a face picture of the same animal when it is calm as opposed to when it has been stressed through isolation or shearing and can also be trained to discriminate between calm and anxious versions of the same face. Interestingly, while they have a preference for the faces of familiar individuals they prefer to chose the face of a calm unfamiliar animal to that of a stressed familiar one (Tate et al. 2006; see Fig. 4.5 for examples of faces).

4.3.4 What are the Welfare Implications of Face Recognition and Attraction in Sheep?

The first and most obvious point is that sheep need to have unimpeded vision if they are going to be able to use this sense to negotiate successfully their social and physical environments. As we can see from a picture taken at a UK Agricultural

Fig. 4.6 Picture of a sheep at an agricultural show with fleece covering its eyes

Show this is not always appreciated (Fig. 4.6). Rapid treatment of eye infections and trimming of overgrown horns is also important.

The fact that sheep can recognise and remember large numbers of sheep and human faces also tells us two important things about them. In the first place no species would have developed such sophisticated individual recognition skills unless they had a need for them due to living in a highly complex social environment. In the second place having a long-term memory for faces both shows that sheep do indeed have relatively advanced cognitive skills and may have the capacity to think about individuals missing from their social environment (this possibility will be discussed more below). Both of these observations provide strong arguments for keeping sheep in a stable social environment.

Since faces are undoubtedly a source of attraction for sheep, in the same way as they are for humans, it occurred to us that being exposed to them might actually help alleviate the stress of isolation. We have found that just the sight of faces of familiar or unfamiliar members of the same breed does indeed have a profound calming influence on sheep experiencing the psychological stress of a brief period of isolation (da Costa et al. 2004). Seeing pictures of a familiar type of face reduces behavioural expressions of stress (vocalisations and increased activity), autonomic indices (heart rate), hormonal indices (adrenalin and cortisol) and activation of areas of the brain controlling stress and fear responses (Fig. 4.7). It would seem therefore that under conditions where sheep need to be isolated from the flock that the presence of just a picture of a sheep face of the same breed could significantly reduce the effects of isolation stress. Indeed in another context it has been shown that pictures of sheep can help encourage animals to enter raceways in stock yards (Franklin & Hutson 1982).

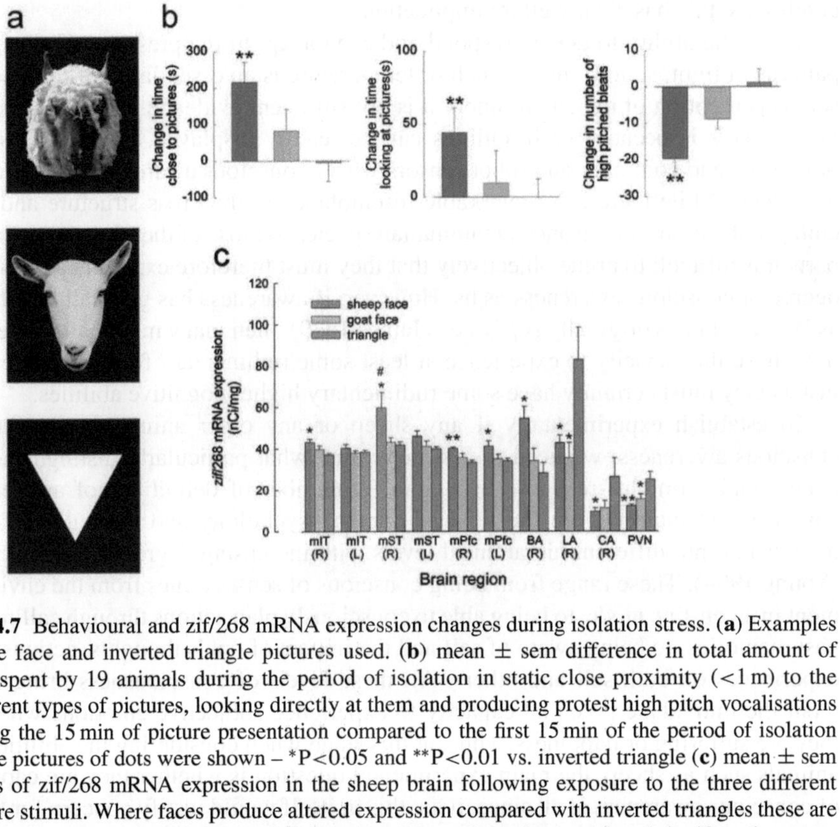

Fig. 4.7 Behavioural and zif/268 mRNA expression changes during isolation stress. (**a**) Examples of the face and inverted triangle pictures used. (**b**) mean ± sem difference in total amount of time spent by 19 animals during the period of isolation in static close proximity (<1 m) to the different types of pictures, looking directly at them and producing protest high pitch vocalisations during the 15 min of picture presentation compared to the first 15 min of the period of isolation where pictures of dots were shown – *$P<0.05$ and **$P<0.01$ vs. inverted triangle (**c**) mean ± sem levels of zif/268 mRNA expression in the sheep brain following exposure to the three different picture stimuli. Where faces produce altered expression compared with inverted triangles these are indicated by *$P<0.05$ and **$P<0.01$ (Tukey test). Where sheep faces produced significantly greater expression that either goat faces or triangles these are indicated by #$P<0.05$. Brain areas: IT (inferotemporal cortex), ST (superior temporal cortex), mPFC (medial prefrontal cortex), BA (basal amygdala nucleus) (these are involved with face recognition), LA (lateral amygdala nucleus), CA (central amygdala nucleus) (both involved in fear responses) and the PVN (hypothalamic paraventricular nucleus) (involved in mediating stress responses). (R) and (L) refer to right and left brain hemisphere respectively and where not indicated measurements were made bilaterally (6–7 animals were used per condition with optical density readings made from 4–8 sections per region per animal) (adapted from Da Costa et al. 2004)

4.4 Are Sheep Conscious of their Surroundings and Can They "Think" about Individuals or Objects in Their Absence?

So sheep are attracted to the faces of different individuals and can remember them for several years or more. Does this mean they are conscious of these individuals when they perceive them and are capable of 'thinking' about them in their absence?

These are very difficult questions to answer in any animal species and in humans it is only through our own experiences and language that we can answer them for ourselves. This does not mean however that we cannot try to provide at least some answers since were sheep and other animals to have even limited capacities for consciousness, and be able to suffer as a result of "missing" absent individuals or conditions, this has clear welfare implications.

While the ability to detect, respond and even adapt to the presence of changing patterns of light, sound, smell, touch or temperature is an essential first step in conscious perception of the environment it is not sufficient evidence for its occurrence *per se.* This is because such abilities can be readily displayed by simple micro-organisms and computer and robot sensors where conscious awareness clearly does not occur. While there is a remarkable resemblance in the gross structure and circuitry of the brains of advanced mammalian species to those of their human counterparts it is difficult to argue objectively that they must therefore experience the same degree of conscious awareness as us. However, if awareness has gradually evolved, as William James originally suggested (James 1879), then many mammalian species must have the capacity to experience at least some rudimentary form of awareness just as they must certainly have some rudimentary higher cognitive abilities.

To establish experimentally if any sheep or any other animal is capable of conscious awareness, we need to first determine what particularly distinguishes it from simple stimulus-response behaviour. A number of definitions of awareness have evolved mainly from the field of human psychology and normally involve its division into different hierarchical levels with increasing degrees of complexity (Young 1994). These range from being conscious of sensory cues from the environment in an on-line mode, to being able to consciously plan actions through calling up past memories, to being aware of self and the impact of one's thoughts, actions and experiences on both self and others. All these levels of consciousness bring with them the immediate potential capacity to experience subjective emotions whether they are suffering or happiness. Thus, at this stage when considering the abilities of animals such as sheep, the main experimental question is whether they are capable of conscious awareness at all since, if so, this in itself would justify concern for their welfare.

A number of complex behaviours are highly suggestive of consciousness, such as self-recognition, social communication, individual recognition of members of their own, or other, species, deceit and empathy, and complex rule learning (Kendrick 1997). However, reasonable evidence for many of these behaviours has only been provided in higher primates and in even in these cases experiments are often open to re-interpretation due to limitations in the experimental paradigms used (Heyes 1994; Kendrick 1997).

The most tractable area for research in this field is the ability to form and use mental images to guide behaviour. While this capacity by itself does not necessarily imply consciousness it is one of the important component parts. One way forward in trying to establish the presence and use of mental imagery in sheep, or other animals, is to combine behavioural assessments (which are often open to problems of interpretation) with a consideration of how an animal's brain is organised to process sensory information from the environment. Where possible this can then be contrasted with what is known about how the human brain functions under similar circumstances (see Kendrick 2007). Such a neurobiological approach to understanding consciousness has also been proposed by others (Crick & Koch 1990).

Recent advances in functional brain imaging techniques using magnetic resonance imaging (MRI) and positron emission tomography (PET) have allowed studies to be conducted in humans aimed at understanding which brain regions are functionally active during actual perception of objects and whether these are the same or different from those which are active when an individual forms a mental image of them. Results from these studies, together with those from the neuropsychological literature derived from brain damaged human patients, have shown considerable overlap between brain regions which are activated during direct perception of objects and when mental images are formed of them (Farah 1995; Farah & Feinberg 1997; Kanwisher et al. 1996, 1997; Koch & Braun 1996). There is not, of course, complete overlap as illustrated by the phenomenon of 'blindsight' in humans, where brain damaged patients can still respond appropriately to objects without actually being aware of them (Farah & Feinberg 1997; Kentridge et al. 1999; Milner & Goodale 1995).

The implications of this research are that if we can show that the sheep brain processes complex sensory information from objects in the same way that the human brain does then it should have at least some capacity to form mental images of them in their absence and hence potentially be consciously aware. One of the most important areas to focus on in this respect is social recognition and in particular recognition using visual cues from the face since sheep appear to have very similar specialised systems within the brain for recognising faces as we do.

4.4.1 Can Sheep Form and Use Mental Images of Faces?

We now have some preliminary evidence to suggest that sheep can indeed form and use mental images of faces. The first experiments we carried out addressed the issue of whether sheep could recognise different views of faces to those they were trained to recognise in a choice Y-maze. Thus we asked the question of whether they could immediately recognise profile views of the same sheep faces having been trained only using frontal views of them, or *vice versa*. While some of the same facial features are present in both frontal and profile views, their orientation, and therefore appearance, are very different. In all cases, face images are viewed against a neutral uniform black background to eliminate any common peripheral visual cues. The ability to determine that a particular profile view belongs to the same animal viewed from the front should therefore involve some capacity to mentally rotate the face image. Sheep can indeed perform this task with both familiar and unfamiliar sheep faces, and even with human faces (Kendrick et al. 2001c). There are also cells within the temporal cortex which respond equivalently to frontal and profiled views of faces (Peirce 2000) and presumably these may assist in the process of mental rotation. While we cannot entirely rule out the ability of the animals to use some sophisticated form of stimulus generalisation to determine that a profile view of a face belongs to the same individual that has previously been viewed only from the front (or *vice versa*) it seems unlikely that this is happening. The main reason for supposing this is that features that are common to both frontal and profile views of faces look very different in the two cases due to being turned by 90°. One could also argue that if they were able to use stimulus generalisation in this context then they should also be able to recognise inverted views of the same face which they cannot. We also have evidence that many animals can even match-up rear head views with frontal ones, which is even more impressive.

A second task we have developed which should require the use of mental imagery is matching to sample using face images. In this task the animal views a single face and, after a variable delay, is shown two faces, one of which is the original face. The animal is required to identify the face it has just seen (matching to sample), or alternatively the face that it has not previously seen (non-matching to sample) in order to obtain a food reward. This is a common working memory task in human experimental psychology and successful performance is generally thought to involve the individual having the capacity to form and hold a mental image of the training stimulus up until the point where the recognition test is presented. As such, it is a very difficult task and not one that is easily demonstrated (in terms of visual object recognition) in non-primate animals. Nevertheless goats have been found to be able to do this with pictures of shapes (Soltysik & Baldwin 1972) and there is some evidence in sheep that they can do this too with visual stimuli after short delays (5–10 s). However, further work is needed to assess their overall abilities more precisely and considerable amounts of training are required for them to acquire this task.

4.4.1.1 Is there Evidence from the Brain for Sheep Forming Mental Images of Faces?

We have made two approaches to addressing this question. The first has been to establish whether non-visual cues can evoke activation of regions of the temporal and frontal cortices that respond to visual cues from faces. The model chosen was maternal ewes responding to the bleats of their lambs after a period of separation. This was chosen since by one month post-partum ewes readily recognise their lambs from their faces and it was felt that hearing the lambs bleat alone after a period of separation might evoke a visual mental image of its face. Previous electrophysiological studies had also failed to find any evidence that face-sensitive neurones in the temporal cortex could respond to auditory stimuli. Under these circumstances it was found that bleats did indeed produce similar levels of activation (as evidenced by quantifying altered messenger RNA expression of the immediate early gene, c-*fos*, which is widely used as a neuroanatomical marker of increased neural activity in the brain – see Broad et al. 2000) in the temporal and frontal cortices as seen following exposure to pictures of the lambs faces. As a control it was shown that exposing the ewes to the same lamb bleat vocalisations but re-organised in a random sequence (i.e. sections of the sound spectrogram were cut out and reassembled in a different order to provide a sound containing identical acoustic components but in the wrong sequence) failed to produce the same level of activation (Fig. 4.8).

A similar approach has also provided some evidence that odour cues from an absent lamb can provoke responses from cells in the medial frontal cortex of a maternal ewe which respond to pictures of her lamb's face. Although difficult to prove, this may be due to the odour of the absent lamb evoking a visual mental image of its face. This is suggested by the fact that combining the lamb odour and the sight of its face did not have any additive effect (it would be expected to if there was convergence of visual and odour cues) and that an unexpectedly high number of face sensitive cells were also influenced by the lamb's odour. Interestingly, the only face-sensitive cells that showed this effect were those that were view independent (they responded equivalently to different views of the face). Those that only responded to a specific view (view dependent) showed no responses to odours. So it is possible that it is only the view-independent cells that are involved in mental imagery in this case (Tate et al. 2006).

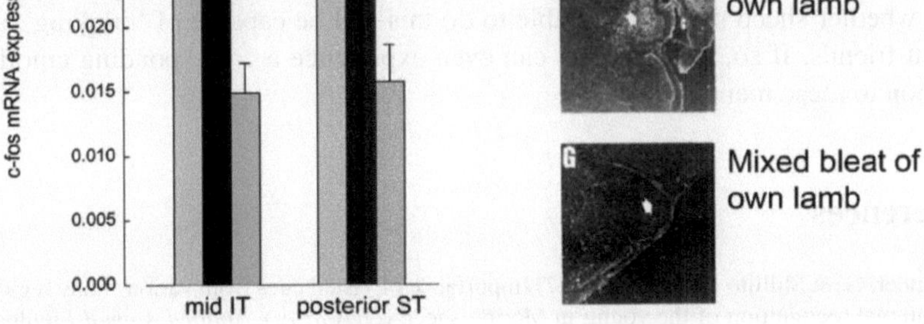

Fig. 4.8 Histograms and brain sections showing patterns of neural activation (using a specific gene marker c-*fos* and indicated by white in the brain sections on the *right*) in the part of the sheep temporal cortex that is particularly responsive to the sight of faces. When a mother ewe is separated from her lamb and hears its bleat but does not see it, the same level of activation is seen in this visual area. This does not happen if the bleat is made unrecognisable by scrambling its sequence, implying that the sound of the bleat may have caused the mother to form a mental visual image of her lamb's face.

The second method we have started to employ is use of digitised video sequences as stimuli while recording from cells in the temporal cortex that are responsive to the faces of specific familiar individuals. Face-sensitive cells in this region can often respond selectively to only one or two socially familiar individuals. We have constructed video sequences which are shot from a sheep's eye level and which eventually reveal the presence of the specific socially familiar individual in its home pen. The animals are repeatedly exposed to this video sequence while recordings are made from cells that respond selectively to pictures of that sheep's face. Randomly interposed in these repeated video sequences are similar ones which, when the home pen is finally revealed, no longer contain the familiar individual. The cells continue to show altered responses to parts of the film sequence where the familiar sheep is expected to be there but is not (Kendrick et al. 2001c). This shows that neural systems responding to the actual sight of faces are indeed active at times when a sheep expects to see a particular individual and may therefore be the result of the animal forming a mental image of them. Clearly other paradigms that are designed to evoke mental images of specific individuals, such as them being obviously hidden behind screens or in the delay period during a matching to sample task, will be needed to provide more supporting evidence.

4.5 General Conclusions

Overall it can be seen that sheep make very sophisticated use of their senses for both social and non-social purposes and rely much more on their visual sense than one might have expected. They have remarkable abilities for recognising and remembering large numbers of different individuals from their faces and use the same specialised system for doing this in their brains just as in humans. It would appear that they can even learn to recognise human facial expressions and visual cues indicating emotional state on the faces of other sheep. Since we also use this specialised system in our own brains to imagine faces this does beg the question as to whether sheep may also be able to do this and be capable of 'thinking' about absent friends. If so, perhaps they can even experience a corresponding emotional reaction to these mental images.

References

Alexander, G. & Shillito Walser, E. (1977) Importance of visual cues from various body regions in maternal recognition of the young in Merino sheep (*Ovis aries*). *Applied Animal Etho*logy, 3, 137–143.

Alexander, G. & Shillito Walser, E. (1978). Maternal responses in merino lambs to artificially coloured lambs. *Applied Animal Ethology*, 4, 141–152.

Alexander, G. & Stevens, D. (1979) Discrimination of colours and grey shades by merino ewes: Tests using coloured lambs. *Applied Animal Ethology*, 5, 215–231.

Alexander, G. & Stevens, D. (1981) Recognition of washed lambs by merino ewes. *Applied Animal Ethology*, 7, 77–86.

Backhaus, D. (1959) Visual acuity in ungulates. *Zeitschrift fur Tierpsychologie*, 16, 454.

Baldwin, B. A. (1981) Shape discrimination in sheep and calves. *Animal Behaviour*, 29, 830–834.

Baldwin, B. A, McLaughlin, C. L. & Baile C. A. (1977) The effect of ablation of the olfactory bulbs on feeding behaviour in sheep. *Applied Animal Ethology*, 3, 151–161.

Baldwin, B. A. & Meese, G. B. (1977) The ability of sheep to distinguish between conspecifics by means of olfaction. *Physiology and Behaviour*, 18, 803–808.

Baldwin, B. A. & Shillito, E. E. (1974) The effects of ablation of the olfactory bulbs on parturition and maternal behaviour in Soay sheep. *Animal Behaviour*, 22, 220–223.

Bazely, D. R. (1988) '*Foraging behaviour of sheep* (*Ovis aries L.*) *grazing on swards of perennial ryegrass* (*Lolium perenne L.*)' Doctoral thesis. University of Oxford.

Blair, R. J. R. (2003) Facial expressions, their communicatory functions and neuro-cognitive substrates. *Philosophical Transactions of the Royal Society London B*, 358, 561–572.

Blissett, M. J., Boland, K. P. & Cottrell, D. F. (1990). Discrimination between odours of fresh oestrous and non-oestrous ewe urine by rams. *Applied Animal Behaviour Science*, 25, 51–59.

Broad, K. D., Mimmack, M. L. & Kendrick, K. M. (2000) Is right hemisphere specialization for face discrimination specific to humans? *European Journal of Neuroscience*, 12, 731–741.

Carroll, J., Murphy, C. J., Neitz, M., Ver Hoeve, J. N. & Neitz, J. (2001) Photopigment basis for dichromatic color vision in the horse. *Journal of Vision*, 1, 80–87.

Clarke, P. G. H. & Whitteridge, D. (1976) The cortical visual areas of the sheep. *Journal of Physiology* (*London*), 276, 497–508.

Clarke, P. G. H., Donaldson, I. M. L. & Whitteridge, D. (1976). Binocular visual mechanisms in cortical areas I and II of the sheep. *Journal of Physiology* (*London*), 256, 509–526.

Crick, F. & Koch, C. (1990) Towards a neurobiological theory of consciousness. *Seminars in the Neurosciences*, 2, 263–275.

Da Costa, A. P. C., Leigh, A. E. Man M-S & Kendrick, K. M. (2004) Face pictures reduce behavioural, autonomic, endocrine and neural indices of stress and fear in sheep. *Proceedings of the Royal Society B*, 271, 2077–2084.

Darwin, C. (1872) *The Expression of the Emotions in Man and Animals*, London, Albermarle.

Fabre-Nys, C., Ohkura, S. & Kendrick, K. M. (1997) Male faces and odours evoke differential patterns of neurochemical release in the mediobasal hypothalamus of the ewe during oestrus: An insight into sexual motivation. *European Journal of Neuroscience*, 9, 1666–1677.

Farah, M. J. (1995) Current issues in the neuropsychology of image generation. *Neuropsychologia*, 33, 1455–1471.

Farah, M. J. & Feinberg, T. E. (1997) Consciousness of perception after brain damage *Seminars in Neurology*, 17, 145–152.

Ferreira, G., Terrazas, A., Poindron, P., Nowak, R., Orgeur, P. & Levy, F. (2000) Learning of olfactory cues is not necessary for ealy lamb recognition by the mother. *Physiology and Behavior*, 69, 405–412.

Franklin, J. R. & Hutson, G. D. (1982). Experiments on attracting sheep to move along a laneway. III. Visual stimuli. *Applied Animal Ethology*, 8, 457–478.

Geist, V. (1968) On the inter-relation of external appearances, social behaviour and social structures of mountain sheep. *Zeitschrift fur Tierpsychologie*, 25, 199–215.

Geist, V. (1971) '*Mountain sheep: A study in behaviour and evolution*'. University of Chicago Press, Chicago.

Heffner, H. E. & Heffner, R. S. (1992) Auditory perception. In: '*Farm Animals and the Environment*', (Eds. C. Phillips & D. Piggins) CAB International, Oxford pp. 159–184.

Heyes, C. M. (1994) Social cognition in primates. In: '*Animal Learning and Cognition*'. (ed. J. Mackintosh), Academic Press, San Diego, pp. 281–305.

James, W. (1879). Are we automata? *Mind*, 4, 1–22.

Kanwisher, N., McDermott, J. & Chun, M. M. (1997) The fusiform face area: A module in human extrastriate cortex specialized for face perception. *Journal of Neuroscience*, 17, 4302–4311.

Kanwisher, N., Chun, M. M., McDermott, J. & Ledden, P. J. (1996). Functional imaging of human visual recognition. *Cognitive Brain Research*, 5, 55–67.

Kendrick, K. M. (2007) Quality of life and the evolution of the brain. *Animal Welfare*, 16(S), 9–15.
Kendrick, K. M. (1992) Cognition. In: '*Farm Animals and the Environment*', (Eds. C. Phillips & D. Piggins) CAB International, Oxford, pp. 209–231.
Kendrick, K. M. (1994) Neurobiological correlates of visual and olfactory recognition in sheep. *Behavioural Processes*, 33, 89–111.
Kendrick, K. M. (1997) Animal awareness. *British Society for Animal Science Occasional Publication*, 20, 1–8.
Kendrick, K. M. (2001) Oxytocin, motherhood and bonding. *Experimental Physiology*, 85S, 111S–124S.
Kendrick, K. M., Atkins, K., Hinton, M. R., Broad, K. D., Fabre-Nys, C. & Keverne, E. B. (1995) Facial and vocal discrimination in sheep. *Animal Behaviour*, 49, 1665–1676.
Kendrick, K. M., Atkins, K., Hinton, M. R., Heavens, P. & Keverne, E. B. (1996) Are faces special for sheep – evidence from facial and object discrimination-learning tests showing effects of inversion and social familiarity. *Behavioural Processes*, 38, 19–35.
Kendrick, K. M. & Baldwin, B. A. (1987) Cells in temporal cortex of conscious sheep can respond preferentially to the sight of faces. *Science*, 236, 448–450.
Kendrick, K. M., Hinton, M. R., Atkins, K., Haupt, M. A. & Skinner, J. D. (1998) Mothers determine sexual preferences. *Nature*, 395, 229–230.
Kendrick, K. M., da Costa, A. P., Hinton, M. R., Leigh, A. E. & Peirce, J. W. (2001a) Sheep don't forget a face. *Nature*, 414, 165–166.
Kendrick, K. M., Haupt, M. A., Hinton, M. R., Broad, K. D. & Skinner, J. D. (2001b) Sex differences in the influences of mothers on the socio-sexual preferences of their offspring. *Hormones and Behavior*, 40, 322–338.
Kendrick, K. M., Leigh, A. E. & Peirce, J. (2001c) Behavioural and neural correlates of mental imagery in sheep using face recognition paradigms. *Animal Welfare*, 10, 89–101.
Kendrick, K. M., Levy, F. & Keverne, E. B. (1992) Changes in the sensory processing of olfactory signals induced by birth in sheep. *Science*, 256, 833–836.
Kentridge, R. W., Heywood, C. A. & Weiskrantz, L. (1999) Attention without awareness in blindsight. *Proceedings of the Royal Society of London B*, 266, 1805–1811.
Knight, T. (1983) Ram induced stimulation of ovarian and oestrous activity in anoestrus ewes – a review. *Proceedings of the New Zealand Society for Animal Production*, 43, 7–11.
Koch, C. & Braun, J. (1996) Towards the neuronal correlate of visual awareness. *Current Neurobiology*, 6, 158–164.
LeDoux, J. E. (1996) '*The Emotional Brain: The Mysterious Underpinnings of emotional Life*'. New York, Simon and Schuster.
McCarthy, G., Puce, A., Gore, J. C. & Allison, T. (1997) Face-specific processing in the human fusiform gyrus. *Journal of Cognitive Neuroscience*, 9, 605–610.
Milner, A. D. & Goodale, M. A. (1995) '*The Visual Brain in Action*', Oxford University Press, Oxford, pp. 25–155.
Ollivier, F. J., Samuelson, D. A., Brooks D. E., Lewis, P. A., Kallberg M. E. and Komáromy A. M. (2004) Comparative morphology of the tapetum lucidum (among selected species) *Veterinary Opthalmology*, 7, 11–22.
Peirce, J. W. (2000) Effects of hemispheric asymmetry and configural coding in sheep face recognition. PhD Thesis, University of Cambridge.
Peirce, J. W., Leigh, A. E., daCosta, A. P. C & Kendrick, K. M. (2001) Human face recognition in sheep: Lack of configurational coding and right hemisphere advantage. *Behavioural Processes*, 55, 13–26.
Peirce, J. W., Leigh, A. E. & Kendrick, K. M. (2000) Configurational coding, familiarity and the right hemisphere advantage for face recognition in sheep. *Neuropsychologia*, 38, 475–483.
Piggins, D. (1992) Visual perception. In: '*Farm Animals and the Environment*', (Eds. C. Phillips & D. Piggins) CAB International, Oxford pp. 131–158.
Piggins, D. & Phillips, C. J. C. (1996) The eye of the domesticated sheep with implications for vision. *Animal Science*, 62, 301–308.

Poindron, P. & Levy, F. (1990) Physiological, sensory and experiential determinants of maternal behaviour in sheep. In: '*Mammalian Parenting*', (Eds. N. A. Krasnegor & R. S. Bridges), Oxford University Press, New York, pp. 133–156.

Sergent, J., Ohta, S. & Macdonald, B. (1992) Functional neuroanatomy of face and object processing – a positron emission tomography study. *Brain*, 115, 15–36.

Sergent, J. & Signoret, J. L. (1992) Functional and anatomical decomposition of face processing: Evidence from prosopagnosia and PET study of normal patients. *Philosphical Transactions of the Royal Society London B*, 335, 55–61.

Shillito, E. E. (1975) A comparison of the role of vision and hearing in lambs finding their own dams. *Applied Animal Ethology*, 1, 369–377.

Shillito, E. E. (1978) A comparison of the role of vision and hearing in ewes finding their own lambs. *Applied Animal Ethology*, 4, 71–79.

Soltysik, S. S. & Baldwin, B. A. (1972) The performance of goats in triple choice delayed response tasks. *Acta Neurobiologica Experimenta*, 32, 73–86.

Tate, A. J., Fischer, H., Leigh, A. E. & Kendrick, K. M. (2006) Behavioural and neurophysiological evidence for face identity and face emotion processing in animals. *Philosphical Transactions of the Royal Society London B*, 361, 2155–2172.

Young, A. W. (1994) Neuropsychology of awareness In: '*Consciousness in Philosophy and Cognitive Neuroscience*' (Eds. A. Revonsuo & M. Kampinnen) Erlbaum, Hillsdale, USA, Chapter 8.

Chapter 5
The Impact of Disease and Disease Prevention on Welfare in Sheep

P.A. Roger

Abstract Disease causes adverse welfare to the individual or the flock. The ethical and legal bases of our involvement in the duties of care that we owe domesticated species are outlined and the relevance of these duties to the implementation of preventive medicine is explored. Sheep are capable of not only feeling pain but also of learning, displaying emotion and memory. That this represents a level of sentience is unarguable so that the duties we owe to these animals are of importance in a moral context. The ethical approach to the control of disease and its impact on the individual and the flock is a major determinant of the standard for the generally accepted treatment of animals. The impact of specific diseases on the welfare of the individual and the flock is discussed and preventive measures are outlined. The cardinal points of flock health programmes are discussed and the key elements of maintaining a healthy flock and maximising welfare are raised. Continuing vigilance and early recognition and diagnosis of disease is advocated. Measures to prevent disease should be prioritised, coupled with the proper strategic use of licensed products, where these are required, to prevent the development of drug resistance. The importance of a multi-disciplinary approach to the exploration and understanding of welfare and the impact of disease is paramount

Keywords Sheep · Health · Welfare · Disease · Measurement · Flock · Planning · Management · Preventive medicine.

5.1 Introduction

Disease is a state of disturbance to the health status of an animal and can be caused by any factors that alter this status. Diseases can be species specific, shared with other species, or zoonotic (i.e. can be transmitted between animals and humans). Disease threats occur at three levels: (1) threats to the national flock; (2) threats to individual flocks; and (3) threats to individual sheep. The spread of disease can be

P.A. Roger
Veterinary Consultancy Services, Victoria Cottage, Reeth, Richmond, North Yorkshire, DL11 6SZ, UK

C.M. Dwyer (ed.), *The Welfare of Sheep*,

very varied, from airborne or food borne or through other fomites (vectors of disease that are not changed by their contact with disease i.e. tyres, vermin, skin or clothes of handlers etc.), by horizontal transmission in groups, or by vertical transmission of infection from dam to foetus. Predisposition to disease may be inherited or the disease may be genetic in origin. There may be nutritional excess or deficiency leading to disease states. In short, the understanding of the underlying cause and the mechanisms by which disease spreads (the aetiology of disease) is a vital part of the preparation to control disease.

Diagnosis of disease does not alleviate welfare problems but successful treatment does. Treatment necessarily means some shepherding intervention, which may not always be beneficial, particularly for unaffected members of the flock (see discussions in Chapter 8). Fisher and Mellor (2002) explore the influence of well-intentioned intervention by shepherds to provide obstetric assistance at lambing time with the development of easy-care lambing (where animals are rigorously selected on the basis of not requiring shepherding assistance). They suggest that this may indicate the way forward in producing systems more in keeping with both the biology of the sheep in an extensive environment and with the demands placed on the animal by man. Although the easy-care concept was initially directed at reducing shepherding inputs at lambing, there are now moves to extend this to encompass resistance to disease (e.g. footrot), with the aim of reducing other interventions (such as foot-trimming) (see also Chapter 10). Appleby (1995) points out that human contact is going to alter behavioural and physiological reactions and that this in itself may alter our observations, and thus our perception of disease states. These comments are made about experimental data collection but are equally relevant to normal commercial practice and the diagnosis and treatment of disease states.

The relationship between disease states, mechanisms for disease prevention and treatment and the welfare of the sheep will be the focus of this Chapter. I will not provide an exhaustive list of the disease conditions that can affect the sheep, although specific diseases will be mentioned to illustrate specific welfare issues. In addition, the major disease conditions that are likely to have a considerable impact on sheep welfare will be discussed. For other specific disease information the reader is referred to additional texts, for example, Aitken's The Diseases of Sheep (2007).

5.1.1 Disease Prevention

Disease prevention relies on the understanding of the aetiology of disease and on methods where either specific resistance can be effected in the flock or individual, or where interruption to the development and spread of disease within the flock or individual is brought about. Prevention can be relatively precise and implemented by a number of very reliable tools such as:

1. vaccination against specific diseases, e.g. for clostridial disease (see Section 5.3.6.4)
2. use of less specific vaccines;

3. previous exposure to low levels of disease, which can allow an immunity to develop; for example, allowing exposure of young lambs to low levels of coccidia[1] can establish a natural immunity;
4. specific medication that can prevent spread of infectious disease; for example, antibiotic footbaths can aid in control of contagious ovine digital dermatitis (CODD)[2]

Sensible management practices can also be effective in disease prevention. For example:

1. Management can reduce incidence of hypothermia in lambs. This can occur from either too much heat loss, such as through the provision of inadequate or absent shelter at lambing time, or from a lack of intrinsic heat production, as might occur with a low intake of colostrum in the first few hours after birth. Hypothermia can also be a combination of the two, e.g. a lamb born outside in poor weather where there is inadequate shelter to a dam in poor condition that has an inadequate supply of colostrum.
2. Proper nutrition can almost abolish metabolic diseases like hypocalcaemia (low blood calcium level often seen in the period around lambing. This lowering of the blood calcium level can cause collapse and eventual death as muscle contractility is compromised) and twin lamb disease (a disease caused by the lack of glucose or glucogenic materials in the diet, requiring the breakdown of fat to provide nutrients, with a resulting accumulation of ketone bodies in the bloodstream causing a toxaemia and leading to coma and death).
3. Planned breeding strategies may improve ease of lambing and lamb viability (Dwyer & Lawrence, 2005), thus reducing lamb losses from hypothermia and bacterial infection. Sawalha et al. (2007) suggest that viability is a selection criterion and that optimal rather than maximal lamb birth weights should be our aim. The survival of lambs is greatly influenced by both management and genetic factors. Multi-trait selection is possible for resistance to disease (Conington et al., 2001) and is potentially advantageous (Scobie et al., 1999; Lawrence et al., 2004). These methods may, for example, produce animals that are more resistant to footrot or gastrointestinal parasitism, or less likely to be afflicted by flystrike.

The mechanisms for disease prevention and control will be discussed in more detail in Section 5.4.

---

[1] Coccidia are protozoan gastrointestinal parasites that establish in the lining of the intestine destroying the finger-like villae that project into the open lumen of the gut and thus decreasing the area available for absorption of nutrients and causing a severe scour and weight loss to the animals infected. See also Section 5.3.5

[2] CODD is a virulent disease of the sheep's hoof causing the wall of the hoof to separate from the rest of the foot allowing infection to reach sensitive inner structures of the hoof and to markedly alter the sheep's gait due to the perceived pain that this causes. See also Section 5.3.1.

5.2 Disease and Welfare

Conditions that favour spread of disease are also conditions that have a potential adverse effect on animal welfare (e.g. overcrowding, poor hygiene, rapid changes in temperature or weather conditions, inadequate shelter). Thus poor welfare may facilitate disease. However, the presence of disease may also impact on an animal's welfare state.

5.2.1 Welfare and its Relationship to Disease

Implicit to the understanding of the welfare of sheep is the idea that disease has an adverse effect on welfare. Although freedom from disease is specifically listed in the Five Freedoms (the most widely accepted definition of welfare, see Chapter 1), a diseased animal may, for example, also experience pain, be unable to feed adequately and experience hunger, and be prevented from behaving normally. The ways in which disease can be manifested are many and in all disease situations there are measurable deviations from the norm in the various indicative parameters (see Tables 5.1 and 5.2). Equally disease is an entity that cannot be wholly avoided and disease resistance may be a beneficial trait conferring important adaptive benefit in the evolution of a species. Domesticated species, though, have an extrinsic agency (i.e. man) selecting development traits and improving genetics for specific purposes which may not have as their goal the continuation of the species but rather their commercial viability as a product (Fisher, 2001). A sick animal is frequently an indication of a failure to foresee problems and take avoiding action (Henderson, 1990). Thus, planning for health and production should account for potential problems and aid in the development of proper preventive strategies.

Health is an integral part of the welfare of the individual. As Lynch et al. (1992) show, health is part of the balance that leads to transformation into the affective state of suffering. The links between cognition, perception and behaviour lead to the expression of that suffering. The balance that allows this state to exist is relatively easily influenced by a number of external factors as well as by the inherent status

Table 5.1 Basic clinical observations of sheep and range of normal values[1]

Clinical observation	Units	Normal range
Temperature	Degrees centigrade	39.0
Pulse	Beats/min	70–90
Rumen contractility	Contractions/min	1–3
Respiration	Breaths/min	19

[1]In addition to watching the animals for the assessment of numerical values for these parameters an assessment of the character of the action should also be conducted. For example, although the pulse may fall within the normal range the variability of other character may be abnormal. Similarly neonates will show different ranges to adult animals and these observations must be taken as a guide not as defined accurate and exact measurements.

Table 5.2 Laboratory values for some parameters measured in sheep blood taken from three different commercial laboratories on the same day. The ranges are similar, but not identical, suggesting that interpretation at the limiting values could be a cause of differing interpretation

Parameter	Units	Laboratory 1	Laboratory 2	Laboratory 3
Haemoglobin	g/dl	9–15	9–15	9–15
Red Blood Cells	$\times 10^{12}/l$	9–15	5–10	9–15
White Blood Cells	$\times 10^{9}/l$	4–12	4–12	4–12
Neutrophils	$\times 10^{9}/l$	0.6–4.5	0.7–6	0.7–6
Lymphocytes	$\times 10^{9}/l$	1.5–8	2–9	2–9
Eosinophils	$\times 10^{9}/l$	< 0.5	< 0.75	< 1
Packed Cell Volume	l/l	0.26–0.42	0.27–0.45	0.25–0.45
Total Protein	g/l	60–80	60–72	60–79
Albumin	g/l	25–41	28–34	24–34
Globulin	g/l	20–45	32–43	32–45
Urea	mmol/l	2.5–6.7	2.7–6.6	2.6–6.6
Glucose	mmol/l	2.8–4	1.9–2.8	2.7–4.4
Beta-hydroxybutyrate	mmol/l		< 1	< 1.2
Calcium	mmol/l	2–3	2.1–2.8	2.3–3
Magnesium	mmol/l	0.80–1.20	0.7–1.2	0.7–1.3
Copper	μmol/l	9–20	9–31	9–19
Cobalt/B12	pmol/l	> 300	> 400	> 221
Creatine Kinase	IU/l	10–90	< 68	0–200
Gamma glutamyl transferase	IU/l	10–35	< 58	0–30
G Lactate dehydrogenase	IU/l	80–1635	< 2	0–25

of the individual. The factors that tend to unbalance the individual's health could be referred to as negative stressors whilst those that aid the maintenance of this balance could be said to be beneficial stressors. Necessarily all living beings experience stress and it is only when a stressor exceeds the animal's ability to cope that distress is deemed to be present (Broom and Johnson, 1993). Duncan (2004) warns that it is not helpful to lump together different states as 'distress' as, for example, fear, frustration and boredom can warrant very different remedies.

Stress levels can be anticipated in most husbandry systems. The animal's welfare is therefore in a constant flux and the balance is maintained by the animal's reaction at behavioural, physiological and biochemical levels. Examples of some indicators are given with biochemical parameters sourced from three different laboratory groups (Table 5.2). Although base levels for individual groups of sheep are still recognised as important, standard reference levels are hard to achieve with breed, age, sex and condition causing variety amongst others (Dubreuil et al., 2005).

5.2.2 Legal Requirements

In specific disease situations, primary and secondary legislation direct the action of the government agencies and its deputies (Anon, 1992; Radford, 2001). The major acts involved in welfare legislation in the United Kingdom are the Protection of

Animals Act 1911 and the Agriculture (Miscellaneous provisions) Act 1968 and associated regulations under statutory instruments such as the Welfare of Farmed Animals (England) Regulations 2000 (S.I.2000 No. 1870). Public perception can be manipulated by the media (e.g. this is explored by Rollin, 1995) and the sheep industry needs clear and deliberate presentation, such as Wilmut et al. (1997). These define legal limits in the ways we can treat animals. There is a difficulty in the definition of some legal terms as animals are defined as property in the 1911 act. Thus legal interpretation of unnecessary suffering is not necessarily the same as a welfarist's view. Recent legislation, namely the Animal Welfare Act (2006), has attempted to amalgamate the diversity of welfare legislation under one Act, rather as the Protection of Animals Act 1911 tried to do for the preceding 70 odd years, but has added intentionality into the legislative framework. Parallel to the development of this Act in the United Kingdom, the European Union has recently passed legislation recognising that animals are sentient beings. That this recognition is now on the statute book should give a solid foundation to move welfare priorities forward.

Further to this, Veterinary Surgeons are governed in the United Kingdom by a Code of Practice controlled by the Royal College of Veterinary Surgeons (Anon, 2004). There are similar codes in place in Denmark, New Zealand, Australia, Canada and the United States, although the wording may vary slightly. A review of codes of conduct, through the Federation of Veterinarians in Europe, may lead to the establishment of a pan-European code. The need for an interdisciplinary approach to any system of applied ethics is argued convincingly for by Beauchamp and Childres (1994). We need to build upon the input of ethicists, the legal profession, scientists and the various veterinary disciplines to create a widely acceptable ethical code for the protection of the animals, the veterinary surgeons and their clients. The present Code of Practice is based upon such input, and details ethical and accepted practice as well as etiquette and includes, for example, strict guidance on certification and on the overarching duty of the veterinary surgeon to uphold the welfare of animals under his/her care. The term 'welfare' in this context has not been defined by the Royal College of Veterinary Surgeons (Porter, 1991). Ethical theory will have a vital role in the continuing development of such concepts and so should have a major influence on future revision of the code.

Other legal requirements for those who care for or keep animals exist to impose certain duties upon them and should encourage the search for reasonable parameters to assess their welfare (Ingram et al., 2002), for example, in the transport of animals, treatment in markets and reporting of notifiable diseases. These all act to reinforce the duties that we owe to the animals we keep, although reports suggest that these requirements may not always be met (Warriss et al., 2002).

5.2.3 Pain and Disease

Another concept in the panoply of welfare that needs consideration is that of pain (Hellebrekers, 2000). Pain is part of welfare but not the only determinant needed to assess welfare conditions. An animal that suffers from poor welfare is not the same as an animal that suffers pain, whereas an animal that suffers pain necessarily means that its welfare is compromised, at least in the short term (Rutherford, 2002). Understanding and assessing the experience of pain is difficult in all species, but particularly so in non-human animals where verbal witness is not available (Wall, 1992). However, attempts to assess pain (for more information see http://www.vet.ed.ac.uk/animalpain/) have used objective measures (physiological, biochemical and behavioural responses) and subjective assessment (including verbal descriptors, visual analogue scales and numerical rating scales). Duncan (2004) argues that there can be no simple physiological determinant of welfare because of the emotional dimension (which means that there can be no simple measurement of welfare). In addition, pain experience may show differences amongst species as well as individuals. For example, Mellor et al. (1991) suggested that lambs showed the most distress following castration when compared to kids and calves. In addition, there is significant variation between veterinary practitioners in their assessment of the pain associated with different procedures or conditions (Harkins et al., 2002; Huxley and Whay, 2006). In attempts to reduce this variation, Fitzpatrick et al. (2006) describe the methodology necessary for ascribing a single value for animal welfare, based on Quality of Life assessment developed in human medicine extended to animals. This uses individual dimensions of the impact of pain on the physical ability of the animal to carry out normal behaviours and activities, the social interactions the animal can engage in and the psychological wellbeing of the animal, scaled to achieve a single score.

It is arguable that pain is the greatest evil for an animal and that the most fundamental right is to have that pain alleviated or terminated (Rollin, 2000). A counter to this would be that pain can have value for an animal by alerting it to potentially dangerous situations, e.g. ingestion of vegetable toxins may lead to abdominal pain or the presence of a sprain alerts an animal to damage to a particular joint (Broom, 1998), so reducing its use and the potential to increase tissue damage. In sheep medicine in the United Kingdom, there are few licensed products available to address the problem of pain control. The use of non-steroidal anti-inflammatory drugs, morphine-based drugs or alpha2 agonists has been recorded (Kent et al., 2004). The use of regional anaesthesia, in particular, the use of caudal epidural block techniques including local anaesthetic and an α-agonist in combination is reported to act for up to 36 h post-injection and has been a valuable improvement in the alleviation of suffering in the ewe. All practitioners should consider the use of epidural blocks for any hindquarter procedure in the sheep. Most United Kingdom practitioners have responded positively to this technique and would seek to promote pain relief from other sites by judicious clinical use of other pharmaceutical agents.

In the United Kingdom and the European Union, various veterinary initiatives to increase the availability of drugs used in pain control in farmed animals have met with a number of predictable responses:

1. Added and unnecessary expense.
2. Pain aids healing by promoting disuse of the affected body-part.

3. In the wild, only those able to take the pain or condition would survive.
4. No licensed products available.
5. Statutory licensing authorities prohibit the use of unlicensed medicines in food producing animals.

The first three arguments fail as they are based on false conceptions. We farm domestic animals – they are not bred to exist in the wild. We control their husbandry and production, and pain control should be part of the moral responsibility we take on in return for the products we use. The argument regarding immobilisation is not rigorous in that pain relief is an expected part of modern medicine and if the pain is not present then the healing process will proceed more rapidly as resources are not redirected to deal with pain issues (Broom, 1998). Further we can also distinguish between pain that alters an animal's physiology and behaviour to reduce or avoid tissue damage (where the perception of pain might be functional) and non-functional (non-useful) pain where the intensity or duration of the pain is not appropriate for the damage sustained and where physiological and behavioural responses are unsuccessful at alleviating the pain (Molony and Kent, 1997). The evolutionary role of pain can be an area where cognitive dissonance is sometimes met (cognitive dissonance [Festinger, 1957; Festinger and Carlsmith, 1959] is the feeling of uncomfortable tension which comes from holding two conflicting thoughts in the mind at the same time). Flockmasters, shepherds, animal owners and even veterinary surgeons can all be subject to this and it is a relatively common occurrence in the consulting room of the small animal practice. Effectively the cognitive approach is altered to suit the circumstance. By doing this the owner/person responsible for decision-making avoids a certain outcome. If however the circumstance was happening to someone else and the owner/responsible person was advising a third party, they would not try to avoid this particular outcome. This is often the basis behind the 'we want to do everything possible to save the animal' approach where the focus is not on the animal but on the emotional concepts of the owner/responsible person. This is often a difficult situation for the clinician and a similar situation can occur on the farm. This could arise where there is a reluctance to accept the degree of deterioration that has occurred or to accept responsibility for a particular condition. For example, a high percentage of lameness in a sheep flock may go unrecognised or unreported (Clements et al., 2002). This is not an unusual circumstance and I have been on farm walks where the farmer has commented that lameness was not a problem in his flock despite a 15% incidence being visible in the flock we were walking through. A similar situation has been reported in dairy cows, where farmer reports of lameness prevalence are considerably lower than the recorded prevalence (Main et al., 2003).

Individual conditions are important in the context of an overall assessment of welfare on any one unit. This does not mean that the occurrence of one animal with one condition is necessarily unreasonable and is unnecessarily causing suffering to that animal. The individual condition needs to be considered and then compared to national or local incidence rates if known and to the presence or absence of preventive measures. For some conditions, avoidance is impossible as we do not fully understand the aetiology of the condition.

5.2.3.1 An Example: Prolapse of the Vagina and Cervix

Prolapse of the vagina and cervix is a condition of late pregnancy, generally occurring in the last three weeks of pregnancy (Fig. 5.1). This can be a relatively minor clinical incident with part of the vaginal wall appearing at the vulval lips as the ewe lies down and disappearing as she stands or it can lead to chronic inflammation of the vaginal wall and cervix with local infection from traumatic damage and, in the worst cases, actual rupture of the vaginal wall and prolapse of the intestine through it. The bladder may commonly be involved and becomes reflected thus obstructing urination. The inflammatory changes can make the ewe strain further and the condition is the potentially more severe as it may not easily be seen in the early stages.

There are a number of potential outcomes from this condition, which can affect not only the welfare of the ewe but also that of her lambs. Hosie (1989) summarised the condition, estimating each case cost the farmer *circa* £41. Scott and Gessert (1998) estimated costs between £21–£90 and point out that this could exceed the value of the ewe. In both studies these costs were based on the assumptions that:

a. some ewes will die
b. some will abort
c. some will show reduced fertility
d. some will be culled
e. stillbirth and neonatal mortality are increased
f. the risk of dystocia is increased.

The differential list of predisposing factors is illustrated in Table 5.3. These predisposing conditions can be grouped into those which we can influence such as condition, exercise, amount of dietary fibre; those which are related to shepherding practice such as previous dystocia, short tail docking and vaginal irritation; those that are more animal related and therefore more difficult to limit such as foetal load, inherited disposition and hormonal imbalances; and environmental triggers such as dietary oestrogens and their precursors (Hosie, 1989).

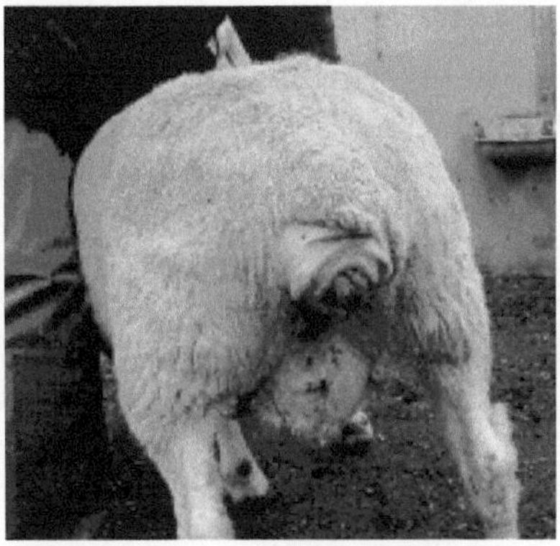

Fig. 5.1 Vaginal prolapse of the ewe
Photo: Paul Roger

Table 5.3 Factors predisposing to prolapse of the vagina and cervix in late pregnancy (after Hosie, 1989)

Management in pregnancy
Ewes overly fat in pregnancy (above BCS 4)
Ewes overly thin in pregnancy (below BCS 2)
Ewes inadequately exercised, e.g. housed at high stocking density
Ewes fed bulky foods (e.g. root crops)
Ewes fed excessive dietary fibre
Shepherding practice
Previous dystocia and treatment
Short tail docking
Vaginal irritation (e.g. previous treatment for prolapse)
Animal-based factors
Large foetal load (e.g. twin or greater litter sizes)
Hormonal excess or inbalance
Hypocalcaemia
Inherited predisposition (e.g. more common in lowland ewes)
Ewe age (more common in older ewes)
Environmental factors
Dietary oestrogens and their precursors
Sloping terrain

In practical terms this condition is very distressing for the ewe and can lead to severe local infection and inflammation, which is most irritant, and eventually to death causing considerable suffering for the individual. However, a dramatic reduction in incidence is possible using a number of simple approaches, particularly if a number of prolapses happen within the same flock. In the field the first steps to take are to reduce indigestible fibre intake whilst trying to maintain nutrient throughput, which is a difficulty in itself, and to increase exercise by, for example, walking the ewes around the outer perimeter of the sheep shed twice daily. Batching the ewes according to body condition score, lamb numbers and due date aids in controlled feeding and helps to avoid this problem. Monitoring beta-hydroxybutyrate (BHOB) levels through regular blood sampling to check the adequacy of energy in the diet is also beneficial (Hosie, 2007). Target figures should be between 0.8 mmol/l to 1.1 mmol/l.

Treatment under caudal epidural anaesthesia and the use of aids to retention is described by Noakes (2001). The plastic spoon and the Buhner suture method are described, but the author has had better success with the use of webbing harnesses and epidural. My opinion is that use of the plastic spoons must be avoided as the aggravation of an inflamed cervix by placing a solid object against its surface must irritate and tends to promote the resumption of straining. The Buhner stitch technique can be useful in neglected cases but the major factor in rapid and lasting clinical recovery in the use of effective caudal epidural anaesthesia including xylazine to prolong the effect and allow the tissues a chance to commence healing. The lay use of nappy pins, safety pins, stitching with twine and sacking needles is wholly inappropriate.

5.2.4 Disease and Welfare Assessment

Losses in sheep flocks from disease and associated problems can be high, relative to other farmed livestock, and can occur relatively rapidly from some causes. These issues provide welfare concerns due to their high incidence and morbidity rates. Compared with other farmed species, these rates can be exceptionally high and reduction of these rates generally requires a concomitant injection into cost, which may create a negative margin. However, with proper management procedures in place, it is possible to avoid the conditions that lead to these high losses. This would be the conventional approach to preventive medicine.

Sheep would be better served if the approach we took was removed from this resource-based analysis and took an approach considering total welfare indices. Total welfare indices can be derived by considering the number of animals within a flock affected by the disease or condition related to the number of animals in the national flock with the disease or condition. This leads to a better understanding of the impact of disease on the sheep rather than looking at resource-based measures or performance targets that are based on economic and clinical parameters. Wemelsfelder (1997) suggests a subject-based approach to try to recognise the active role of the individual animal in regulating its interaction with its environment. The role and validity of qualitative judgements of sheep behaviour (such as fear, timidity, agitation, pain, calmness, curiosity/play) could be considered (Wemelsfelder and Farish 2004) rather than the quantitative observations often used. Qualitative interpreatation may help clarify any ambiguity of quantitative measurements. Thus if these total welfare indices could be tempered by qualitative ratings, then perhaps a more sheep-centred system could be developed.

What does this mean to sheep in terms of health and disease? It leads to the recognition of the importance of disease states to the sheep, which are not necessarily the same goals as those dictated by economic parameters. In order to implement welfare standards, a beneficial economic return is needed to reinforce the importance of these standards with the sheep farming industry. For example, there are a number of important diseases which manifest by causing abortion. In some cases this may be a single occurrence in the productive life of the ewe and not be transmissible to other females or to man; in other cases the cause of abortion can readily be transmitted to both other sheep and to man. In order to implement control measures it is important that all incidents of abortion are investigated. As a rule of thumb, this should involve investigation of any abortion incidence above 2% and then a sample of any continuing cases (Mearns, 2007a). More than one cause may be present; this is becoming more common as the diseases that cause abortion in ewes usually cause some degree of immunosuppression.

5.3 Major Disease Concerns

Welfare issues of concern to the sheep industry are well documented. The five major issues that provide most concern are lameness, lamb losses, reproductive failure, sheep scab and anthelminthic resistance. These major concerns, and the particular welfare issues associated with them, will be discussed below.

5.3.1 Lameness

Lameness is a major health and welfare problem in all sheep-producing countries throughout the world. It is a significant cause of discomfort and pain and is a major source of economic loss to the sheep industry (Defra, 2002). Lameness is also one of the conditions that has a number of infective causes that can be controlled and in some cases eradicated by use of current treatments and management practices. In the United Kingdom the Royal Veterinary College (1997) carried out a postal survey of 547 farms to identify the causes of lameness. These were identified as:

- Inter-digital dermatitis (scald): 43%
- Footrot: 39%
- Foot abcesses: 4%
- Post-dipping lameness: 4%
- Swollen joints: 2%
- Soil balling: 2%
- Fibroma: 2%
- Other: 1%

The vast majority of lameness cases can be attributed to scald (infection with *Fusobacterium necrophorum,* a naturally occurring environmental pathogen), and footrot (infection with *Dichelobacter nodosus*).

In the Royal Veterinary College survey 92% of farms responding had a lameness problem and reported an incidence of between 6–11% of their sheep were affected annually. In more recent surveys over 90% of sheep flocks (SAC, 2002) or over 80% of flocks (Winter, 2004a), contained lame sheep with prevalence in some flocks of over 9% for footrot and over 15% for scald. The clinical symptoms of footrot (Fig. 5.2) are varying degrees of lameness (from transient and mild to persistent and severe), recumbency and reluctance to move, reduced feed intake, low body weight and reduced wool growth (Marshall et al., 1991; Egerton, 2007). With virulent foot rot the clinical signs are more severe, resulting in angry, red and swollen feet, complete detachment of the horn and spread of the infection to above the hoof (Winter, 1997). Some animals, particularly young stock, were left with permanent damage and no regrowth of horn. Thus the scale of the problem cannot be overestimated, although the high incidence and frequency of lameness in sheep flocks may mean that the pain and suffering associated with these conditions are under-emphasised.

Fig. 5.2 (**a**) Lamb suffering from lameness, and (**b**) hoof affected with footrot
Photos: Paul Roger; SAC Image Bank

Lame sheep are less able to graze and compete for feed. The adverse effects of lameness on flock productivity therefore include (Nieuwhof and Bishop, 2005):

- inadequate body condition (affecting ovulation rate in ewes and sperm production in rams);
- hence leading to reduced fertility;
- reduced conception rates (as lame rams will serve fewer ewes in addition to the effects on fertility);
- increased predisposition to disease (including metabolic diseases in lame pregnant sheep);
- increased lamb mortality of the offspring of lame ewes (due to reduced birth weights and milk production);
- reduced growth rates and delayed fattening;
- reduced wool growth.

Thus the pervasive influence of this condition on the flock can be seen. In addition to the effects on productivity, sheep with footrot and other causes of lameness show physiological responses of pain and stress. Sheep with footrot have elevated vasopressin and prolactin (Ley et al., 1991a) and elevated plasma cortisol with severe lesions (Ley et al., 1994). Sheep with both mild and severe footrot show elevated plasma adrenaline and noradrenaline (Ley et al., 1992) suggesting activation of the sympathetic adreno-medullary system and the hypothalamic-pituitary-adrenal axis with footrot (see Chapter 1 for details of stress responses). Sheep with severe footrot also have a significantly reduced threshold for nociceptive stimuli compared to healthy controls, indicating an increased sensitivity to acute pain (Ley et al., 1989; Ley et al., 1995), and the analgesic effectiveness of xylazine is also reduced with chronic foot rot (Ley et al., 1991b). However, for all the effects seen in animals experiencing chronic foot rot, treatment and an apparent resolution of the clinical symptoms is not accompanied by an alteration in the physiological effects by 3 months after treatment (Ley et al., 1989; Ley et al., 1995). In addition, the reduction in mechanical nociceptive threshold is still present in animals that were apparently sound. Thus although the sheep are not judged to be lame they still have increased sensitivity to acutely painful stimuli.

All authorities agree that a precise diagnosis is the key to correct treatment and control of lameness. As described above, the two major conditions of the foot are scald (also known as strip or ovine interdigital dermatitis) and footrot (also called hoof rot, pretin, pedero or rotkreupel). Scald occurs on wet pasture due to the invasion of F. *necrophorum* into the skin of the interdigital cleft. It is a relatively mild condition and its importance is that it predisposes to the development of footrot if D. *nodosus* is present in the environment. D. *nodosus* is a more labile bacterium and will only survive in the environment for about 7 days so this knowledge makes a control programme possible. The treatment and control of the different causes of lameness differ (Table 5.4), thus knowledge of both the diagnostic features and the appropriate treatment regimens is important to prevent further suffering. For example, foot trimming or paring is effective in the treatment of footrot, but not the control, and is not indicated for either the treatment or control of scald.

Footrot may be benign or virulent and maintaining a footrot-free status requires attention to detail. Eradication schemes in Australia were more successful where the number of infected sheep had been reduced to a low level prior to the start of the programme (Plant, 2007). Due to the ubiquitous nature of footrot the eradication of this disease would alleviate a great amount of suffering, and a target programme is given in Fig. 5.3.

Lameness is an indicator, not a specific condition, and a lame ewe seen in June or July may not have footrot but may, for example, be suffering from mastitis (Fig. 5.4). A good shepherd or flockmaster is aware of conditions that may predispose to lameness (Wassink et al., 2003; 2004). Proper diagnosis is the first step in alleviating adverse welfare.

5.3.2 Reproductive Failure

Reproductive failure encompasses losses from mating to parturition. This includes a failure to show signs of oestrus and early embryonic loss, as well as the more

Table 5.4 Suitability of different methods for the treatment and control of scald (OID) and footrot (after Winter, 2004b)

	Scald		Footrot	
	Treat	Control	Treat	Control
Foot trimming	x	x	yes, with care	x
Antibiotic spray	yes (but not with footbath)	x	yes (but not with footbath; not adequate for CODD)	x
Antibiotic injection	x	x	yes	x
Footbath	yes (but not with spray)	yes	yes (but not with spray)	yes
Vaccination	x	x	yes	yes

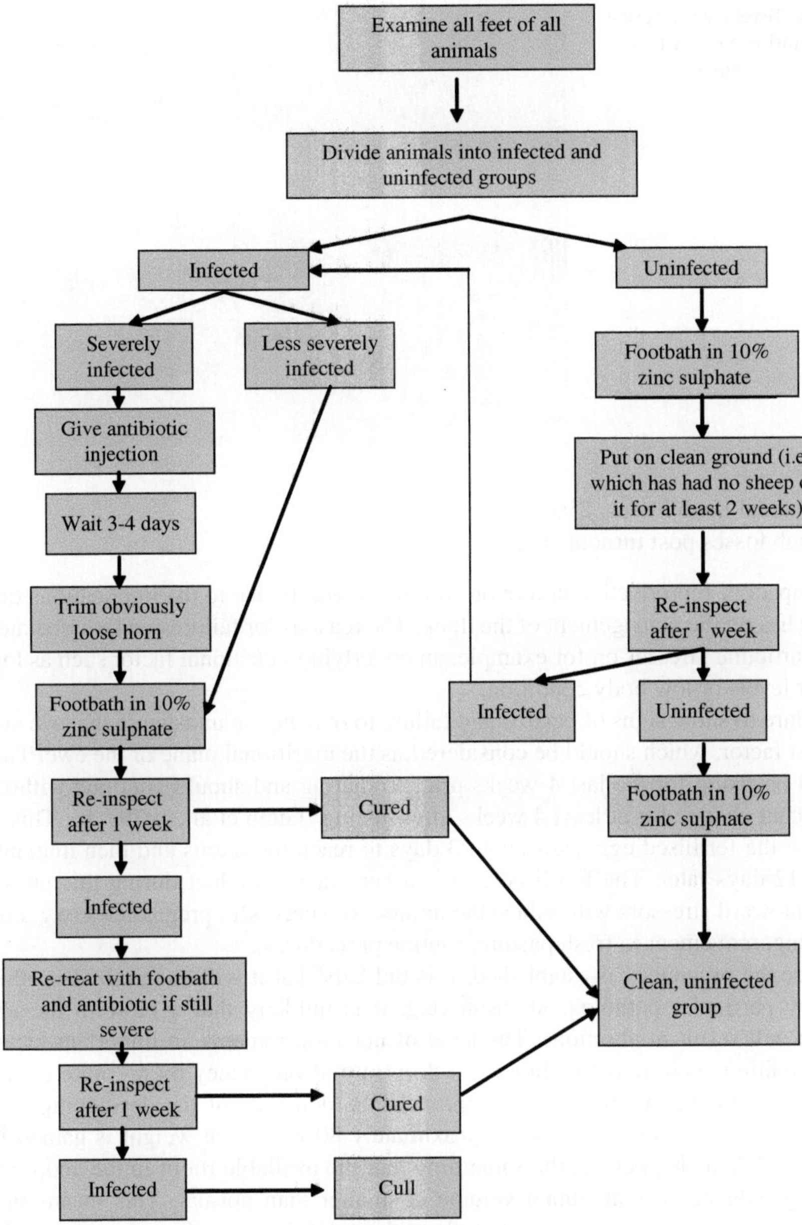

Fig. 5.3 Scheme for the eradication of footrot (after Winter, 2004b)

obvious signs of infectious abortion. Unpublished work by ADAS has partitioned reproductive losses for the average United Kingdom lowland farm as:

- Embryonic losses from mating to scanning: 33%
- Foetal losses from abortion and stillbirth: 30%

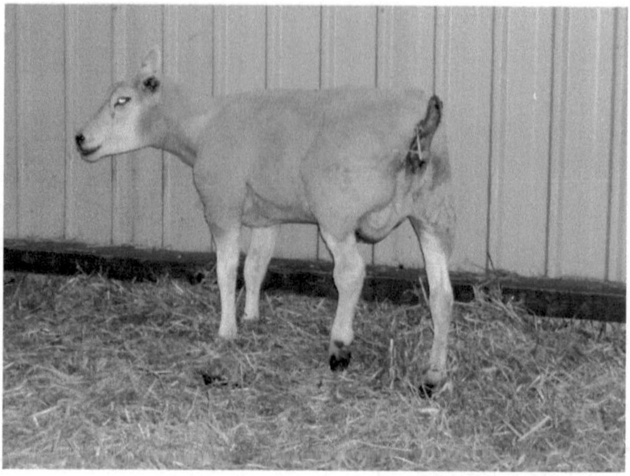

Fig. 5.4 Texel ewe suffering from arthritis and mastitis
Photo: Brian Hosie

- Neonatal lamb losses: 25%
- Lamb losses post turnout: 12%

The impact of reproductive failure on welfare is largely due to the implications this failure has on the management of the flock. The reasons for failure can be associated with particular stresses or, for example, an underlying nutritional factor such as low copper levels or low body condition.

Failure to show signs of oestrus and failure to ovulate are unusual in the ewe and the first factor, which should be considered, is the nutritional plane of the ewe. This should be rising for the last 4 weeks prior to mating and should continue without significant change for at least 4 weeks after mating (Gunn et al., 1979a, b). This is because the fertilised egg takes about 3 days to reach the uterus and then implants about 12 days later. The fertilised egg is a very labile product during this phase. Any untoward stressors will reduce the number of successful pregnancies (e.g. dog worrying, movement to fresh pasture, routine procedures).

Once the pregnancy is established, it is unlikely that it will be terminated other than by particular pathological insult (e.g. it is unlikely that a blow to the abdomen will result in abortion). The level of nutrition remains an important factor and condition score may be allowed to drop in mid-pregnancy by no more than a half score. As the last 6–8 weeks approach, the demands of the developing foetus(es) increase dramatically and approximately 80% of birth weight is gained in this period. This happens at the same time that the available room in the abdomen is being reduced so that rumen volume is smaller than normal. This means that rumen turnover must be increased so that adequate amounts of nutrient are available to the ewe and the rapidly developing lamb. The quality of the protein input is important and a higher level of undegradable protein is required to enhance reproductive efficiency (Robinson et al., 2002). Robinson et al. (2002) also point out that improved protein intake can enhance immunity to gastro-intestinal parasites in the late pregnant and lactating ewes. If there is a breakdown in delivery of feed or the diet is inadequate, this can lead to metabolic disease which can lead to reproductive failure. Overfeeding and the development of overfat ewes can lead not only to problems with metabolism but also increase the risk of dystocia through the physical narrowing of the pelvic outlet by fat deposits, and delivery of oversized lambs.

Infectious causes of abortion are well documented and almost all have zoonotic implications. The main causes of infectious abortion in ewes are:

- Chlamydia (enzootic abortion, EAE): 52%
- Toxoplasma: 24%
- Campylobacter: 9%
- Salmonella: 3%
- Listeria: 2%
- Other: 10%

Infectious abortion can result in ewes returning to the ram at unexpected intervals, empty ewes at term, abortion, stillbirth, weakly lambs or apparently healthy lambs at term that may, nonetheless, spread infection in the flock. The level of abortion in any flock should be low and an investigation of the cause is better entered into early than awaiting the incidence level to rise. For instance in the case of chlamydial (enzootic) abortion, introduction to the flock may go unnoticed in the first year if all cases of abortion are not investigated and can be followed in the second year by a high level of abortion in the flock, typically of the order of 25% or more. This may have little effect on the ewe in terms of physical illness but may have behavioural effects if the event takes place close to term, as the full udder and the maternal instinct will be well prepared at this time, although the impact of this on the ewe has never been fully explored. In addition, infected but surviving lambs may be delivered at term which, apart from spreading infection until they have dried, may be underweight and thus more weakly than if the infection was absent.

Agents of infectious abortion are found throughout the sheep producing countries of the world but different areas have different prevalence and importance. For example, Chlamydial abortion is notably absent in Australia and New Zealand and is of greater importance in the countries of Northern Europe (Aitken and Longbottom, 2007), whereas *Salmonella abortus ovis* ranks highly as a cause of infectious abortion in Western Asia and in some European countries. With the opening of boundaries in the European Union, *Brucella melitensis* is a potential problem. This is not presently found north of 45° latitude. It is a disease of primarily the Mediterranean and Middle-East but has tended to spread eastwards into the former USSR, Mongolia and Northern China. Where vaccination is possible, then health programmes should consider the use of preventive treatment to avoid the unnecessary losses and the distress that this condition affords ewes.

Twin-lamb disease occurs in this last part of pregnancy and is a preventable condition. The condition arises as a consequence of inadequate energy intake. This condition causes signs of blindness, inappetance and ketosis (detected by the typical smell of pear-drops on the breath) due to the breakdown of body fat as there is no other readily available source of energy. This can be for a number of reasons:

1. Inadequate dietary constitution
2. Inadequate amount of diet
3. Inadequate supply of diet
4. Interrupted supply of diet

All of these predispositions can be avoided by correct feeding.

Hypocalcaemia (low blood calcium) is another condition seen near or at term causing collapse and leading to coma and eventually death if untreated. It occurs in the last part of pregnancy as the sudden increase in demands by the developing foetus cannot be matched by the ewe's ability to supply. The ewe needs to mobilise the body reserves through a complex hormonal system and this is not directly linked to dietary content of calcium (that is, calcium deficiency cannot be overcome purely by including high levels of calcium in the diet). There is a complex interrelationship between magnesium and calcium metabolism and currently advice is to feed low calcium and normal phosphorus levels before lambing to stimulate bone mobilisation and absorption followed by an increase at parturition. Vitamin D may be useful in management of intractable problems at a level of 2000 I.U./kg live weight. Stressors such as movement can precipitate episodes of hypocalcaemia. Response to treatment is rapid and ewes change in demeanour from moribund to bright and active within 30 min to one hour of treatment. They should be carefully monitored as occasionally a relapse will occur. Intercurrent disease can complicate these conditions and may act as a trigger for their development so that, where these diseases are seen, they may act as markers for compromised welfare not only of the affected individuals but also the whole flock.

5.3.3 Lamb Losses

World wide high levels of lamb mortality are a significant welfare concern (Table 5.5). The average lamb mortality in developed countries is 15–20% (Table 5.5), with nearly 50% of these lamb deaths occurring within the first 3 days of life (Fig. 5.5). Mellor (2007) categorises 70–90% of these losses to be due to functional disorders, birth and neonatal adaptation to the external environment with 10–30% due to infectious agents. The principal causes of lamb deaths are:

- Pre- or periparturient disorders (e.g. due to birth difficulty or dystocia): 30–40%;
- Weakly lamb/exposure/starvation: 25–30%
- Infectious disease and gastrointestinal problems: 20–25%
- Congenital disorders: 5–8%
- Predation, misadventure and unknown causes: 5%.

The behavioural contributions to lamb mortality have been discussed in Chapter 3, thus this chapter will focus on the contribution of other factors to lamb mortality.

The risks of lambs succumbing to any of the causes of death will vary somewhat by management. For example, outdoor lambing systems may have higher deaths from dystocia (as the risks of a ewe experiencing difficulties and not being assisted

Table 5.5 *Summary of published lamb mortality figures for various countries*. Values given are preweaning mortalities unless otherwise stated. Ranges are used where several values are given, reflecting variation by litter size, breed and/or season

Lamb mortality (%)	Location	Breed	Comments	Reference
Europe				
10	United Kingdom	Not specified	60% lowland, 73% housed	Binns et al., 2002
9.7	United Kingdom	Not specified	All housed	Green & Morgan, 1993
11.8	United Kingdom	Scottish Blackface	Hill/upland	Sawalha et al., 2007
25.8	United Kingdom	Scottish Blackface, Cheviot, Welsh Mountain	Upland	Wiener et al., 1983
11.4	Norway	Unknown	Rangeland grazing	L Grøva, pers. Comm.
22.9*	Norway	Predominantly Steigar	Rangeland with high predator density	Warren et al., 2001
6.9	Turkey	Akkaraman, Merino and crosses	Only 4.8% of lambs were twins	Thieme et al., 1999
Australia/ New Zealand				
14.0	New Zealand	Merino and New Zealand breeds	Range on different farms was 1.4% to 43.5%	Forrest et al., 2007
5.4–24.4	New Zealand	New Zealand breeds		Summarised in Fisher 2003
14.5–33	New Zealand	Romney x Suffolk	Figure to 25 days only	Scales et al., 1986
17.8	New Zealand	Romney, Border Leicester x Romney		Hight & Jury, 1970
10.9–29.8	Western Australia	Merino		Kelly 1992
20	Australia	Merino	Paddock and pen studies	Jordan & Lefeuvre, 1989
Americas				
14.7	USA	Terminal sire composite: 50% Columbia, 25% Hampshire, 25% Suffolk		Southey et al., 2004
18.3	Tobago	Barbados Blackbelly		Rastogi, 2001
8.2–12.2	Colorado, USA	Not specified	Housed lambing	Rowland et al., 1992

(continued)

Table 5.5 (continued)

Lamb mortality (%)	Location	Breed	Comments	Reference
Overall 22 16.7–32.8	USA	Finnsheep, Dorset, Rambouillet, Suffolk, Targhee, Composite	Housed lambing	Gama et al., 1991
25.9	USA	Not specified		Yapi et al., 1992
18.0	USA	Targhee		Huffman et al., 1985
Asia				
6.6 (12.6 by 1 year)	Uttar Pradhesh, India	Muzaffarhagni (mutton breed)		Mandal et al., 2005
14	India	Nali, Lohi, crosses with Nellore		Malik & Acharya, 1972
9 (neonatal only)	Lebanon (Bedouin)	Awassi	Nomadic management	Bhattacharya & Harb, 1973
Africa				
11.5 (27.5 by 1 year)	Ghana	Sahelian	Only 5% of lambs were twins	Turkson & Sualisu, 2005
21	Ghana	Sahelian crossbred		Kabuga & Akowuah, 1991
53 by 1 year	Ghana	Djallonke		Turkson, 2003
19.6	Guinea	Djallonke		Mourad et al., 2001
8.8–25.3 (28–59 by 1 year)	Ethiopia	Menz and Horro		Mukasa-Mugerwa et al., 2000
19.3 (first 4 days only)	Ethiopia	Menz		Mukasa-Mugerwa et al., 1994
46.3–51.5	Ethiopia	Local breeds (not specified)		Bekele et al., 1992
12.6–23	S. Africa	Not specified		Quoted in Haughey, 1993
10.3–19.9	S. Africa	Dormer Merino, SA Mutton Merino		Cloete et al., 1993
28.1 by 1 year	Cote d'Ivoire	Improved village flocks		Armbruster et al., 1991
17.6–31.3 by 1 year	Morocco	Local breeds (not specified)		Chaarani et al., 1991
5–10 (to 60 days)	Morocco	D'Man, Sardi		Berger et al., 1989
23	Mali	Sahelian crossbred		Wilson et al., 1985
20–25	Nigeria	Sahelian crossbred		Abu & Ngere, 1979

*For lambs released onto range at ages greater than 2 weeks.

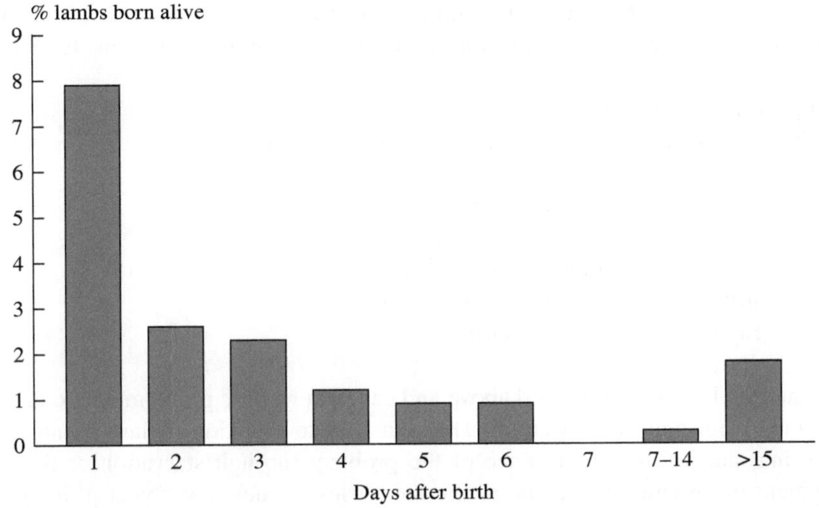

Fig. 5.5 Distribution of lamb mortality by lamb age (after Huffman et al., 1985). Total mortality was 18%, 14.3% of lambs died in the neonatal period

are greater) and exposure/starvation, whereas indoor lambing systems face greater risks of infectious diseases and abortions (Binns et al., 2002). Litter size is also a contributory factor with almost half twin lamb deaths attributed to starvation, whereas less than a quarter of singleton lamb deaths occur due to starvation (Nowak et al., 2000). Mortality in twin born lambs peaks on day 1 after birth rather than on the day of birth when most singleton deaths occur. These data come predominantly from studies in the developed world and lamb deaths in the developing countries are generally higher, appear to occur at older ages, and a greater number can be attributed to infectious disease (see also Chapter 6, Section 6.4.7.2).

Although many studies have categorised the causes and risk factors for lamb mortality, particularly from the viewpoint of improving productivity, the welfare implications for the neonate have less frequently been considered. The review of Mellor and Stafford (2004) has concluded that the main issues for the welfare of the neonate are breathlessness during neonatal adaptation (or failure to adapt) to postnatal life, hypothermia, hunger, sickness and pain. Of these, they consider that hunger, sickness and pain are the most severe insults to welfare experienced by the neonatal lamb. However, Duncan (2005) argues that the welfare of a sick animal (which would include any moribund lamb) should be considered to be compromised without the necessity of demonstrating that the animal 'feels' ill. Thus lamb mortality at the levels currently seen around the world must be considered a significant welfare problem.

Good shepherding, condition scoring, pregnancy scanning, proper feeding, housing or shelter provision, and good preparation and planning can keep these losses below 10%. Most causes of lamb loss in the first week of life are preventable.

Maintaining optimal body condition throughout the reproductive cycle of the ewe can be considered as the pivotal factor in preventing lamb mortality. This leads to:

- Improved conception rates
- Minimal risk of twin lamb disease and other metabolic disorders
- Lower risk of prolapse and dystocia
- Production of strong, vigorous and viable lambs
- Better onset and quality of maternal care
- Lower mortality of lambs at or before birth
- Good milk and colostrum production by the ewe
- Reduced lamb mortality after birth

Hypothermia has been described above and can be a welfare problem where not all the lambs from one ewe succumb. This can lead to rejection of lambs and mis-mothering, and cause a recurrence of the problem through starvation or the development of an enterotoxaemia, as the lamb tries to suck any object in its quest for food. Watery mouth or rattle belly is another condition seen in young lambs where they ingest environmental contaminants before they have taken colostrum from the dam. The ubiquitous *Escherichia coli* is the commonest isolate from these infections and colonises the stomach as successive waves of bacterial growth die and an endotoxin is released. This causes the typical terminal picture seen in this condition of a hunched-up lamb, often drooling saliva and staggering, unable to properly raise its head then descending into collapse and death (Fig. 5.6). This condition is preventable. The shepherd should ensure the intake of colostrum at around 50 ml/kg and then the condition will not occur, providing that the antibodies from the colostrum are in place before the infectious challenge arrives. If feeding by stomach tube it should be cleaned and disinfected between lambs as a dirty stomach tube can introduce the disease easily. Routinely many farms use oral antibiotics to control this condition, however blanket prophylactic use of antibiotics can lead to an increase in resistance in the bacterial population. The condition is associated with intensification and should properly be controlled by colostral immunity. There are

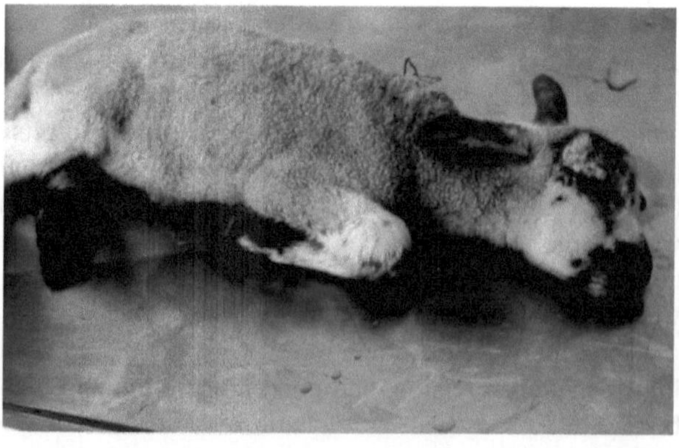

Fig. 5.6 Neonatal lamb with watery mouth
Photo: Brian Hosie

vaccines available that boost maternal antibody levels to coliforms but their effective use relies on the ability of the shepherd to ensure the lambs have sucked.

Most causes of lamb loss can be avoided by correct nutrition of the ewe, good lambing hygiene and attention to the new born lamb. This emphasises the role of correct nutrition in the promotion of the welfare of the flock.

5.3.4 Ectoparasites

Ectoparasites live on the body surface or the wool of the sheep and cause economic problems by their effect on productivity, reducing milk and meat yields, the downgrading of wool and leather and the requirement for expensive control programmes (Bates, 2007). They may be permanent such as mange mite, keds and lice or seasonal such as blow flies, head flies and nasal bot flies. They tend to be host specific and so are rarely found on other species and these are thought to be unlikely to be an aid to transmission. In addition to their effects on productivity, ectoparasites can have major effects on sheep suffering and distress. Fly strike, or cutaneous myiasis, may frequently go unnoticed in extensive farming conditions (Bates, 2007), but it is most common in lowland farms, particularly in warm and wet areas, and where sheep are kept at high stocking density (French et al., 1994). Fly strike causes a depression in feed intake, weight gain and wool growth (Heath et al., 1987; Walkden-Brown et al., 1999a). This is accompanied by elevations in body temperature, plasma ACTH and cortisol, and decreases in plasma glucose and β-endorphin (Shutt et al., 1988; Walkden-Brown et al., 1999b). Physiologically, therefore, fly strike is associated with indicators of distress and suffering in sheep. Initial infestations are accompanied by behavioural indicators of distress including agitation and dejection (Bates, 2007). Stamping, vigorous shaking of the tail and gnawing or rubbing at the infested areas develop as the infestation continues. Fly strike occurs in waves and untreated animals may suffer primary, secondary and tertiary attacks before death.

Sheep scab is an intensely irritating condition caused by the allergic reaction to the sheep scab mite, *Psoroptes ovis* (Fig. 5.7). Initially small lesions coalesce to larger areas of wool loss and an intense irritation develops. Early infestations may take some days before symptoms are manifest as the host sheep becomes increasingly sensitised to the mites. Sheep in this stage of the disease may show the development of abnormal behaviour patterns: restlessness, intense rubbing of areas of the fleece, biting at the flanks and head tossing (Sargison, 1995; Bates, 1997). These behaviours appear to be related to the development of the scab lesions. As the disease progresses, the infected sheep becomes increasingly distressed and agitated by the presence of the allergens, with increased rubbing and head tossing. This is also accompanied by a stereotypic nibble or mouthing response, in the absence of stimuli, characterised by lip-smacking and tongue protrusion (Sargison, 1995; Corke and Broom, 1999). In a study of the behaviour of sheep infected with sheep scab, bouts of maintenance behaviours (grazing, idling, ruminating) were frequently interrupted by rubbing, scratching and biting bouts (Corke and Broom, 1999), although

Fig. 5.7 (**a**) Gathered sheep infested with sheep scab and (**b**) close up of a sheep scab lesion
Photo: Paul Roger

the total amount of time spent in maintenance behaviour did not increase when the sheep scab was treated. In sheep infested with a virulent form of sheep scab, ewes were reported to show a hypersensitive reaction to handling or movement (Bygraves et al., 1993). This was characterised by the typical mouthing or nibble reflex, progressing rapidly to head tossing, biting and chewing of the lesions, involuntary and increasingly frenetic scratching of the lesions with the limbs and 'epileptiform seizures'. Seizures lasted for 10–20 min with loss of voluntary control, champing of the jaws and convulsions.

Sheep can also become infested with psoroptes species mites within the ear canal (Bates, 2007). These infections may be in the absence of infections with sheep scab. Subclinical infestations are found in up to 25% of sheep in infected flocks with higher incidences in rams (Bates, 1996). Behaviourally sheep showed signs of head tossing or shaking, ear rubbing, and scratching the ears that can lead to aural haematomas or cauliflower ears. Clearly, sheep scab causes intense distress to the sheep and early indicators of infestation are essential to prevent excessive suffering. As allergic symptoms are related to the formation of lesions and may persist for the duration of the lesion, after treatment has killed the mites, techniques for the early detection and treatment of the infestation are required.

Transmission is usually sheep to sheep but contact with wool, tags or scab can transmit the mite. The condition is found in most sheep producing countries of the world with the exception of Australia, Canada, New Zealand and the USA. Control is by plunge dipping (Fig. 5.8) in a synthetic pyrethroid-type (SP) or organophosphorus (OP) dip or by use of injectable macrocyclic lactones. Most SP dips require a second dip after about 14 days and injectable treatments are perceived by practitioners to be more effective if repeated after 10 days. OP dips are recognised to be the most effective way of treating this debilitating condition but strict attention needs to be paid to the safety of the operator and the handling of the sheep. In the United Kingdom, a fashion has developed for the treatment and prevention of scab by the use of showers despite the lack of effective chemical to add to these. The use of showers is widespread in Australia and New Zealand where sheep scab is not a problem and the major concerns are fly strike and surface living ectoparasites. There is no licensed product for the control of scab by showers and reports of their widespread use may indicate inadequate control of the disease. Bates et al. (2005) report a limited trial in the United Kingdom to investigate the efficiency of spray applications of SP dip products. The concern is that inadequate penetration of the fleece is achieved by this topical application of chemical and that if adequate levels

Fig. 5.8 Plunge dipping of sheep for the control of ectoparasites
Photo: SAC Image Bank

are not penetrating into the follicles then control can not be effected. They raise concerns about resistance development and, on the limited study that was carried out, a reasonable level of efficacy was achieved where ewes had recently been shorn. Further information is needed before this technique can be endorsed.

Sarcoptic mange (caused by *Sarcoptes scabei*) is seen in Europe, the Middle East, the Balkans, India, South and Central America, but has never been recorded in the United Kingdom. It causes a mild irritation on the face and ears and can become more serious if it spreads. In Australia and New Zealand, *Chorioptes ovis* causes foot mange or scrotal mange at a low incidence but when it occurs it can be a problem in rams as it can be irritant and severe scrotal mange can cause seminal degeneration and reduced fertility. The re-emergence of this condition and the increased incidence of lice infestations in the United Kingdom may be due to a reduction in the use of dipping (Cranwell et al., 2002).

Both blood-sucking and chewing lice have a world wide distribution. In the United Kingdom, the use of injectable macrocyclic lactones to control scab has an effect only on the blood-sucking lice. Topical treatment is needed to rid the fleece of the chewing lice. All of these conditions are irritant to the sheep and therefore warrant inclusion in health planning. Following the widespread use (and misuse) of anthelminthics, and the ensuing resistance development, the spectre of resistance may broaden to include ectoparasiticides. Reports of resistance to the SPs are well documented and, in the interests of the future welfare of sheep, proper use of the chemicals available coupled with a sensible quarantine strategy for the introduction of fresh sheep to the flock is an imperative for the industry.

The promotion of welfare relies not only on the education and training that is available to shepherds but also on having effective remedies available when the preventive strategies are stretched and fail. The potential welfare benefits of breeding sheep that could be more resistant to ectoparasites, and to endoparasites, has great potential benefits for sheep welfare (Scobie et al., 1999; Chapter 10).

5.3.5 Endoparasites

Endoparasites include the gastrointestinal parasites and liver fluke. Gastrointestinal helminths are major contributors to reduced productivity in ruminants. Parasitic infections may cause acute and chronic disease but, with the use of anthelminthics, subclincial infections may be common even if sheep are apparently healthy (Jackson and Coop, 2007). The clinical symptoms are diarrhoea, dehydration, loss of appetite, failure to gain weight and anaemia in *Haemonchus* infections. Sheep that are outwardly healthy but have subclinical parasitism still display signs that suggest their welfare may be compromised. Coop (1979) suggests that animals infected with parasites have a reduction in their voluntary feed intake and utilise nutrients less efficiently than non-parasitised animals. Various production effects have been reported including reduced growth rates, reductions in wool production, increased mortality, especially in young sheep, reductions in milk production and reproductive success, and an increase in the frequency that sheep succumbed to fly strike (Coop, 1982; Waller et al., 1987a, b; Anderson et al., 1987; Festa-Bianchet, 1988). Some of these effects may be secondary to the reduction in voluntary feed intake. Sheep infected with either *Haemonchus* or *Trichostrongylus* have elevated plasma cortisol and low thyroxine (Prichard et al., 1974; Fleming, 1997). These effects could not, however, be entirely attributed to their reduced feed intake, as sheep experimentally fed a similar low feed intake did not show the same decrease in thyroxine, and the elevation in cortisol was much less marked (Prichard et al., 1974).

Studies of the grazing and diet selection behaviour of subclinically infected lambs demonstrate that, although all animals tended to avoid areas of the sward contaminated with faeces, parasitised animals are more selective than uninfected animals (Hutchings et al., 1998), and would reject even high quality swards if contaminated (Hutchings et al., 1999). Sheep with large internal parasite burdens, although not showing signs of disease, spend less time grazing and are less active than uninfected sheep (Hutchings et al., 2000). Parasitised sheep have the same number of grazing bouts as control sheep but graze for shorter periods and have a reduced herbage intake. These data suggest that subclinical infections are sufficient to alter the behaviour of sheep with worm burdens, in particular reducing activity and feed intake. These behaviours may be indicative of animals suffering some discomfort from the worm burdens or general malaise, but could serve to indicate animals that may go on to show clinical symptoms of parasitic infection.

There has been an increase world-wide in the importance of endoparasitism to the sheep industry. This has become manifest due to the spread of resistant worm populations and the failure to develop proper control strategies and to encourage their uptake by the industry. In the United Kingdom alone Hindson (1985) estimated that £10 million was wasted annually on anthelminthic treatments using the wrong product on the wrong class of animal at the wrong time. Although the routine use of anthelminthics together with grazing management has controlled worms very successfully in the United Kingdom over the last 30 years the prevalence of anthelminthic resistance (AR) has risen sharply and an increasing number of flocks are finding that one or more of the three available chemical groups used to treat helminths are no longer effective (Abbott et al., 2004a; Sargison et al., 2007). Sargison et al. (2007) record the difficulty in managing a multiply resistant flock and warn that economic sheep farming could become unsustainable in the face of multiple anthelminthic resistance. The routine use of batch faecal egg counts and the planning of worming strategies coupled with good pasture management have to point the way forward as the development of vaccines against gut worms is still a long way off. Love and Coles (2002) have reported on the levels of AR seen in sheep in New South Wales, Australia, and the importance of objective information and planning combined with effective use of the anthelminthics that are available.

Fascioliasis is caused by the liver fluke, *Fasciola hepatica*. It presents a flock problem where it occurs and is seen in mild wet climates where the intermediary host, a dwarf pond snail, *Limnea Truncatula*, is found. Disease is seen as acute, subacute or chronic and the syndromes may overlap. Disease caused by other flukes such as F. *gigantica* (seen in Africa, Asia, southern USA, Spain, southern Russia and the Middle East) can be controlled in a similar way to *F hepatica* but *Dicrocoelium dendriticum*, another fluke which is found globally (but restricted in the United Kingdom to the western Isles of Scotland) cannot be controlled as easily, as snail sites cannot be defined adequately and conventional flukicides are not effective at normal dose rates (Mitchell, 2007). In the acute disease, acute haemorrhagic anaemia and hypoalbuminaemia result from the large numbers of fluke migrating through the liver causing sudden death. Abdominal pain and a palpably enlarged liver may be shown by other flock members. The subacute form may lead to slow deterioration over 1–2 weeks ending in death. In the chronic form of the disease weight loss and anaemia are clinically evident, and hypoalbuminaemia and eosinophilia can be seen on biochemical and haematological examination. Death may occur due to the progressive anaemia and cachexia or the condition can precipitate metabolic disease or concurrent infection. Control is by improving pasture and removing snail habitat through drainage and local fencing, as well as by strategic use of flukicides.

Programmes for parasite control should include an assessment of fluke risk to the flock and the use of combined fluke and worm products should be discouraged as control programmes need to be targeted at specific times of the year and according to individual farm needs. Resistance to flukicides has been recorded in the United Kingdom (Mitchell et al., 1998).

5.3.6 Other Important Health Issues

5.3.6.1 Respiratory Disease

Two chronic progressive pulmonary virus infections that produce disease throughout the world are sheep pulmonary adenomatosis and Maedi-Visna. These diseases are both caused by viruses and are found in many countries. They have both been imported into the United Kingdom and Sharp and De Las Heras (2007) and Pritchard and McConnell (2007) describe these diseases in detail.

Sheep pulmonary adenomatosis (SPA) is diagnosed on clinical signs and confirmed on post mortem. This contagious lung tumour causes marked respiratory effort and, due to the slow insidious onset of the disease, leads to a slow deterioration and eventual death. The classic clinical sign is the 'wheel barrow' test which delivers fluid from the nose when the hindquarters are raised (Fig. 5.9). This condition can also cause severe economic loss, apart from the obvious welfare implications to the flock and the individual. There are no laboratory tests currently available. It is a disease that is recognised in commercial sheep flocks and culling of affected individuals is the usual control. There is no treatment recognised or advised.

Maedi-Visna (M-V) is a viral infection with a long incubation period and is characterised by a progressive pneumonia and wasting, culminating in multi-organ failure and death. A variety of clinical symptoms can be seen from Central Nervous System (CNS) signs through to mastitis. Exposure to the virus seems to be widespread and the maintenance of an M-V free status is a goal that many countries, including the United Kingdom, would like to achieve. The exact incidence is uncertain as many flocks do not undertake the expense of investigations or

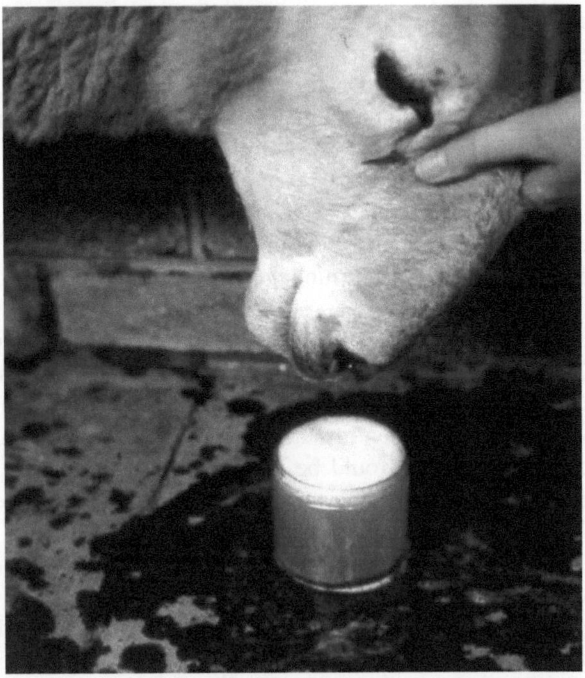

Fig. 5.9 Response of ewe with SPA to the 'wheel barrow' test
Photo: Brian Hosie

continuing survey. However, the presence of at least one sheep seropositive for M-V has been found in nearly 70% of Canadian sheep flocks (Campbell et al., 1994), and more than 100,000 United Kingdom sheep are known to have been infected. The signs of M-V are twofold: Maedi is a progressive pneumonia resulting in nasal discharge and laboured breathing and resulting in Visna, a wasting condition so weight loss, exercise intolerance, inappetance and general depression are part of the adverse welfare consequences of this disease. M-V can also be involved in intractable mastitis so needs consideration where this is found. Scottish Agricultural College figures show that, in one lowland flock an outbreak of M-V resulted in 68% of the flock being infected, and the deaths of 14% of adult sheep and 30% of lambs (Watt et al., 1992).

The M-V virus is very closely related to the *Caprine arthritis and encephalitis (CAE) Lentivirus* in goats and cross infection can occur (Oliver et al., 1988), therefore goats need to be tested for this disease before being brought in as foster mothers.

Pasteurellosis is the most commonly diagnosed pneumonic condition in sheep in the United Kingdom by the Veterinary Laboratories Agency (VLA) (formerly the Veterinary Investigation Service); it is also the most commonly misdiagnosed on gross post-mortem outside the Veterinary Laboratories Agency. It can have a very significant economic impact on a flock but often the disease precipitates suddenly and disappears as quickly, leaving a discernible mortality in its wake. Pasteurellosis occurs in two forms: pneumonic and systemic. Both are caused by *Mannheimia haemolytica*. *Pasteurella multocida* rarely causes pneumonia and is not recognised as important in that context. Although *P. multocida* is a relatively uncommon pathogen of sheep in temperate climates, it can on occasion cause incidents of severe disease. Watson and Davies (2002) describe an outbreak of septicaemia in neonatal lambs, particularly the increased risk to susceptible lamb populations, and compare this to the recognised susceptibility of Bighorn sheep exposed to domestic sheep. In peracute disease, the presenting sign is often the discovery of a number of dead sheep in a flock which has been subject to a recent stressor, such as movement to a new pasture, change in temperature or weather conditions. Avoidance of sudden environmental and nutritional changes should be part of the management protocol for all sheep flocks.

In the acute phase, the disease is often seen in the early months of the sheep year (spring and early summer) and a very effective vaccine is available. This vaccine is killed so that 2 doses are needed at a 4–6 week interval followed by booster vaccination at 6 monthly intervals to ensure maximum protection at peak times of challenge. These would be in the spring and autumn, although in flocks housed prior to lambing, a booster dose prior to housing is recognised as good practice. The vaccine is also available combined with clostridial components, either suitable for the breeding flock or the commercial growing young stock. Rams should be included in this programme.

Treatment for pasteurellosis is of limited efficacy and, as preventive measures are available, these should be instituted as soon as the disease is recognised, but preferably as standard practice, as the organisms involved are normal commensals of the upper respiratory tract. The use of a bovine parainfluenza 3 vaccine has been advocated to limit the invasion of Mannheimia species. This is an effective supplementary treatment where apparent breakdowns to vaccine are occurring (Rodger, 1989).

5.3.6.2 Scrapie

Scrapie is a chronic, progressive and invariably fatal degeneration of the CNS of sheep caused by an infection agent that is virus-like but remains characterised as a prion (Kitching, 1997). It is one of the spongiform encephalopathies that include chronic wasting disease of mule deer and Rocky Mountain Elk, Kuru and Creutzfeldt-Jakob disease of man and similar conditions such as transmissible mink encephalopathy and bovine spongiform encephalopathy (Fraser, 2000). The pattern of disease is complicated by the major effect that host genetic factors have on the development of the disease following infection.

Many cases of scrapie probably go unrecognised, especially in poorly supervised flocks, as an affected sheep may rapidly become uncoordinated and die from exposure. The signs of scrapie include an intense itchiness and a nibbling reflex and there is often a noticeable tremor that develops with the course of disease. This can manifest as a lack of coordination or slight lameness in the early stages of the disease. In the later stages of the disease, it would be arguable that the sheep is unaware of its plight but the developing signs of the disease cause welfare problems to the individual. Until the mechanism of spread is fully understood, these sheep must be treated as presenting a risk to the flock.

There is no treatment for this condition and the current United Kingdom eradication programme is based on selection of genetically-resistant breeding stock. The selection for resistance based on genotype is a sensible move, but it needs to be used as fine tuning not as a blunt weapon. That is, although it is apparently a very small gene and therefore unlikely to cause concern through translocation, there may be other genetic attributes that are being lost by the continual and purposeful selection of only the strongest levels of resistance. Research into the potential links between valuable breeding characteristics and scrapie-resistant genotype is currently being pursued. Freezing of semen and fertilised ova may help to reduce these concerns. This is partly in response to market pressure to ensure that all meat is derived from animals that show partial resistance to scrapie. Recently the emergence of atypical scrapie has complicated this issue as it appears that the genotypes that endow resistance to the development of classical scrapie do not offer the same protective umbrella to atypical scrapie. The development of scrapie in an imported, scrapie-free, population of sheep has also given rise to concerns that the aetiology of the disease may need yet further investigation, and that perhaps there are aspects of prion disease that need to be reassessed. So, although the results have been encouraging nationally in creating a largely dominant genotype modelled to help prevent or reduce a specific disease, the recent legislation may well drive the disease underground as government authorities may require the culling of infected flocks, whilst other genotypes that may be lost to the industry may be recognised for their importance too late. Confirmatory tests for scrapie can only be done at post mortem on CNS material at present, though tests are being validated on lymphatic tissue taken trans-rectally in the live animal.

5.3.6.3 Caseous Lymphadenitis

Caseous lymphadenitis (CLA) is present in all the major sheep-rearing areas of the world and was introduced to the United Kingdom in 1987 (Baird, 2003). As many as 18% of terminal sire sector flocks may be infected. Surface abscesses in regional lymph nodes are seen in the disease cases presented in the United Kingdom, particularly in the parotid region but also affecting other sites. Baird et al. (2004) showed a widespread net of seropositive regions covering most of the U.K. In Australia, the signs are more commonly found on post mortem involving extensive infiltration of the chest and pneumonic lesions and high levels of carcase condemnation.

Abscesses tend to be painful and when this is coupled with an impaired respiratory system then there are obvious adverse effects on welfare. There is currently no vaccine available in the United Kingdom but autogenous vaccines have been used with mixed success. These do not prevent the shedding of the infective organism, *Corynebacterium pseudotuberculosis* (A. Jones, pers. comm., 2003). Autogenous vaccines are not used any longer in other parts of the world (Baird, 2003). Without effective vaccine production, control strategies are fraught, particularly with no validated serological test presently available in the United Kingdom, although an ELISA test is close to release with an estimated sensitivity of 80% (Baird et al., 2004).

5.3.6.4 Clostridial Diseases

Clostridial diseases are a group of soil-borne organisms, which cause a number of different diseases through the very rapid release of toxins into the body. These diseases include tetanus, lamb dysentery and botulism. Clostridial diseases remain an threat to sheep flocks (Table 5.6). Clostridial disease announces its presence by causing sudden death in most cases and diagnosis is by post-mortem techniques. A number of types of disease are seen, all are the reaction to the various toxins produced by the clostridial bacteria (Table 5.6).

Vaccination is the only protection against clostridial disease. The vaccines are very effective and can be limited or contain protection against up to 10 different forms of clostridial disease. It is generally considered by veterinary clinicians that failure to vaccinate against clostridial disease is tantamount to an admission of poor welfare standards on any farm. The fact that flocks get away without use of vaccine is often remarked on but when that flock succumbs to clostridial disease, which it inevitably will, the immediate losses provide a convincing argument that basic farm economics should guide all sheep farmers to use these vaccines. The vaccine regime is similar for all killed vaccines but clostridial vaccines are very strongly immunogenic and so an initial two-dose primary course can be boosted yearly. The rams need to be kept in this regular protocol as well.

5.3.6.5 Thin Ewe Syndrome

As mentioned elsewhere, the body condition of an adult sheep reflects the recent feeding history and whilst body condition may vary, target scores for different types of animal are set for different times during the production year (see Section 5.4.2.1). These figures are guides and not to be taken as an end in themselves, as there will be variation around any mean score given. However body condition score (BCS) is a useful management technique and, if a significant number of ewes are found at the low end of an expected score, then the possible causes should be investigated.

Thin ewes represent a potential welfare problem. The relationship between a low body condition score and hunger in ewes is not known. Nevertheless it is probable that a ewe without an adequate feed intake for any period of time, whether they have the ability to search for food or not, is not in a good welfare state. Even in the absence of feelings of hunger the presence of thin ewes (generally defined as a ewe at of below BCS 2.0), particularly if they are bred, will lead to other welfare problems, being prone to metabolic disorders in pregnancy, and having a higher incidence of lambing difficulty, prolapse, low birth weight lambs and higher lamb mortality than ewes with BCS above 2.0 (and below 4.0).

As Clarkson and Faull (1990) observe, the list of possible causes is short and thin ewes are a common occurrence so that this should encourage us to do something about them. The causes are: lack of food, inadequate dentition and/or chronic disease. A lack of food can be established by biochemical parameters as well as post-mortem estimation of rate of loss of body fat and presence or absence of fat reserves. Correct feeding or improved quality of diet may help return the sheep

Table 5.6 A brief outline of clostridial diseases affecting sheep (after Lewis, 2007)

Organism	Disease	Precipitating factor	Signs	Comment	Prevention
Enterotoxaemias					
Clostridium perfringens A	Entertoxaemia	Poor hygiene	Death rapid over 2–12 h initially	Rare	
C. perfringens B	Lamb dysentery and haemorrhagic enteritis	Soil contamination, poor hygiene	Sporadic losses usually found dead	Lambs less than 21 days old United Kingdom (UK) and Europe Haemorrhagic enteritis worldwide	Vaccination
C. perfringens C	Struck	Abrupt changes in diet		SA and AUS Uncommon in UK	Vaccination
C. perfringens D	Pulpy kidney	Dietary change	Peracute disease, death often occurring within 2 h	Cosmopolitan	Vaccination
C. novyi B	Black disease	Often precipitated by fluke		Cosmopolitan Generally adults	Vaccination
C. haemolyticum C	Bacillary haemoglobinuria	Migrating young fluke or other hepatic insult	Jaundice leading to recumbency and death over 2–3 days	Sporadic UK and Ireland	Vaccination
C. sordellii	Abomasitis and toxaemia	Possibly associated with increased passage of carbohydrates		UK, NZ. All age groups	Vaccination
C. septicum	Braxy	Ingestion of frozen forage, often after first autumn frosts	Usually found dead	UK, Scandinavia. Mainly weaned lambs and shearlings in autumn	Vaccination

(continued)

Table 5.7 (continued)

Organism	Disease	Precipitating factor	Signs	Comment	Prevention
Myonecrosis and toxaemia					
C. chauvoei	Blackleg Post parturient gangrene	Wounds particularly on contaminated pasture	Usually found dead. Carcase decomposes very rapidly	Cosmopolitan	**Vaccination**
C. septicum	Malignant oedema	Wound or injection site	Death over 1–2 days	Rare in Europe	Good management of fighting risks as well as good hygiene, experienced shearing teams and avoidance of contaminated pastures
C. novyi A	Big head, Malignant oedema	Wound or injection site	Death over 1–2 days	Rams esp. hot,arid conditions	As above
C. perfringens A	Malignant oedema	Wound or injection site	Death over 1–2 days	Rare	As above
C. sordellii	Malignant oedema	Wound or injection site	Death over 1–2 days	USA,NZ	As above
Neurotropic disorders					
C. perfringens D	Focal symmetrical encephalomalacia	Poor condition coupled with anthelminthic treatment and move to new pasture	Drift into coma	Cosmopolitan	**Vaccination**
C. tetani	Tetanus	Wounds particularly docking and castration	Generalised stiffness. Death over 4–7 days. May be up to 3 week lapse before appearance of signs	Cosmopolitan. Mainly lambs	**Vaccination**
C. botulinum C and D	Botulism	Ingestion of toxin from feedstuff usually carcasses	Stiffness and incoordination leading to deathover a period of days	SA and Australia, especially in draught conditions. UK from poultry litter (type C)	Specific vaccination available in SA and Australia

to an acceptable condition. The inability of the dentition to deal with the diet can be assessed by physical examination of the mouth and dental arcades; missing or uneven teeth or abscesses, gaps, balling of food and impaction can make grazing and chewing difficult. These animals will be unable to acquire sufficient nutrients without provision of special diets or long grass swards.

Chronic disease may be more difficult to address. The particular diseases involved would include:

1. chronic endoparasitism
2. chronic fascioliasis
3. chronic pneumonia
4. chronic lameness
5. johnes disease

The first four syndromes have been discussed above. Control is possible, or the flock may have to live with a low level of disease (as is the case with SPA, see above). Johnes disease is relatively uncommon, but may occur with incidence levels in a flock approaching 10% or more. The disease is essentially a chronic enteritis found throughout the world in all ruminants. It is important because of the overt clinical disease, but also due to the reduced productivity of the prolonged preclinical stages of disease. Sheep become inappetant and start to lose condition and deteriorate. Classically, clinical Johnes disease can progress fairly rapidly and, in advanced cases, faeces may become soft and unformed giving rise to faecal soiling around the perineum. The disease can occur in any sheep over about the age of 1 year. Diagnosis can only be made definitively on post-mortem although blood tests may be useful on a flock basis. Control is difficult as monogastric animals as well as ruminants may be affected, and this complicates development of effective control strategies.

Sharp (2007) discusses the aetiology and importance of the disease and its uncertain potential as a zoonotic threat. When considering the effect of disease on welfare, the presence on any unit of a planned preventive strategy should address the steps to be taken to investigate any increase in disease incidence rates, and the establishment of local disease prevalence can only help to maintain and improve the welfare of all local flocks.

5.3.6.6 Injuries

Injuries are largely preventable but accidents happen. To reduce the chances of injury, all handling facilities should be regularly checked for sharp edges and state of repair. Moving sheep rapidly through handling facilities means that these facilities need to be in good working order and preferably simple to use as complex operations require more concentration on the mechanics than the animal. The facilities need to be well sited so that there is no shadow at critical points and so that the shepherd has easy, quick and clear access to the sheep. Stockman skills and attitude can also reduce injuries as animals that at driven aggressively, or are highly fearful, are more likely to panic and collide with handling facilities.

Common injuries seen on sheep units include damage to the new born lamb from an inexperienced or overenthusiastic dam (e.g. prolapse of intestines through mechanically created umbilical hernia), as well as distal limb fractures of lambs investigating new surroundings and drowning of lambs in troughs soon after birth. Injuries to adult sheep include limb fractures, road traffic accidents and shearing injuries through either speed or inattention at a critical moment. These are all accidental to some extent and can be reduced by improved environments, or more rigorous shepherding, but can never be completely eliminated.

Other commonly-seen injuries that are non-accidental may due to predation or attack from predators, especially domestic dogs. The prevalence and severity of predation depend greatly on the local frequency of large carnivores, for example in the United Kingdom foxes, badgers and avian scavengers are the only significant predators of sheep, with most attention being directed to lambs. In other countries predation can be a more significant threat to sheep farming with coyotes, bears, lynx, mountain lions, wolverines, dingoes, wolves, foxes, feral pigs, eagles and other predatory birds as potential threats to both adult sheep and lambs (see Chapter 1). Crows are attracted by the bright sheen of the eyes and will often peck out the eyes of a comatose or freshly dead lamb. If the lamb is alive but unable to defend itself, this must be an extremely painful experience. Foxes tend to kill lambs out-right but often seem to do so for reasons other than for the necessity of food.

Dogs are responsible for increasing numbers of reported attacks on sheep particularly on flocks in areas around centres of population (towns or cities). These attacks can be vicious and leave many injured and suffering sheep behind. Commonly, dogs attack not on their own, but in a number so that they can work the flock like their ancestral wolf-packs would have done. Once a dog has attacked sheep, it is probable that the dog will attack again if given the opportunity. It is legally permitted in the United Kingdom for a farmer to shoot a dog that is worrying sheep. It is not permissible to catch the dog and then to shoot it. However, it may be difficult for the farmer to shoot a dog cleanly without also endangering his sheep.

These types of problem do not take account of the effect that the stress may have on the flock, particularly in the last weeks before the start of lambing when worrying may lead to abortion and the precipitation of metabolic diseases. Increased incidence of disease due to acute distress including abortion, metabolic disease or physical injury is an accepted sequel to attack or predation and should be taken into account in viewing the economic benefits of flock security.

5.3.6.7 Continual Monitoring and Recognition of Disease

These diseases are important examples of the need to keep the spread of information about disease, its diagnosis, preventive measures and treatments continually updated and to share discussion about the needs for delivery of new technology and new ideas for improving the ways in which we can implement positive change in the health and welfare of our flocks. Improved welfare can be assessed using indices from clinical observation (Webster et al., 2004). The systems to allow this have been established in some farmed species, such as poultry (Kestin et al., 1992), using behavioural indices with repeatable success. Dawkins (2004) advocates the use of behavioural assessments as they can not only show a useful guide to the physical health of the animal but could also act as an early warning system. This will be recognised by many experienced shepherds.

5.4 Disease Control and Welfare

Welfare issues arise with disease control in a number of areas:

1. The response to disease control should be proportionate to the threat. For example: Foot and Mouth Disease (FMD) control of hefted flocks (see Chapter 6) should consider risk assessment of spread of FMD within the flock – it is unusual to find a large number of sheep showing overt clinical symptoms of the disease at the same time – and the risk to contiguous groups needs to be weighted against the difficulty of re-establishing a hefted flock. The risk of carrier status in these extensively managed sheep is probably very low (Kitching et al., 2005).
2. The response should result in a lower incidence of overt disease, e.g. the prevention of spread and maintenance of high health status flocks that can be certified free of Maedi-Visna is enabled by blood sampling and serological testing. The benefits of preventing Maedi-Visna outweigh the aversive effects of gathering the flock and blood sampling the sheep. Grandin (1989) describes a system that was devised to allow sheep to volunteer for phlebotomy which, whilst it allows a nearer stress-free measurement of haematological and biochemical parameters, still restrains the animal's autonomy and arguably provides an immeasurable constraint on the cognitive resolve of that individual.
3. The response should reduce the use of therapeutic agents to the minimum needed to maintain efficacy and to prolong the useful life of those agents. For example, the use of zinc sulphate foot baths is ineffective without the use of a wetting agent to enhance penetration of the horny hoof, and the footbath must allow for standing time in the solution. Proper controlled use of footbaths and conditioning of the flock to routine procedures lessens the aversive welfare impact of human intervention. Similarly the general use of antibiotics in foot baths should be discouraged, except in extreme instances, in order that resistant strains of bacteria do not develop and compromise use of these compounds in the national flock.

5.4.1 Flock Health Plans

There are accepted general categories of control within an approach to flock health programmes (Spedding et al., 2007):

Statutory schemes. These would include the control of notifiable disease such as those on the Organisation Internationale des Epizooties (OIE) category A list of infectious diseases. The control schemes are put in place nationally but are monitored by this body. This statutory control has raised questions about the welfare consideration given to animals when looking at national animal disease control

(Crispin et al., 2002) and also about the international accreditation of these schemes and the use and validation of technical advances and their methods of implementation (Marshall and Roger, 2004). The decision to implement mass slaughter as a disease control policy when other countries have shown the efficacy of vaccination policy is an example of the poor priority welfare is sometimes accorded when dealing with commercial interests. The use of stakeholders (a word coined to mean all those with an interest in any one particular area) in helping to formulate contingency plans for dealing with exotic disease may help to ensure that welfare interests are given a better priority, although trade perspectives may argue on economic imperatives and regard welfare as an incidental benefit or detraction from their aims. Commercial interests of the large supermarkets and the major retail outlets for food must shoulder some of the responsibility for this due to their reluctance to use products from vaccinated animals. Community action in the European Union on FMD may have global impact on disease control. Preparation for dealing with exotic disease is a priority for state veterinary medicine (Cawthorne, 2002).

Certified national/regional schemes (non-statutory). National accreditation schemes for diseases, such as Maedi-Visna and enzootic abortion-free schemes, exist in some countries. These help to stem the spread of disease and accredited status is seen as an important goal for pedigree breeders. Enzootic abortion of ewes is less of a problem now with the advent of effective vaccines, but freedom from the disease is still a better option for the sheep, and for farmers.

Coordinated flock health programmes. Lovatt (2005) has described the developing importance of health planning to the sheep industry and highlights the benefits of flock health plans and a proactive approach to preventive medicine. This needs to include a review of previous performance and to provide targets for the production goals of the unit. On many sheep farms, particularly extensive systems, record-keeping may be minimal and this can allow unknown chronic disease problems to affect both welfare and productivity for many years. Thus the health planning approach can bring about positive benefits for the sheep and for the farmer. This type of planning involves a significant input of time from the veterinary surgeon and cost effectiveness needs to be shown to the farmer. Spedding et al. (2007) have said that programmes are self limiting as they are success terminated, but, if that moves the improvement of welfare forward and on to the next farm, that is a successful outcome. Lewis (2002) emphasises the importance of an up-to-date written health plan to not only comply with the welfare code for sheep but also to record the proposed procedures for all involved with the flock.

Planned unsupervised schemes. The generation of flock health programmes by computer may apparently satisfy the legal requirements of the codes of welfare but to reward the sheep (and farmer) with the level of welfare improvement that can be expected from a good scheme, supervision is imperative. The implementation and the joint effort between farmer, shepherd and veterinary surgeon is vital to the continuing success of a properly constructed and instituted health scheme.

Basic response to specific episodes. The emergency veterinary work, that never goes away and for which planning is difficult, can at least be brought under control by starting to plan preventive strategies. This will not avoid the lambing problems or caesarean sections seen on intensive lowland farms, and during the season on upland units, or the sudden and explosive incidents of disease that break out from time to time, but may reduce the number of deaths on easy-care or extensive holdings.

Spedding et al. (2007) also emphasise the importance of well-defined end points for such schemes. These schemes have been welcomed by the industry, with the usual reservations as to cost and application. It is often felt that these schemes are most relevant to the pedigree sector, which produces a more valuable product for breeding and can spread the cost of these schemes into their product more easily. Unfortunately the reluctance to adopt an open stance on health issues is often most noticeable in this sector. This is because of the better prices they can command so that, for example, were CLA to become a problem in a particular breed, that would affect the price and thus the return that producers expected. In such a case, the temptation not to declare the infection must be present, caveat emptor!

However the emphasis in such schemes has been more on disease prevention and control rather than on welfare promotion. It can be argued that these issues go hand in hand but without reference to the welfare of the individual, these schemes can represent a mechanistic approach to animal husbandry. In the United Kingdom the current Department for Environment, Food and Rural Affairs (Defra) Sheep Codes of Recommendation for the Welfare of Livestock (2000) recommend a written flock health programme.

5.4.1.1 Flock Health Plan Requirements

The primary requirement for a basic flock health programme is a diary of events for the flock recording dates for mating, lambing, weaning and around which strategic preventive veterinary plans can be instituted. This could take the form of vaccination programmes and foot-bathing regimes, as well as suggested times for batch sampling of faeces for faecal egg count monitoring. All events should be recorded, including ewe and lamb losses, and the reasons for these, as well as records of any investigations. Planning for feeding at critical times of the flock year should also be included.

To be an intrinsic part of any shepherd's approach to husbandry, it is vital that science provides a fuller understanding of the complex requirements of livestock. In his discussion on Flock health programmes, Lewis (2004) quotes the following poem:

Go in search of your people,
Love them,
Learn from them,
Plan with them,
Serve them,
Begin with what they know,
Build on what they have.

Kwame N'Krumah

This summarises the approach needed to consolidate the uptake and implementation of flock health planning and the reasonable ethical and consensual atmosphere that will help drive this goal. It is also important as the industry needs to consider sheep within the context of the moral community as they possess intrinsic qualities deserving of this consideration. The Five Freedoms provide a basic ethical framework which can aid this but the powers of legislation and statutory control do not prevent abuses of the system.

5.4.2 Management Tools for Disease Control

5.4.2.1 Preventive Management Strategies

A number of preventive management strategies exist that can help to prevent, or lessen the impact of, the most common and predictable disease conditions.

Condition scoring: This is one of the most important tools for evaluating nutritional performance within a flock. It involves handling the adult breeding stock and assessing the amount of muscle and fat cover over the transverse processes of the lumbar vertebrae and around the tail head (Russel, 1984). It is useful as a guide to the performance of a group of sheep and cannot be used as a predictive indicator for any single individual.

Pasture management: This is part of the planning regime and the regulation of sward height is a useful way to assess the contribution that can be expected from grazing (Mayne et al., 2000). Movement of sheep should be planned, deliberate and careful.

Nutrition: The role of nutrition is central to successful production management and a full consideration of the feeding programme is a necessary adjunct to any planned health and welfare programme. This needs to consider availability and quality of forage, the constituents of concentrate feed as well as amounts needed by the flock, grazing availability and rotation of classes of livestock. The role of nutrition of small ruminants is of major economic importance worldwide and manipulation of diet is as important as the development of sheep genotypes and new production systems (Freer and Dove, 2002).

Blood-sampling: The regular use of blood sampling to check nutritional adequacy and at certain times of the cycle to check trace element status (particularly for example, copper, vitamin B12/cobalt, vitamin E/selenium and iodine levels in mid pregnancy) is a useful strategy (Suttle, 2005; Milne and Scott, 2006). This permits corrective action to be taken to prevent the adverse effects of deficiencies or to ensure that toxic levels are not reached. This technique should be planned and discussed to implement a pro-active approach to problem solving.

Shelter provision: Housing or shelter needs should be organised in advance of bad weather, and any specific preventive measures prior to the use of this should be considered (Eales et al., 2004). For example, housing will bring animals into closer contact than at pasture, therefore ensuring that incidences of footrot or other infectious diseases have been eliminated before housing will prevent the spread of infection.

Hygiene: Hygiene and cleaning regimes should be planned well in advance of need. This is particularly important, for example, at lambing when poor hygiene practices can be a significant contributor to lamb mortality (Binns et al., 2002).

5.4.2.2 Vaccination

For commercial flocks, vaccination against clostridial diseases and pasteurella pneumonia is a basic requirement for all those whose concern is the welfare of their sheep. For many flocks vaccination of all replacements against *Chlamydophila abortus* (the cause of enzootic abortion) and *Toxoplasma gonddii* (the cause of toxoplasmosis, another common cause of abortion in the United Kingdom) is standard practice. These vaccines are perceived as giving lasting protection over the reproductive life of the ewe, although the manufacturers' recommend revaccination after 3 years. As part of a control programme against footrot, vaccination is a useful tool but the frequency of booster doses needs to coincide with the peak periods of expected incidence of the disease. Vaccinations against *E. coli* and against *Erysipelas rhusiopathiae* (a cause of arthritis) are available and can be useful where disease is caused specifically by these organisms. These diseases and their welfare implications have been discussed above. The innovative use of a cattle vaccine to help protect sheep against parainfluenza-3 virus (Rodger, 1989) is an example of the cooperation needed to progress disease control between farmer, clinician and pharmaceutical industry. Vaccination against orf, a parapox virus causing skin surface lesions particularly around the mouth and nose, is not as straight forward as the vaccines mentioned above, as the mechanism by which immunity arises is cell-mediated rather than humoral (blood-born antibodies). This means that the immunity wanes faster than with other vaccines and, in order to get the best effect from the vaccine, it should be used just prior to expected peak incidence of the disease or about 4–6 weeks prior to sale time.

Vaccines are also available against Salmonellosis, a bacterial infection and Louping ill, a tick-born viral infection. Salmonella species have been associated with abortion and deaths in most parts of the world. *S. abortus ovis* is a major cause of abortion in parts of Europe and Western Asia; other Salmonella species, such as *S. typhimurium* and *S. Dublin* can cause illness as well as abortion and *S. Montevideo* is becoming a more common isolate in the United Kingdom. In other parts of the world different serotypes may predominate (Mearns, 2007b). Louping-ill was thought to be restricted in occurrence to the upland grazings of the United Kingdom but indistinguishable disease has been reported in Bulgaria, Turkey, Norway and Spain. It is an acute disease of the central nervous system and can affect most domestic animals and man. Reid and Rodger (2007) describe methods of control and emphasise the danger of this potent zoonosis.

Autogenous and emergency vaccines have been used in the face of disease (e.g. CLA) but these are variable in quality and effectiveness and do not necessarily prevent the continuing spread of infection. Autogenous vaccines are those manufactured under licence by private laboratories from infective material submitted from a particular farm. These vaccines are only for use on the farm or in the flock from which the samples originate. Strictly, autogenous vaccines are prepared from material from one animal for use in that animal whereas emergency vaccines may be used on the whole flock.

Outside the United Kingdom, vaccines are available against CLA, Campylobacteriosis, Salmonellosis and Johnes disease and have proved effective in control strategies against these diseases.

5.4.2.3 Specific Treatment Regimes

These are specific management practices designed to reduce the incidence of specific diseases. The examples of preventive treatment and control for FR have been given earlier. Other examples include early effective treatment or preventive treatment, such as navel dipping of newborn lambs in a 10% spirituous solution of iodine. These regimes are often specific to the farm (for example, copper supplementation in mid-pregnancy to prevent swayback in lambs could be fatal if the level of copper in the flock is not monitored properly) and need to be the product of discussion between farmer and clinician. Another useful example would be the continuing analysis of faecal egg counts to determine the correct dosing strategy for the control of worms in the flock in different classes and ages of sheep.

The use of action plans linked to target production levels, in flock health plans, can flag up intervention levels. These raise the mechanistic side of health programmes. Welfare should be an overriding consideration and not depend on the acquisition of target production figures. However the setting of production targets can include reducing morbidity and mortality from disease and can focus the shepherd's attention on areas of the management strategies that need improvement.

5.4.2.4 Breeding

Selection for production traits may lead to increasing problems (e.g. selection for increased milk yield is linked to increasing incidence of mastitis in dairy cattle; Ouweltjes et al., 2007), but the potential for improvement of resistance to certain diseases and traits continues. The danger of pursuing one line of attack has been discussed above (see scrapie). The benefits to welfare from planned breeding and selection are discussed in Chapter 10.

5.4.3 Husbandry and Disease Control

Different husbandry methods lead to different disease risks. Different production goals also can alter the perception of disease risks. For example, sheep dairying regions may breed for milk yield and milk composition and may have an increased incidence of mastitis, a condition perceived as very painful as well as problematic to treat (Cokrevski et al., 2001). The intensive lowland farm offers different welfare perspectives to those offered by the extensive ranch or hill flocks as the stocking levels range from a number of ewes per hectare to a number of hectares per ewe. The welfare issues surrounding different husbandry practices will be discussed further in Chapter 6 and only specific remarks relating to disease issues will be made here.

Perhaps the main factor influencing disease control and associated suffering is the frequency of inspection under different systems. Shepherding levels are recognised as appropriate for the type of system which is being shepherded. In the United Kingdom, for example, it would be normal practice to see all the sheep on a holding daily in lowland situations, but on extensive hill systems it may not be possible to see all the sheep every week, indeed they may only all be seen at gathering times, a few times a year. The Farm Animal Welfare Council (FAWC, 1994) has suggested that an upper limit of 1000 ewes, with extra help at peak times such as lambing and shearing and dosing, is a reasonable number for one shepherd. In Australia and New Zealand, these numbers are often much higher (Fisher, 2001). The main risk with infrequent inspection is that disease or injury may go unnoticed or untreated for prolonged periods of time, leading to extended suffering in affected animals. These concerns are partly ameliorated by the types of animals managed under low input systems, e.g. sheep breeds, such as Romneys, specifically bred for hardiness and disease resistance in New Zealand, where the frequency of disease is lower than in more intensive lowland systems.

Extensively farmed sheep are at a lower risk from infectious diseases but are more at risk from nutritional and meteorological effects. Water requirements may also be a limiting factor in some areas of the world, particularly where sheep are milked, as the requirements of these sheep are at least double that of the non-lactating ewe. Intensive systems, on the other hand, mean risk of infectious disease increases but regular inspections and treatments or preventive actions are easier to undertake. A further welfare consideration for extensive systems is the stress involved in handling animals unused to human contact for the purposes of preventive treatments, and balancing the welfare of the individual, that may require treatment, against the stress imposed on the rest of the flock in gathering to provide individual care. The ability to provide care to individual sheep, with minimal disruption to the rest of the flock, is much greater in a more intensive system.

It would seem important that health planning is more detailed for the flocks not regularly inspected, as remedial actions can more easily be instituted for sheep that are regularly inspected. For example, Coccidiosis is often called a disease of intensification. It is a protozoan parasite and causes disease by invasion of the intestinal wall of young lambs and is insidious in its onset and causes much of the damage that affects the lamb prior to clinical evidence of the disease. The disease is preventable if its incidence can be accurately anticipated. It is difficult to use any clinical markers to do this as oocyst counts are notoriously variable and do not necessarily reflect peak infection. Presence or absence of the parasite does not necessarily equate with the disease situation in the live lamb. The disease is reported as mainly affecting young lambs between 4–6 weeks old (Wright and Coop, 2007). The author has experienced outbreaks regularly in the autumn in lambs 24–26 weeks old. If the pattern of disease is similar in sheep previously unexposed to infection, then it could be argued that this is a disease of overproduction: too many lambs leading to increased contamination, disease rather than immunity. Equally it may be that changing husbandry methods have lessened the exposure of lambs to an oocyst burden so that whenever naïve lambs meet challenge they are less equipped to respond. Many farmers are coming to the view that feeding the lambs properly reduces disease incidence. This is a part-truth. Proper nutrition requires that lambs are equipped to mount an immunological response to disease and if they are lacking in nutrients they will be unable to do this. This is a good reason to check flock health status using biochemical and haematological indicators prior to lambing.

Regardless of farming system, stockperson skills and abilities are an important part of disease control, for example the ability to deal appropriately with different lameness conditions, and to recognise the early signs of disease. There is a lack of trained stockpersons in many sheep-producing countries, as there is a political climate that seems not to value livestock agriculture and food production and has seen the demise of a large number of agricultural colleges and training schemes for educating shepherds and initiating the transfer of new technologies.

5.4.4 Quarantine

Transboundary disease such as Foot-and-mouth disease (FMD), Bluetongue, Peste des petits Ruminants, Sheep and goat pox and Rift Valley fever are well known. The continuing monitoring of the epidemiology and changing distribution of these diseases across the world is of importance to all sheep producers (Rashwan et al., 2001). There are continuing changes in populations of insect vectors (viz. Bluetongue, Sarto I Monteys et al., 2005) and so, with climate change, diseases more commonly associated with Tropical or Mediterranean climates are becoming recognised in Temperate zones. Continued vigilance is needed. Increase in globalisation may enable the free movement of livestock across ever-widening frontiers and vast numbers of relatively unpoliced travellers pass through airspaces and airports along with their luggage, waste and occasionally contraband. For example, the amount of illegal bushmeat seized at Heathrow Airport is measured in tons and outbreaks of Classical Swine Fever and the United Kingdom FMD epidemic of 2001 may have had their origins from this type of source.

In maintaining biosecurity, a consideration of the introduction of purchased sheep into an existing flock is vital. The maintenance of the health status of the existing flock depends on the effort that is made to keep this. Isolation facilities are not a luxury; all sheep farms should have adequate self-contained isolation facilities. The flocks from which introduced stock are purchased should be of a known health status and it is useful to purchase direct so that no contact is made with third parties or their problems on the way.

A quick summary of the issues presented by the diseases described above will usefully remind a sheep purchaser of the questions that the vendor should be able to answer

1. Does the flock have accredited M-V free &/or enzootic abortion free status?
2. What is the Scrapie status of the flock?
3. Has the farm a written health scheme?
4. What were the dates of vaccination, worming and last foot-care treatments?

5. What were the products used for these procedures?
6. Is there a disinfection point close to the point of departure from the vendor's premises?

The purchased animals should be isolated for not less than 21 days. During this time routine treatments can be carried out and common diseases observed for. Any sheep entering should be assumed to be bringing sheep scab and lice, as well as resistant intestinal worms, and to be carrying footrot bacteria on their feet (Lewis, 2001). The Sustainable Control of Parasites in Sheep (SCOPS) group in the United Kingdom was convened to explore strategies to address the increasing problem faced by the sheep industry of parasitic resistance to treatment. Resistant intestinal worms are an increasing problem and the recommendations of the SCOPS group (Abbott et al., 2004b) should be adhered to. These include treatment on arrival with a combination of anthelminthics and isolation. This is a difficult message to deliver to farmers. It clashes with widely held concepts and, with major sale times for breeding sheep, especially rams, occurring close to the breeding season, may prevent adequate quarantine. The industry needs to reconsider these timings to allow proper quarantine procedures to be followed.

The ACME principle (Jackson et al., 2004) lists the following:

Adopt an effective quarantine and treatment strategy using a macrocyclic lactone and an imidazothiazole to treat animals and minimise the risk of importing resistant worms

Check the efficiency of your drenches

Monitor to ensure that you use the appropriate parasiticide and minimise the number of treatments

Ensure that you follow best practice advice.

To which should be added:

Talk to your veterinary surgeon

Implement these ideas now

Plan your preventive strategy

Use this **TIP** to achieve the **ACME.**

5.5 Conclusions

There are major challenges to the industry from the disease perspective particularly in terms of parasite control. The establishment of basic welfare parameters and sheep-based indicators of welfare are needed to allow proper assessment of the needs of our sheep. The Farm Animal Welfare Council (2004) note the urgent need to develop a farm welfare surveillance system to give robust and reliable information on the prevalence of a range of health and welfare traits for different species of livestock and further recommends significant new investment in molecular genetics in cattle and sheep. This needs to be balanced with a practical understanding of the difficulties faced in farming sheep whether on an extensive or intensive system, and the benefits that well-run flocks offer the larger community in terms of environment and leisure as well as from production. This also needs to be balanced by the problems that may be faced in parts of the world with human welfare needs.

The development of an objective animal welfare index which is applicable at farm level by individual assessment would seem to be an area for priority and this should involve integration of the presently recognised contributors to animal welfare measurement, i.e. behavioural, physiological, biochemical, pathological, emotional and legal indicators. The relationship between veterinary clinicians practising in the field and the leading research institutes and universities needs to be encouraged and developed to allow a more immediate transfer of technical advances and to help shape the practical development of measurable welfare attributes. Collaboration in this area should be rewarding for both groups – on one hand allowing good practical clinical access and experience and on the other providing an extra intellectual stimulus and a broader understanding of research targets and goals.

The difficulties faced by thc sheep industry globally are encapsulated in two different scenarios: (1) the emergence of the global threat of anthelminthic resistance has led to areas where sheep husbandry is no longer possible or is fraught with difficulty; and (2) the development of the perception of risk of TSEs to the human population has shown the difficulty in proving a negative (that sheep do not suffer from BSE). The development of risk reduction strategies has aimed at increasing the genetic resistance of the population, whilst proscribing certain genetic lines. Recently atypical scrapie has suggested that this may not have been as sound an idea as first thought. These difficulties demonstrate the need for a continual reassessment of the threats facing the sheep industry and the importance of maintaining a scientific community that can embrace, debate and respond to change rapidly and effectively.

The promotion of sheep health and welfare through the implementation of flock health planning on a local level, backed by veterinary surgeons with an interest in developing and supporting these plans working with flockmasters is an important goal. This initiative, including the help of others to effect technology transfer and to increase our understanding of the sheep's welfare requirements, deserves support not only from all those involved in the industry but from the consumer and from central government.

References

Abbott, K.A., Taylor, M.A. & Stubbings, L.A. (2004a) Anthelminthic resistance management in sheep. *Veterinary Record* **154**: 735–736.

Abbott, K.A., Taylor, M.A. & Stubbings, L.A. (2004b) Sustainable worm control strategies for sheep. A Technical Manual for Veterinary Surgeons and Advisers, March 2004, SCOPS (Sustainable Control of Parasites in Sheep), National Sheep Association, Malvern, United Kingdom.

Abu, I.F. & Ngere, L.O. (1979) The indigenous sheep of Nigeria. *World Review of Animal Production* **15**: 51–62.

Aitken, I.D. (2007) *Diseases of Sheep.* Blackwell Science, Oxford, United Kingdom.

Aitken, I.D. & Longbottom, D. (2007) Chlamydial abortion. In *Diseases of Sheep* (ed. I. D. Aitken), pp. 105–112. Blackwell Science, Oxford, United Kingdom.

Anderson, N., Barger, I.A. & Waller, P.J. (1987) Impact of gastrointestinal parasitism on pasture utilization by grazing sheep. In *Temperate pastures: their production, use and management* (eds. J. L. Wheeler, Pearson, C. J. and Robards, G. E.), pp. 555–566. Australian Wool Corporation and CSIRO, Melbourne, Australia.

Anon (1992) Animal Health Act (1981) as amended by Animal Health Act Regulations1992 (Statutory Instrument (SI) No.3923, 1992) Her Majesty's Stationary Office, London, United Kingdom.

Anon (2000) *Codes of recommendations for the welfare of livestock: sheep*. Her Majesty's Stationary Office, London, United Kingdom.

Anon (2004) *Guide to professional conduct*. Royal College of Veterinary Surgeons, London, United Kingdom.

Appleby, M. (1995) Farm Animal (mammal) enrichment. In *Environmental enrichment information resources for Laboratory Animals* 1965–1995, pp. 63–67, Animal Welfare Information Centre, USDA, USA.

Armbruster, T., Peters, K.J. & Hadji-Thomas, A. (1991) Sheep production in the humid zone of West Africa. 3. Mortality and productivity of sheep in improved production systems in Cote D'Ivoire. *Journal of Animal Breeding and Genetics* **108**: 220–226.

Baird, G. (2003) Current perspectives on caseous lymphadenitis. *In Practice* **25**: 62.

Baird, G., Synge, B. & Derckson, D. (2004) Survey of caseous lymphadenitis seroprevalence in British terminal sire sheep breeds. *Veterinary Record* **154**: 505–506.

Bates, P.G. (1996) Epidemiology of subclinical ovine psoroptic otoacariasis in Great Britain. *Veterinary Record* **138**: 388-393.

Bates, P.G. (1997) The pathogenesis and ageing of sheep scab lesions. Part 1. *State Veterinary Journal* **7**: 11–15.

Bates, P. (2007) Sheep scab. In *Diseases of Sheep* (ed. I. D. Aitken), pp. 321–325. Blackwell Science, Oxford, United Kingdom.

Bates, P., Rankin, M., Clifford, D. & Stubbings, L. (2005) Shower dipping in diazinon or cypermethrin dipwash to control ovine psoroptic mange (sheep scab). *Veterinary Record* **156**: 655.

Beauchamp, T.L. & Childres, J. (1994) *Principles of Biomedical Ethics* 4th Edition, pp. 3–40, Oxford University Press, Oxford, United Kingdom.

Bekele, T., Woldeab, T., Lahlou-Kassi, A. & Sherington, J. (1992) Factors affection morbidity and mortality on-farm and on-station in the Ethiopian highland sheep. *Acta Tropica* **52**: 99–109.

Berger, Y.M., Bradfrod, G.E., Essaadi, A., Johnson, D.W., Bourfia, M. & Lahlou-Kassi, A. (1989) Performance of Dman and Sardi sheep on accelerated lambing 3. Lamb mortality, growth and production per ewe. *Small Ruminant Research* **2**: 307–321.

Binns, S.H., Cox, I.J., Rizvi, S. & Green, L.E. (2002) Risk factors for lamb mortality on UK sheep farms. *Preventive Veterinary Medicine* **52**: 287–303.

Bhattacharya, A.N. & Harb, M. (1973) Sheep production on natural pasture by roaming Bedouins in Lebanon. *Journal of Range Management* **26**: 266–269.

Broom, D.M. (1998) Welfare, Stress and the Evolution of Feelings. *Advances in the study of Behaviour* **27**: 371–403.

Broom, D.M. & Johnson, K.G. (1993) *Stress and Animal Welfare*. Chapman and Hall, New York, USA.

Bygraves, A.C., Bates, P.G. & Daniel, N.J. (1993) Epileptiform seizures in ewes associated with sheep scab mite infestation. *Veterinary Record* **132:** 394–395.

Campbell, J.R., Menzies, P.I., Waltner-Toews, D., Walton, J.S., Buckrell, D.C. & Thorsen, J. (1994). The seroprevalence of maedi-visna in Ontario sheep flocks and its relationship to flock demographics and management practices. *Canadian Veterinary Journal* **35**: 39–44.

Cawthorne, R.G. (2002) Developing a control strategy for bluetongue. *Veterinary Record* **151**: 518.

Chaarani, B., Robinson, R.A. & Johnson, D.W. (1991). Lamb mortality in Mekenes Province (Morocco). *Preventive Veterinary Medicine* **10**: 283–298.

Clarkson, M.J. & Faull, W.B. (1990). *A Handbook for the Sheep Clinician,* 4th Edition. Liverpool University Press, Liverpool, United Kingdom.

Clements, A.C.A., Fitzpatrick, J.L. & Mellor, D.J. (2002). Reporting of sheep lameness conditions to veterinarians in the Scottish Borders. *Veterinary Record* **150**: 815–817.

Cloete, S.W.P., Van Halderen, A. & Schneider, D.J. (1993) Causes of perinatal lamb mortality amongst Dormer and SA mutton merino lambs. *Journal of the South African Veterinary Association* **64**: 121–125.
Cokrevski, S. Crcev, D., Loria, G.R. & Nicholas, R.A.J. (2001) Outbreaks of contagious agalactia in the Republic of Macedonia. *Veterinary Record* **148**: 21–667.
Coop, R.L. (1979) Production loss in subclinical helminth infections. *Veterinary Record* **105**: 189.
Coop, R.L. (1982) The impact of subclinical parasitism in ruminants. In *Parasites – their world and ours* (eds. D. F. Mettrick and S. S. Desser), pp. 439–450. Proceedings of the 5th International Congress of Parasitology, Toronto, Canada.
Conington, J., Bishop, S.C., Grundy, B., Waterhouse, A. & Simm, G. (2001) Multi-trait selection for sustainable UK hill sheep production. *Animal Science* **73**: 413–423.
Corke, M.J. & Broom, D.M. (1999) The behaviour of sheep with sheep scab, *Psoroptes ovis*, infestation. *Veterinary Parisitology* **83**: 291–300.
Cranwell, M.P., Rose, M.A. & Blake, M. (2002) Chorioptic mange of sheep. *Veterinary Record* **151**:14–427.
Crispin, S.M., Roger, P.A., O'Hare, H. & Binns, S.H. (2002) The 2001 foot and mouth disease epidemic in the United Kingdom: animal welfare perspectives. *Revue Scientifique et Technique: Office International des Epizooties* **21**: 877–883.
Dawkins, M.S. (2004) Using behaviour to assess animal welfare. *Animal Welfare* **13**: S3–S7.
DEFRA (2002) *Lameness in Sheep*. Her Majesty's Stationary Office, London, United Kingdom.
Dubreuil, P., Arsenault, J. & Belanger, D. (2005) Biochemical reference ranges for ewes of different ages. *Veterinary Record* **156**: 636–638.
Duncan, I.J.H. (2004) Pain, fear and distress. *Proceedings of Global Conference on Animal Welfare: an OIE initiative.* Feb 23–24, 2004, Paris, France, pp. 163–172.
Duncan, I.J.H. (2005) Science based assessment of animal welfare: farm animals. *Revue Scientifique et Technique: Office International des Epizooties* **24**: 483–492.
Dwyer, C.M. & Lawrence, A.B. (2005) Frequency and cost of human intervention at lambing: an interbreed comparison. *Veterinary Record* **157**: 101–104.
Eales, A., Small, J. & Macaldowie, C. (2004) Practical Lambing and Lamb Care: a Veterinary Guide. 3rd Edition, Blackwell Publishing, Oxford, United Kingdom.
Egerton, J.R. (2007) Diseases of the feet. In *Diseases of Sheep* (ed. I. D. Aitken), pp. 273–281. Blackwell Science, Oxford, United Kingdom.
Farm Animal Welfare Council (2004) *Report on the Welfare Implications of Animal Breeding and Breeding Technologies in Commercial Agriculture*. Defra, London.
Farm Animal Welfare Council (1994) *Report on the Welfare of Sheep*. MAFF, London.
Festa-Bianchet, M. (1988) Nursing behaviour of bighorn sheep: correlates of ewe age, parasitism, lamb age, birthdate and sex. *Animal Behaviour* **36**: 1445–1454.
Festinger, L. (1957) *A theory of cognitive dissonance*, Stanford University Press, Stanford, USA.
Festinger, L. & Carlsmith, J.M. (1959) Cognitive consequences of forced compliance, *Journal of Abnormal and Social Psychology* **58**: 203–211.
Fisher, M.W. (2001) Lambing management in New Zealand: ethics and welfare considerations. *Surveillance (Wellington)* **28**: 16–17.
Fisher, M.W. (2003) New Zealand narratives of the benefits of reduced human intervention during lambing in extensive farming systems. *Journal of Agricultural and Environmental Ethics* **16**: 77–90.
Fisher, M.W. & Mellor, D.J. (2002) The welfare implications of shepherding during lambing in extensive New Zealand farming systems. *Animal Welfare* **11**: 157–170.
Fitzpatrick, J., Scott, M. & Nolan, A. (2006) Assessment of pain and welfare in sheep. *Small Ruminant Research* **62**: 55–61.
Fleming, M.W. (1997) Cortisol as an indicator of severity of parasitic infections of *Haemonchus contortus* in lambs (*Ovis aries*). *Comparative Biochemistry and Physiology* **116B**: 41–44.
Forrest, R.H., Hickford, J.G.H. & Frampton, C.M. (2007) Polymorphism at the ovine β_3-adrenergic receptor locus (ADRB3) and its association with lamb mortality. *Journal of Animal Science* **85**: 2801–2806.

Fraser, H. (2000) Phillips report and the origin of BSE. *Veterinary Record* **147**: 724.

Freer, M. & Dove, H. (2002) Preface In: *Sheep Nutrition* (eds. M. Freer and H. Dove) pp. ix–x. CABI publishing, Wallingford, United Kingdom.

French, N., Wall, R., Cripps, P.J. & Morgan, K.L. (1994) Blowfly strike in England and Wales: the relationship between prevalence and farm and management factors. *Medical and Veterinary Entomology* **8**: 51–56.

Gama, L.T., Dickerson, G.E., Young, L.D. & Leymaster, K.A. (1991) Effects of breed, heterosis, age of dam, litter size, and birth weight on lamb mortality. *Journal of Animal Science* **69**: 2727–2743.

Grandin, T. (1989) Voluntary acceptance of restraint by sheep. *Applied Animal Behaviour Science* **23**: 257–261.

Green, L.E. & Morgan, K.L. (1993) Mortality in early born, housed lambs in South-West England. *Preventive Veterinary Medicine* **17**: 261–271.

Gunn, R.G., Doney, J.M. & Smith, W.F. (1979a) Fertility in Cheviot ewes. 2. The effect of level of pre-mating nutrition on ovulation rate and early embryo mortality in North and South Country Cheviot ewes in moderately-good condition at mating. *Animal Production* **29**: 17–24.

Gunn, R.G., Doney, J.M. & Smith, W.F. (1979b) Fertility in Cheviot ewes. 3. The effect of level of nutrition before and after mating on ovulation rate and early embryo mortality in North and South Country Cheviot ewes in moderately-good condition at mating. *Animal Production* **29**: 25–32.

Harkins, L.S., Fitzpatrick, J.L., Nolan, A.M., Barrett, D.C. & Scott, E.M. (2002) Perception of pain in sheep. *Sheep Veterinary Society* **57**: 17–20.

Haughey, K.G. (1993) Perinatal lamb mortality – it's investigation, causes and control. *Irish Veterinary Journal* **46**: 9–28.

Heath, A.C.G., Bishop, D.M. & Tenquist, J.D. (1987) The effects of artificially-induced fly-strike on food intake and liveweight gain in sheep. *New Zealand Veterinary Journal* **35**: 50–52.

Hellebrekers, L.J. (2000) Pain in animals. In *Animal Pain: A practice oriented approach to effective pain control in animals*, pp. 11–16. Van der Wees, Blackwell Science, Oxford, United Kingdom.

Henderson, D.C. (1990) *The Veterinary Book for Sheep Farmers*. Old Pond Publishing, Suffolk, United Kingdom.

Hight, G.K. & Jury, K.E. (1970) Hill country sheep production 2. Lamb mortality and birth weights in Romney and Border Leicester x Romney flocks. *New Zealand Journal of Agricultural Research* **13**: 735.

Hindson, J. (1985) Flock Health Control – Where next? *Proceedings of the Sheep Veterinary Society* **9**: 44–48.

Hosie, B.D. (1989) Prolapse of the vagina and cervix in the ewe. *In Practice* **11**: 215–217.

Hosie, BD (2007) Prolapse and hernia. In *Diseases of Sheep* (ed. I. D. Aitken), Blackwell Scientific, Oxford, United Kingdom.

Huffman, E.M., Kirk, J.H. & Pappaioanou, M. (1985) Factors associated with neonatal lamb mortality. *Theriogenology* **24**: 163–171.

Hutchings, M.R., Gordon, I.J., Robertson, E., Kyriazakis, I. & Jackson, F. (2000) Effects of parasitic status and level of feeding motivation on the diet selected by sheep grazing grass/clover swards. *Journal of Agricultural Science* **135**: 65–75.

Hutchings, M.R., Kyriazakis, I., Anderson, D.H., Gordon, I.J. & Coop, R.L. (1998) Behavioural strategies used by parasitized and non-parasitized sheep to avoid ingestion of gastro-intestinal nematodes associated with faeces. *Animal Science* **67**: 97–106.

Hutchings, M.R., Kyriazakis, I., Gordon, I.J. & Jackson, F. (1999) Trade-offs between nutrient intake and faecal avoidance in herbivore foraging decisions: the effect of animal parasitic status, level of feeding motivation and sward nitrogen content. *Journal of Animal Ecology* **68**: 310–323.

Huxley, J.N. & Whay, H.R. (2006) Current attitudes of cattle practitioners to pain and the use of analgesics in cattle. *Veterinary Record* **159**: 662–668.

Ingram, J.R., Cook, C.J. & Harris, P.J. (2002) The effect of transport on core and peripheral body temperatures and heart rate of sheep. *Animal Welfare* **11**: 103–112.

Jackson, F., Coop, R.L. (2007) Gastrointestinal helminthosis. In *Diseases of Sheep* (ed. I. D. Aitken), pp. 185–195. Blackwell Science, Oxford, United Kingdom.

Jackson, F., Coop, R.L. & Bartley, D. (2004) To dose or not to dose? Prophylactic anthelminthic treatments. *Proceedings of the Sheep Veterinary Society* **27:** 97–100.

Jordan, D.J. & Lefeuvre, A.S. (1989) The extent and cause of perinatal lamb mortality in 3 flocks of Merino sheep. *Australian Veterinary Journal* **66**: 198–201.

Kabuga, J.D. & Akowuah, F. (1991) Reproductive performance of Djallonke x Sahellain ewes in Ghana. *Small Ruminant Research* **5**: 245–254.

Kelly, R.W. (1992) Lamb mortality and growth to weaning in commercial Merino flocks in Western Australia. *Australian Journal of Agricultural Research* **43**: 1399–1416.

Kent, J.E., Thrusfield, M.V., Molony, V., Hosie, B.D. & Sheppard, B.W. (2004) Randomised, controlled field trial of two new techniques for the castration and tail docking of lambs less than two days of age. *Veterinary Record* **154**: 193–200.

Kestin, S.C., Knowles, T.G., Tinch, A.E. & Gregory, N.G. (1992) Prevalence of leg weakness in broiler chickens and its relationship with genotype. *Veterinary Record* **131:** 190–194.

Kitching, P. (1997) Notifiable viral diseases and spongiform encephalopathies cattle, sheep and goats. *In Practice* **19**: 51.

Kitching, R.P., Hutber, A.M. & Thrusfield, M.V. (2005) A review of foot-and-mouth disease with special consideration for the clinical and epidemiological factors relevant to predictive modelling of the disease. *The Veterinary Journal* **169:** 197–209.

Lawrence, A.B., Conington, J. & Simm, G. (2004) Breeding and animal welfare: practical and theoretical advantages of multi-trait selection. *Animal Welfare* **13**: S191–S196.

Lewis, C.J. (2001) Disease prevention when buying in stock. *Moredun Foundation* **3:** 13.

Lewis, C.J. (2002) Developing a healthy flock. *In Practice* **24**: 12–16.

Lewis, C.J. (2004) Farm Health Programmes. *Proceedings of the Sheep Veterinary Society* **27**: 105–107.

Lewis, C.J. (2007) Clostridial diseases. In *Diseases of Sheep* (ed. I. D. Aitken), pp. 156–167, Blackwell Scientific, Oxford, United Kingdom.

Ley, S.J., Livingston, A. & Waterman, A.E. (1989) The effect of chronic clinical pain on thermal and mechanical thresholds in sheep. *Pain* **39**: 353–357.

Ley, S.J., Livingston, A. & Waterman, A.E. (1991a) Effects of chronic lameness on the concentrations of cortisol, prolactin and vasopressin in the plasma of sheep. *Veterinary Record* **129**: 45–47.

Ley, S.J., Livingston, A. & Waterman, A.E. (1992) Effects of clincially occurring chronic lameness in sheep on the concentrations of plasma noradrenaline and adrenaline. *Research in Veterinary Science* **53**: 122–125.

Ley, S., Waterman, A. & Livingston, A. (1991b) The influence of chronic pain on the analgesic effects of the alpha2-adrenoceptor agonists, xylazine, in sheep. *Journal of Veterinary Pharmacology and Therapeutics* **14**: 141–144.

Ley, S.J., Waterman, A.E., Livingston, A. & Parkinson, T.J. (1994) Effect of chronic pain associated with lameness on plasma cortisol concentrations in sheep: a field study. *Research in Veterinary Science* **57**: 332–335.

Ley, S.J., Waterman, A.E. & Livingston, A. (1995) A field study of the effect of lameness on mechanical nociceptive thresholds in sheep. *Veterinary Record* **137**: 85–87.

Lovatt, F (2005) Developing flock health plans. *In Practice* **26:** 290–295.

Love, S.C.J. & Coles, G.C. (2002) Anthelmintic resistance in sheep worms in New South Wales, Australia. *Veterinary Record* **150**: 87.

Lynch, J.J., Hinch, G.N. & Adams, D.B. (1992) *The Behaviour of Sheep: Biological Principles and Implications for Production* CAB International, Wallingford, Oxon, United Kingdom.

Main, D.C., Whay, H.R., Green, L.E. & Webster, A.J. (2003) Effect of the RSPCA Freedom Food scheme on the welfare of dairy cattle. *Veterinary Record* **153**: 227–231.

Malik, R.C. & Acharya, R.M. (1972) A note on factors affecting lamb survival in Indian sheep. *Animal Production* **14**: 123–125.

Mandal, A., Barua, S., Rout, P.K., Roy, R., Prasad, H., Sinha, N.K. & Sharma, N. (2005) Factors affecting the prevalence of mortality associated with pneumonia in a flock of Muzaffarnagari sheep. *Indian Journal of Animal Sciences* **75**: 407–410.

Marshall, D.J., Walker, R.I., Cullis, B.R. & Luff, M.F. (1991) The effect of footrot on body weight and wool growth of sheep. *Australian Veterinary Journal* **68**: 45–49.

Marshall, M. & Roger, P.A.(2004) Policy and science of FMD control: the stakeholder contribution to decision making. A call for integrated animal disease management. Report of the open session of the Research Group of the Standing Technical Committee of the European Union for the Control of Foot-and-Mouth Disease (ECCFMD) Chania, Crete, 12–15 October ECCFMD, Rome, pp. 246–249. Available at: http://www.fao.org/ag/againfo//commissions/docs/greece04/App78.pdf (accessed on 25th September 2007).

Mayne, C.S, Wright, I.A. & Fisher, G.E.J. (2000) Grassland management under grazing and animal response. In *Grass, its Production and Utilization* (ed. A. Hopkins) pp. 247–291. Blackwell Science, Oxford, United Kingdom.

Mearns, R. (2007a) Abortion in sheep 1. Investigation and principal causes. *In Practice* **29**: 40–46.

Mearns, R. (2007b) Other infectious causes of abortion. In *Diseases of Sheep* (ed. I. D. Aitken) pp. 127–136. Blackwell Scientific, Oxford, United Kingdom.

Mellor, D.J. (2007) Welfare of fetal and newborn lambs. In *Diseases of Sheep* (ed. I.D. Aitken), pp. 22–27. Blackwell Scientific, Oxford, United Kingdom.

Mellor, D.J., Molony, V. & Robertson, I.S. (1991) Effects of castration on behaviour and plasma cortisol concentrations in young lambs, kids and calves. *Research in Veterinary Science* **51**: 149–154.

Mellor, D.J. & Stafford, K.J. (2004) Animal welfare implications of neonatal mortality and morbidity in farm animals. *Veterinary Journal* **168**: 118–133.

Milne, E. & Scott, P.R. (2006) Cost-effective biochemistry and haematology in sheep *In Practice* **28**: 454–461.

Mitchell, G.B.B. (2007) Liver fluke. In *Diseases of Sheep* (ed. I. D. Aitken), pp. 195–204. Blackwell Scientific, Oxford, United Kingdom.

Mitchell, G.B.B., Maris, L. & Bonniwell, M.A. (1998) Triclabendazole-resistant liver fluke in Scottish sheep. *Veterinary Record* **143**: 399.

Molony, V. & Kent, J.E. (1997) Assessment of acute pain in farm animals using behavioural and physiological measurements. *Journal of Animal Science* **75**: 266–272.

Mourad, M., Gbanamou, G. & Balde, I.B. (2001) Performance of Djallonke sheep under an extensive system of production in Faranah, Guinea. *Tropical Animal Health and Production* **33**: 413–422.

Mukasa-Mugerwa, E., Said, A.N., Lahlou-Kassi, A., Sherrington, J. & Mutiga, E.R. (1994) Birthweight as a risk factor for perinatal lamb mortality, and the effects of stage of pregnant ewe supplementation and gestation weight gain in Ethiopian Menz Sheep. *Preventive Veterinary Medicine* **19**: 45–56.

Mukasa-Mugerwa, E., Lahlou-Kassi, A., Anindo, D., Rege, J.E.O., Tembely, S., Tibbo, M. & Baker, R.L. (2000) Between and within breed variation in lamb survival and the risk factors associated with major causes of mortality in indigenous Horro and Menz sheep in Ethiopia. *Small Ruminant Research* **37**: 1–12.

Nieuwhof, G.J. & Bishop, S.C. (2005) Costs of the major endemic diseases of sheep in Great Britain and the potential benefits of reduction in disease impact. *Animal Science* **81**: 23–29.

Noakes, D.E. (2001) Prolapse of the vagina and cervix. In *Arthur's Veterinary Reproduction and Obstetrics* (eds. D. E. Noakes, M. W. Parkinson, G. H. Arthur and G. W. England), pp. 1145–154. Saunders, Elsevier Health Sciences, Amsterdam, The Netherlands.

Nowak, R., Porter, R.H., Lévy, F., Orgeur, P. & Schaal, B. (2000) Role of mother-young interactions in the survival of offspring in domestic mammals. *Reviews of Reproduction* **5**: 153–163.

Oliver, R.E., Modha, M.X.J., Phipps, J.C., Poole, W. & Robarts, G. (1988) Experimental infection of caprine arthritis-encephalitis virus in sheep. *New Zealand Veterinary Journal* **36**: 94–95.

Ouweltjes, W., Beerda, B., Windig, J.J., Calus, M.P.L. & Veerkamp, R.F. (2007) Effect of management and genetics on udder health and milk composition of dairy cattle. *Journal of Dairy Science* **90**: 229–238.

Plant, J. (2007) Australia. In *Diseases of Sheep* (ed. I. D. Aitken), pp. 498–504. Blackwell Scientific, Oxford, United Kingdom.

Porter, A.R.W. (1991) Animal Welfare. In *Legislation affecting the Veterinary profession*, pp. 17–45. Royal College of Veterinary Surgeons, London, United Kingdom.

Pritchard, G.C. & McConnell, I. (2007) Maedi-visna. In *Diseases of Sheep* (ed. I. D. Aitken), pp. 217–223. Blackwell Scientific, Oxford, United Kingdom.

Prichard, R.K., Hennessy, D.R. & Griffiths, D.A. (1974) Endocrine responses of sheep to infection with *Trichostrongylus colubriformis*. *Research in Veterinary Science* **17**: 182–187.

Radford, M. (2001) *Animal Welfare Law in Britain.* Oxford University Press, Oxford, United Kingdom.

Rashwan, S.M., Al Khalaf, H.N., Al Hammad, M.I., Abdel-Baky, M.H., Azab, A.H. (2001) Determination of sheep pox circulating neutralizing antibodies in immunized sheep and goats in the North Region of the Kingdom of Saudi Arabia. *Journal of Applied Animal Research* **19**: 233–235.

Rastogi, R.K. (2001) Production performance of Barbados blackbelly sheep in Tobago, West Indies. *Small Ruminant Research* **41**: 171–175.

Reid, H. & Rodger, S.M. (2007) Orf. In *Diseases of Sheep*'(ed. I. D. Aitken), pp. 297–302. Blackwell Scientific, Oxford, United Kingdom.

Robinson, J.J, Rooke, J.A, McEvoy, T.G. (2002) Nutrition for Conception and Pregnancy In *Sheep Nutrition* (eds. M. Freer and H. Dove) pp. 189–211. CAB International publishing, Wallingford, Oxford, United Kingdom.

Rodger, J.L. (1989) Para-influenza 3 vaccination in sheep. *Veterinary Record* **125**: 453–456.

Rollin, B.E. (1995) *The Frankenstein syndrome: ethical and social issues in the genetic engineering of animals*. Cambridge University Press, Cambridge, United Kingdom.

Rollin, B.E. (2000) Veterinary ethics and animal welfare. *Journal of the American Animal Hospital Association* **36**: 477–479.

Rowland, J.P., Salman, M.D., Kimberling, C.V., Schweitzer, D.J. & Keefe, T.J. (1992) Epidemiological factors involved in perinatal lamb mortality on 4 range sheep operations. *American Journal of Veterinary Research* **53**: 262–267.

Russel, A. (1984) Body condition scoring of sheep. *In Practice* **6**: 91–93.

Rutherford, K.M.D. (2002) Assessing pain in animals. *Animal Welfare* **11**: 31–53.

SAC (2002) Sheep lameness www.sac.ac.uk/research/animalhealthwelfare/sheep/lameness.

Sargison, N.D. (1995) Differential diagnosis and treatment of sheep scab. *In Practice* **17**: 3–9.

Sargison, N.D., Jackson, F., Bartley, D.J., Wilson, D.J., Stenhouse, L.J. & Penny, C.D. (2007) Observations on the emergence of multiple anthelmintic resistance in sheep flocks in the south east of Scotland. *Veterinary Parasitology* **145**: 65–76.

Sarto I, Monteys, V., Ventura, D., Pages, N., Aranda, C. & Escosa, R. (2005) Expansion of Culicoides imicola, the main bluetongue virus vector in Europe into Catalonia. *Veterinary Record* **156:** 415–417

Sawalha, R.M., Conington, J., Brotherstone, S. & Villaneuva, B. (2007) Analyses of lamb survival in Scottish Blackface sheep. *Animal* **1**: 151–157.

Scales, G.H., Burton, R.N. & Moss, R.A. (1986) Lamb mortality, birth weight and nutrition in late pregnancy. *New Zealand Journal of Agricultural Research* **29**: 75–82.

Scobie, D.R., Bray, A.R. & O'Connell, D. (1999) A breeding goal to improve the welfare of sheep. *Animal Welfare* **8**: 391–406.

Scott, P. & Gessert, M. (1998) Management of ovine prolapse. *In Practice* **20**: 28–34.

Sharp, J.M. (2007) Mycobacterial infections. In *Diseases of Sheep* (ed. I. D. Aitken), pp. 168–173. Blackwell Scientific, Oxford, United Kingdom.

Sharp, J.M. & Des Las Heras, M. (2007) Contagious respiratory tumours. In *Diseases of Sheep* (ed. I. D. Aitken), pp. 211–217. Blackwell Scientific, Oxford, United Kingdom.

Shutt, D.A., Smith, A.I., Wallace, C.A., Connell, R. & Fell, L.R. (1988) Effect of myiasis and acute restraint stress on plasma levels of immunoreactive beta-endorphin, adrenocorticotrophin (ACTH) and cortisol in the sheep. *Australian Journal of Biological Sciences* **41**: 297–301.

Southey, B.B., Rodriguez-Zas, S.L. & Leymaster, K.A. (2004) Competing risks analysis of lamb mortality in a terminal sire composite population. *Journal of Animal Sciences* **82**: 2892–2899.

Spedding, R.N., Hindson, J.C. & Earl, J.A. (2007) Flock health programmes. In *Diseases of Sheep* (ed. I. D. Aitken), pp. 537–544. Blackwell Scientific, Oxford, United Kingdom.

Suttle, N. (2005) Assessing the needs of sheep for trace elements *In Practice* **27**: 474–483.

Thieme, O., Karazeybek, M., Ozbayat, H.I. & Sozmen, R. (1999) Performance of village sheep flocks in Central Anatolia II. Fertility and productivity of ewes. *Turkish Journal of Veterinary and Animal Sciences* **23**: 175–181.

Turkson, P.K. (2003) Lamb and kid mortality in village flocks in the coastal savannah zone of Ghana. *Tropical Animal Health and Production* **35**: 477–490.

Turkson, P.K. & Sualisu, M. (2005) Risk factors for lamb mortality in Sahelian sheep on a breeding station in Ghana. *Tropical Animal Health and Production* **37**: 49–64.

Walkden-Brown, S.W., Daly, B.L., Crook, B.J. & Colditz, I.G. (1999a) Blowfly strike and wool growth. I Role of anorexia in mediating effects on wool. *Recent advances in Animal Nutrition in Australia* **12**: 2A.

Walkden-Brown, S.W., Daly, B., Colditz, I.G., McFarlan, J.R. & Kauter, K.G. (1999b) Blowfly strike and wool growth. II. Potential mediators of effects, other than anorexia. *Recent advances in Animal Nutrition in Australia* **12**: 3A.

Wall, P.D. (1992) Defining "Pain in Animals". In *Animal Pain* (eds. C. E. Short and A. Van Poznak), pp. 63–79. Churchill Livingstone, New York, USA.

Waller, P.J., Donnelly, J.R., Dobson, R.J., Donald, A.D., Axelsen, A. & Morley, F.H.W. (1987a) Effects of helminth infection on the pre-weaning production of ewes and lambs: evaluation of pre- and post-lambing drenching and provision of safe lambing pasture. *Australian Veterinary Journal* 64: 339–343.

Waller, P.J., Axelsen, A., Donald, A.D., Morley, F.H.W., Dobson, R.J. & Donnelly, J.R. (1987b) Effects of helminth infection on the pre-weaning production of ewes and lambs: comparison between safe and contaminated pasture. *Australian Veterinary Journal* 64: 357–362.

Warren, J.T., Mysterud, I. & Lynnebakken, T. (2001) Mortality of lambs in free-ranging domestic sheep (Ovis aries) in northern Norway. *Journal of Zoology* **254**: 195–202.

Warriss, P.D., Edwards, J.E., Brown, S.N. & Knowles, T.G. (2002) Survey of the stocking densities at which sheep are transported commercially in the UK. *Veterinary Record* **150**: 233–236.

Wassink, G.J., Grogono-Thomas, R., Moore, L.J. & Green, L.E. (2004) Risk factors associated with the prevalence of interdigital dermatitis in sheep from 1999 to 2000. *Veterinary Record* **155**: 551.

Wassink, G.J., Grogono-Thomas, R., Moore, L.J. & Green, L.E. (2003) Risk factors associated with the prevalence of footrot in sheep from 1999 to 2000. *Veterinary Record* **152**: 351–358.

Watson, P.J. & Davies, R.L. (2002) Outbreak of Pasteurella multocida septicaemia in neonatal lambs. *Veterinary Record* **151**: 420–422.

Watt, N.J, King, T.J, Collie, D, McIntyre, N, Sargan, D, McConnell, I. (1992) Clinicopathological investigation of primary, uncomplicated maedi-visna virus infection *Veterinary Rec*ord **131**: 455–461.

Webster, A.J.F., Main, D.C.J. & Whay, H.R. (2004) Welfare assessment: indices from clinical observation. *Animal Welfare* **13:** 93–98.

Wemelsfelder, F. (1997) Life in captivity: its lack of opportunities for variable behaviour. *Applied Animal Behaviour Science* **54**: 67–70.

Wemelsfelder, F. & Farish, M. (2004) Qualitative categories for the interpretation of sheep welfare: A review. *Animal Welfare* **13**: 261–268.

Wiener, G., Woolliams, C. & MacLeod, N.S.M. (1983) The effects of breed, breeding system and other factors on lamb mortality. *Journal of Agricultural Science* **100**: 539–551.
Wilmut, I., Campbell, K.H.S. & Young, L. (1997) Modern reproduction technologies and transgenics. In *The Genetics of Sheep* (eds. L. Piper and A. Ruvinsky), pp. 395–411. CAB International, Wallingford, Oxon, United Kingdom.
Wilson, R.T., Traore, A., Peacock, C.P., Mack, S. & Agyemang, K. (1985) Early mortality of lambs in African traditional livestock production systems. *Veterinary Research Communications* **9**: 295–301.
Winter, A. (2004a) Lameness in sheep 1. Diagnosis. *In Practice* **26**: 58–63.
Winter, A. (2004b) Lameness in sheep 2. Treatment and control. *In Practice* **26**: 130.
Winter, A. (1997) Virulent foot rot in sheep. *Veterinary Record* **141**: 27
Wright, S.E. & Coop, R.L. (2007) Cryptosporidiosis and coccidiosis In *Diseases of Sheep* (ed. I. D. Aitken), pp. 179–185. Blackwell Scientific, Oxford, United Kingdom.
Yapi, C.V., Boylan, W.J. & Robinson, R.A. (1992) Effect of genetic and environmental factors on lamb mortality. *World Review of Animal Production* **27**: 35–41.

Chapter 6
Farming Systems for Sheep Production and Their Effect on Welfare

R.J. Kilgour, T. Waterhouse, C.M. Dwyer, and I.D. Ivanov

Abstract Sheep are raised by humans all over the world for a variety of reasons and in many different management systems. In this chapter, we briefly describe the use of sheep worldwide. We then describe how the welfare of sheep is affected by human activity using three production systems as examples. The three systems chosen are extensively managed sheep raised for wool and meat production, intensive sheep dairying and traditional sheep raising in nomadic systems. The specific examples described are: sheep production in Australia, New Zealand and the British Isles, sheep dairying in Eastern Europe and nomadic pastoralism in Africa and Asia. Under the extensive systems, sheep have the capacity to express the full range of their natural behaviours, although some aspects of their normal social organisation are disrupted. These disruptions include weaning earlier than would occur naturally, segregation of sheep on the basis of age and sex and various husbandry operations, which can cause pain or stress. The main potential source of welfare problems for sheep under these systems come from their interactions with man, which are usually stressful and aversive. Sheep dairying is traditionally pasture-based, often includes a suckling period in addition to milking, and is considerably less intensive in comparison to dairy cattle. However, recent intensification in some countries has resulted in fully housed systems, which have dispensed with a suckling period. In these systems welfare challenges arise from the early weaning of lambs and interactions with the milking machine or handler that can be a source of stress. The potential for increased disease risks in intensive systems may also cause welfare problems. Under the conditions of traditional nomadic pastoralism sheep are perceived as a valuable resource and a feature of this system is the close contact between sheep and their human carers, since sheep are herded during the day and housed at night, often in the main dwelling. Under these systems, however, there is a high degree of unpredictability in the climate, herbage availability and disease risk. These have the potential to create catastrophic welfare problems for both the sheep and their human carers. A diverse range of sheep breeds are managed in the different systems described here, and by working with the adaptations of these breeds generally all the

R.J. Kilgour
NSW Department of Primary Industries, Agricultural Research Centre,
Trangie, NSW 2823, Australia

C.M. Dwyer (ed.), *The Welfare of Sheep*,

systems have the capacity to provide good welfare for the animals kept within them, provided that adequate resources (e.g. supplementary feed, labour) can be given when required. However, some systems (e.g. some in Australia) face exceptional climatic conditions that may make it difficult to maintain good welfare at particular times of the year, without very significant inputs.

Keywords Sheep farming systems · Extensive · Dairy · Nomadic · Human-animal relationship · Mortality · Welfare

6.1 Introduction

Sheep are used extensively throughout the world for the production of wool, meat, milk and skins. The most recent estimate of the world's sheep population is 1.06×10^9 animals (FAO 2001). The regional distribution of sheep worldwide, along with those countries in the various regions with more than ten million sheep are summarised in Table 6.1. This table shows that the greatest numbers of sheep are found in Asia, followed by Africa and Oceania. The uses to which sheep are put are summarised in Table 6.2, from which it can be seen that several countries are significant producers of more than one of these sheep products. Iran, Sudan and Turkey are major producers of all three products, while Australia, China, India, New Zealand and Syria are major producers of two. In many cases, the reason for this is that the sheep are used for their meat once their wool- and milk-producing lives have ended.

The management of sheep will vary depending on the product to be harvested from the animals and the country in which they are raised. For example, milk sheep are managed so that they can be milked twice a day whereas wool sheep only have to be shorn once per year. Within different countries, financial, cultural and climatic differences affect such management factors as the numbers of animals supervised by one person and whether the sheep are kept outdoors all year round or spend some time indoors.

Three major management systems used throughout the world for sheep production, namely, extensive production for wool and meat, intensive dairy production, and traditional pastoralism, will now be briefly reviewed. Particular consideration will be given to how these management systems affect the natural behaviour and welfare of the animals.

6.2 Extensive Management Systems

Extensive management systems for sheep production are the most common in all sheep producing countries, and extend from lowland farming systems where relatively small flocks graze fenced enclosures to rangeland management systems where large flocks live on unfenced pastures. Flock size, the ratio of sheep to shepherds and specific management practices follow local norms and it would be impossible here to give a complete overview of the diverse practices in operation. Instead we will

Table 6.1 Distribution of the world's sheep population in 1999 by region and by those countries where the sheep population exceeds ten million

Region	Number of sheep ($x10^6$)	Country	Number of sheep ($x10^6$)
Africa	250	Sudan	47
		South Africa	29
		Ethiopia	22
		Nigeria	20
		Algeria	19
Asia	407	China	133
		India	58
		Iran	53
		Turkey	29
		Pakistan	24
Commonwealth of Independent States	51		
South-East Asia	8		
Eastern Europe	18		
Western Europe and the Baltic States	129	United Kingdom	37
		Spain	24
		Italy	11
		France	10
Latin America and the Caribbean	83	Brazil	15
		Peru	14
		Argentina	13
		Uruguay	13
North America	8		
Oceania	164	Australia	120
		New Zealand	44

Source: FAO (2001)

concentrate on the sheep farming systems operating in three major sheep-producing countries: rangeland sheep production in Australia, the development of 'easy-care' sheep breeds and management in New Zealand and the stratified sheep management system in the UK. For each a brief description of the history and geographical features underpinning sheep production will be given, as well as specific management practices that may be peculiar to that production system and the associated welfare consequences will be considered. Welfare issues that pertain to all extensive sheep farming systems will be discussed at the end of this section.

6.2.1 Australian Sheep Production Systems

Australia was claimed by the British in 1770 and, eighteen years later, settlement began with the establishment of a penal colony at the site of what is now the city

Table 6.2 Numbers of sheep in the world that are raised for wool, meat and milk production and their distribution in the major producing countries

Wool		Meat		Milk	
Country	Production (Month/year)	Country	Production (Month/year)	Country	Production (Month/year)
Australia	700 000	Australia	663 000	Italy	850 000
China	305 000	New Zealand	562 168	China	800 000
New Zealand	250 000	Turkey	313 000	Turkey	785 000
Iran	703 907	Iran	280 000	Greece	670 000
Uruguay	54 818	United Kingdom	258 000	Sudan	490 000
South Africa	52 761	Spain	240 000	Syria	482 000
India	47 600	India	230 400	Somalia	430 000
Sudan	46 000	Pakistan	195 000	Romania	330 000
Turkey	44 300	Syria	195 000	Spain	306 000
Morocco	40 000	Sudan	144 000	Iran	300 000
World	2330626		7532451		7807799

Source: FAO (2001)

of Sydney. Thus began the history of the pastoral industries in Australia, which is described by Peel (1986). Briefly, sheep were imported into the new colony at first as a source of food and, later, for wool production as the interior of the country was discovered. This meant that Australian sheep husbandry techniques were simply adapted from those in use in Britain at the time. At present, this is still largely true, but with two major exceptions. These are the scale of operations in Australia, particularly the numbers of animals in the care of individual human beings and the use of the mules operation to confer some protection against flystrike in the breech area. The use of the mules operation will be discussed in a later section of this chapter.

As regards the scale of operations, these are highly variable (Table 6.3). One of the principal reasons for this variability is climate, most particularly rainfall. For this reason, Australian sheep production is considered to be practised in three climatic zones, the High Rainfall Zone, the Wheat-Sheep Zone and the Pastoral Zone (Shafron et al. 2002). In the High Rainfall and the Wheat-Sheep zones, farm size is smaller and sheep numbers are considerably lower than in the Pastoral Zone (Table 6.3). Division of the number of sheep by the number of labour units in Table 6.3 shows the ranges between regions for the specialist sheep farms. In the High Rainfall Zone, numbers of sheep per labour unit ranges from 1543 in Western Australia to 3009 in Victoria. In the Wheat-Sheep zone, these figures are 1463 (South Australia) and 3099 (Western Australia) while for the Pastoral Zone they are 3269 in New South Wales and 3760 in Queensland. While these figures are simple averages and take no account of variation, they indicate that each labour unit is responsible for several hundred to some thousands of sheep.

The corresponding figures for the mixed farms show a slightly different picture. In the High Rainfall Zone, numbers of sheep per labour unit range from 1378 (South Australia) to 2051 (Western Australia), in the Wheat-Sheep Zone they range from

Table 6.3 Farm size, number of sheep and number of labour units for Australian farms on which sheep formed a major part of the enterprise mix in 2000 and 2001 by climatic zone and state. Adapted from Shafron et al. (2002)

Zone	State[4]	Specialist farms[1]			Mixed farms[2]		
		Size (ha)	Number of sheep	Number of labour units[3]	Size (ha)	Number of sheep	Number of labour units[3]
High Rainfall	NSW	1035	3296	2.1	1128	3250	2.1
	SA	723	2793	1.6	2005	3169	2.3
	Tas	2007	5875	2.5	2358	4912	3.1
	Vic	867	5416	1.8	1011	4345	2.3
	WA	798	4321	2.8	2336	5948	2.9
Wheat/ sheep	NSW	2701	3657	1.5	2770	3586	2.7
	QLD	7061	2899	1.9	8670	3064	2.6
	SA	5078	1756	1.2	1822	2020	2.3
	Vic	824	3449	1.9	1679	2142	2.5
	WA	4584	9606	3.1	2857	3208	2.6
Pastoral	NSW	22760	5884	1.8	42985	6494	2.5
	QLD	35046	8399	2.5	31114	8294	3.1
	WA				215324	7798	3.3
	SA				30385	2814	2.6

[1]Specialist farms are those where the majority of farm income is from wool.
[2]Mixed farms are those where most of the income came from other enterprises such as cropping or cattle production.
[3]One labour unit is equivalent to one person working a 40-hour week.
[4]States: NSW = New South Wales, Qld = Queensland, SA = South Australia, Tas = Tasmania, Vic = Victoria, WA = Western Australia.

856 (Victoria) to 1328 (New South Wales) and in the Pastoral Zone they range from 1082 in South Australia to 2675 in Queensland. While these figures suggest that labour is more available on mixed farms, the other enterprises will require some of this labour.

6.2.1.1 Welfare Issues in Australian Sheep Production

This section will deal with welfare issues specific to Australian sheep production. Welfare concerns that are common to all extensive sheep production systems will be discussed in Section 6.2.4.

1) *Mulesing*: (see also Section 8.5.7) The Mulesing operation is so named after its inventor, Mr J.H.W. Mules, an Australian sheep breeder and is carried out in order to confer some prolonged protection from breech strike. This operation is particularly applied to Merino sheep where there is excessive folding of the skin. Breech strike is a flystrike where the flies are attracted to the wrinkly breech area, which is often soiled by faeces and urine. The principle is to remove the wool-bearing skin from this area by cutting it off with shears without the use of anaesthetic. The procedure causes considerable stress to sheep as seen in the elevated

levels of cortisol and β-endorphin (Shutt et al. 1987, Fell & Shutt 1989). The study of Shutt et al. (1987) examined lambs that were subjected to tail docking or tail docking and castration as well as mulesing, but that of Fell & Shutt (1989) compared animals that had been mulesed only with sham-operated controls. In that study, levels of cortisol and β-endorphin rose significantly soon after the operation and remained significantly elevated for the following 24 h. The behaviour of the mulesed sheep in their paddock was characterised by abnormal posture and locomotion, which did not return to normal until 72 h after the operation. When tested in a human approach-avoidance test, the sheep avoided the human handler who had held them during the operation. This avoidance was apparent for at least 37 days following the operation. While directly comparing levels of hormones across studies is difficult, the levels of cortisol following the mules operation in the study of Fell & Shutt (1989) were similar to levels found by Shutt et al. (1988b) in sheep with flystrike. The prolonged avoidance of the handler following the operation suggests that a considerable degree of pain and stress was associated with the procedure.

The simplicity of the mules operation is that, in its early form, it merely required the removal of wrinkled skin lateral to the tail. This meant that it could be performed by sheep producers themselves. However, in its more radical form, it now involves the removal of wool-bearing skin on the tail and either side of the perineum. If performed incorrectly, instead of conferring prolonged protection from breech strike, it can confer on the animal a lifetime of problems, particularly related to sunburn. Incorrect mulesing may also render the protection less effective. In one state of Australia, New South Wales, there is an association of livestock contractors who undergo an accreditation for their capacity to mules sheep correctly (Evans & Joshua 2003).

However, recent developments have meant that the current practice of mulesing is to disappear from the Australian sheep industry by the end of 2010. This has followed from a concerted campaign by People for the Ethical Treatment of Animals (PETA) to urge consumers to boycott retailers who sell Australian wool. Australian wool and sheep industry leaders made a commitment to phase out conventional mulesing by 31st December 2010. A training programme will be established to educate Australian woolgrowers of practices to reduce mulesing, a system will be developed to identify, at the retail level, wool from non-mulesed sheep, statistics will be collected on the proportion of mulesed and non-mulesed sheep in Australia, and a genetic research program will be set up alternatives to mulesing (Scobie et al. 2007). There is, however, some concern that this will lead to a decrease in sheep welfare, at least in the short term, with an increase in flystrike without increased flock inspections and the use of chemical preventatives (Lee & Fisher 2007).

2) *Neonatal mortality*: Neonatal mortality has long been recognised as a major issue in the Australian sheep industry, and estimates in lambing flocks range from 20% (Jordan & LeFeuvre 1989) to 72% (Smith 1962). However, many of these estimates were based on small numbers of animals and relied on recovery of lamb carcases. Four large field studies have been conducted on large numbers

of ewes on private properties using more sensitive measurement of lamb mortality. The first of these (Luff 1980) was based on more than 45 000 ewes in which 22.1% of all ewes giving birth failed to rear a lamb. This figure ranged from 3.9 to 61.4% on individual properties. The second study of more than 25 000 ewes from 55 flocks found that 11.4% of ewes failed to rear a lamb (Knight et al. 1975). The third study by Kleemann et al. (1990) showed by laparoscopy that the reproductive potential of adult Merino ewes was 135 lambs per 100 ewes and that the loss of lambs between birth and marking at 6 weeks of age was 24.5%. The final study (Kilgour 1992) measured the number of foetuses *in utero* in more than 55 000 ewes and concluded that the lambing potential of the average commercial flock, comprising 25% of maiden ewes, was 121 lambs per 100 ewes. When this was contrasted with the lamb marking figures of 80 lambs per 100 ewes, it was concluded that one-third of lambs *in utero* die before lamb marking.

3) *Adult mortality*: Because of the size of Australian farms and the fact that there is little capacity to house sheep in buildings, there are often occurrences of death due to natural disasters. One such disaster has been high mortality in newly shorn sheep following high rainfall in summer. Buckman (1983) reported a mortality rate of 27.8% and Holm Glass et al. (1992) reported 11,427 deaths in a total of 44,460 sheep (25.7%). These sheep came from 13 properties, and the range in mortality was 5.7–90.3%. Hutchinson (1968) estimated that the overall mortality for sheep during the approximately one month after shearing was around 0.68%. Bushfires are another common feature of the Australian summer and, in 1977, bushfires were responsible for the deaths of 195,000 sheep which were either burnt to death or had to be euthanased later (Anon 1977).

However, even though losses due to natural disaster can be high, such events are relatively rare. Equally rare are published estimates of levels of mortality. Data for one sheep stud reported mortality rates that increase from 0 to 9% in sheep aged two and five years respectively to 90% in sheep 12 years old (Granger 1944). In an experimental flock, over a period of seven years, the average mortality in ewes aged 1½–7½ years was 2.2% per year while in ewes aged 8½–10½ it was 5.4%. In an eighth year, a drought year, the average mortality rate was 3.8% for ewes up to 6½ years and rose to more than 10% in ewes aged 7½ years and up to 46% in ewes aged 10½ years (Turner et al. 1959). More recent surveys indicate an overall loss of 21% in adult ram flocks and 7% in adult ewe flocks (Harris & Nowara 1995) or an average death rate of 5–6% (Shafron et al. 2002). The major causes of death identified were flystrike and ewe losses in autumn close to lambing, although recent data during drought conditions suggest that malnutrition and disease are important factors in adult ewe mortality (Bush et al. 2006).

4) *Sea transport*: Australia is the biggest transporter of animals by sea in the world and 3.3 million sheep were exported in 2004 (Norris 2005). Most sheep are exported from Western Australia, and nearly all are destined for slaughter in the Middle East, particularly Saudi Arabia. Generally, sheep are collected on a pre-departure feedlot for a period of days or weeks before embarkation. A typical sea journey from Freemantle in Western Australia to Jeddah in Saudi Arabia requires

the sheep to be on board for more than 2 weeks. World attention was particularly focused on the welfare of sheep undergoing these journeys when a consignment of nearly 58,000 sheep was rejected by the Saudi Arabian authorities in August 2003. The sheep remained on board in the port at Jeddah for 2 months, before they were gifted to Eritrea and unloaded 80 days after leaving Australia. During this period 9.6% of the sheep died, mostly from heat stress whilst in the port.

Several studies have quantified the levels of mortality during sea transport, the causes and risk factors associated with sheep deaths undergoing these journeys. Death rates vary across studies from 0.6 to 3.27% (Richards et al. 1989; Norris et al. 1989; Norris & Richards 1989; Richards et al. 1991), although a recent study identified a range of 0–28.2% (Higgs et al. 1999). Deaths in lairage are predominantly from disease, particularly salmonella, whereas the main cause of death on board ship (where most mortalities occur) is inanition (Richards et al. 1989). A significant number of deaths also occur following unloading at destination. Highest mortality occurs in the second half of the year, in sheep sent to the Middle East rather than Singapore, in sheep on enclosed rather than open decks and in animals coming from regions in Australia with higher rainfall and a longer pasture growing season (Norris & Richards 1989; Higgs et al. 1991; Richards et al. 1991; Higgs et al. 1999). These authors suggest that under these conditions sheep may be fatter and more likely to mobilise fat reserves rather than eat the relatively unfamiliar grain feed offered on board ship, so are more likely to show persistent inappetence. Clearly, the welfare of sheep undertaking these journeys can be improved by consideration of these factors, and determining whether sheep from particular regions are suitable for sea transport. Higgs et al. (1999) showed that 50% of all mortalities came from 14% of consignments suggesting that sheep from some areas face higher challenges in adapting to sea transport than others.

6.2.2 Pastoral Sheep Farming in New Zealand

Sheep were first landed in New Zealand in 1773 by Captain Cook who brought four Merinos to the Islands. These animals all died but 60 years later James Wright reintroduced the breed and established a breeding flock near Wellington. Sheep numbers expanded rapidly, reaching over 13 million animals by the end of the 1800s. These were nearly all Merinos, a breed originating in Spain, but brought to New Zealand from Australia. The high quality wool produced by the Merinos was the first pastoral export from New Zealand. British long-wool breeds (such as Leicester and Lincoln) were also imported into New Zealand and crossed with Merinos from the mid 1800s. These crosses dominated in the North Island in particular, until the Romney, introduced in the 1850s, supplanted them in 1900. This dual-purpose breed, producing both meat and wool, originated in the lowland fen regions in Britain (the Romney Marsh), but was found to adapt and thrive in the steep hills. Numbers increased rapidly, with the breed developing into a distinct New Zealand Romney breed until, by the 1960s, three-quarters of the 50 million sheep in New Zealand were

Romneys. Productivity varied under different environments, so Romneys were crossed with other British breeds such as the Cheviot (to produce hardy Perendales) or the Border Leicester (to produce Coopworths that do well in upland and wet country). Merinos continued to dominate the rugged hill country of the South Island due to their ability to survive at high altitudes on native grasses. Sheep numbers in New Zealand increased to 70.3 million animals in 1982. However, following the removal of price support and subsidy in the 1980s, the numbers of animals have declined to current levels of around 43 million.

Sheep are farmed in the hills and in the south of the North Island, and are the main form of pastoral agriculture practised in the South Island. Current estimates suggest that there are approximately 36,000 sheep flocks with an average flock size of 1400 animals. Livestock are rarely housed but may be fed small quantities of supplements (e.g. hay, silage) in the winter. The climate allows grass growth for between 8 and 12 months of the year, with lambing usually managed so that it coincides with spring growth. Good grass management strategies mean that the best sheep farms can carry about 25 sheep per hectare throughout the year.

Sheep management systems in New Zealand are characterised by an abundance of land and a low human population in comparison to European countries, such as Britain, where land is less available and the human population is high. Thus, although sheep farming management originally followed British systems, different breeding goals began to emerge. In particular, sheep farms could be relatively low producing (for example with lambing percentages of around 100% rather than 150% or greater in some parts of the UK), with larger flock sizes. New Zealand is also particularly associated with the development of 'easy-care' sheep flocks, that is flocks that have reduced labour requirements (Bradford & Meyer 1986), such that one shepherd can look after up to 4000 ewes.

6.2.2.1 'Easy-Care' Sheep

Easy-care sheep are characterised as animals able to adapt and survive in adverse climatic conditions, to successfully lamb and rear at least one lamb without assistance and to require less shepherding than other breeds (Kilgour & de Langen 1980). The concept of 'easy-care' sheep is generally associated with the development of the 'Marshall' Romney flock in the 1930s in Wanganui on the west of the North Island (Mackereth 1979), although the origins of easy-care sheep are obscure. This process was characterised firstly by breeding rams on the farm that were suited to the environment rather than by buying in rams from fashionable studs. Secondly, in 1935, a decision was made not to shepherd the ewes at lambing time, in the hopes of breeding sheep that could look after themselves (Mackereth 1979). This policy of no shepherding at lambing and allowing natural selection to cull out undesirable animals has also been described by other farmers (as reported by Fisher 2003). Alternative strategies describe marking and culling animals if assistance has been required, or if the ewe is not seen with a live lamb at docking. In addition, animals are set-stocked, i.e. placed on the same land for a long period of time to become accustomed to the location of shelter and other resources (see also hefting in

Section 6.2.3.3), and lambs returned to the areas they were born on following weaning. The outcome of this selection policy, from farmer narratives, is considered to be a change in the type of sheep, long legged and open faced, coupled with easy lambing, good mothering abilities, hardier, more vigorous sheep requiring low or no shepherding (Fisher 2003).

Lamb survival in the easy-care Marshall Romney ewes compared to control Romney ewes varies from 8.5% to 14.2% higher (Hight 1978/9; Knight et al. 1988), or as much as 26% less mortality on steep slopes (Knight et al. 1989). Lambs sired by Marshall Romney rams also have higher survival than lambs sired by control Romney sires (Knight et al. 1988). Improved survival occurs mainly because Marshall Romney lambs are less likely to suffer deaths from dystocia and starvation than control Romney lambs. Much of this increase in lamb survival is due to improved survivability of Marshall Romney twin lambs, but demonstrates that the easy care selection procedure has resulted in better lamb survival. Several studies have compared lambing of Marshall Romneys with control Romneys to determine what differences exist between the lines as a result of easy-care selection. Marshall Romney ewes are nearly 10 kg heavier than the control Romney ewes, and consequently have larger pelvic dimensions (Knight et al. 1988). However their lambs are only 0.08 kg heavier and have differing skeletal dimensions (e.g. head and chest circumference, pelvic width), which may make the birth process easier in the Marshall Romneys. Marshall Romney ewes were found to be more likely to select sheltered lambing sites than control Romneys and were less likely to abandon a lamb that slipped off a steep birth site (Knight et al. 1988; 1989). Easy care selection has resulted in a longer legged sheep, which may be more agile on steep slopes so making it easier for ewes to reach lambs that have slipped. When their lambs were handled at tagging, a greater proportion of Marshall Romney ewes remained with their lambs than control Romneys. As described in Chapter 3, these behaviours are related to improved lamb survival and are likely to have contributed to the lower number of lambs dying of starvation.

6.2.2.2 Welfare Issues with Easy-Care Systems

1) *Development of easy-care sheep*: The development of easy-care sheep was, primarily, a practical response to the problems of lambing on extensive pastoral farms (Fisher 2003), where effective shepherding was difficult if not impossible, or where cost or availability of labour were important considerations. Although pleas are frequently made for 'easy-care' not to be equated with 'no care', at least in the early stages of the development of easy care sheep, this appears to be exactly what happened to the animals. There are several accounts of sheep being left completely alone at lambing time (Mackereth 1979; Fisher 2003), although in other systems any animal that required assistance, and its offspring, were automatically culled. In the systems where animals were left for natural selection to cull them, deaths would have occurred through injury, starvation, hypothermia and dystocia, as well as other obstetric emergencies such as prolapse or mastitis. These are considerable

welfare costs and, as discussed in Chapter 1, the fact that they can be perceived as ‘natural’ deaths does not mitigate our responsibilities to animals in our care. Fisher (2003) provides several accounts of the level of ewe mortalities that occurred during transition to minimal shepherding. These suggest that the level of ewe deaths was initially between 3 and 7% and then fell to 1–2% in subsequent years. However, lamb deaths are likely to be much higher, as suggested by accounts that, when ewes that had received no shepherding were culled if they did not have a live lamb at docking, culling levels were so high that the flock needed to be supplemented by other ewes. Assisting animals suffering at lambing time, and then either culling or preventing their genes from entering the breeding lines, as has been described by some in the development of easy-care sheep, appears to be a more humane approach to developing animals that are more self-sufficient and able to produce and rear lambs with little need for human intervention. As discussed further in Chapter 8, although the outcome of easy-care selection may be benefits in the improvement of fitness of the flock of animals, welfare is generally considered on an individual basis and the costs, the suffering or deaths of animals, are paid by animals which do not gain any welfare benefit from the selection process.

2) *The welfare of easy-care sheep*: As with more traditionally-shepherded extensive flocks, easy-care sheep, and in particular their lambs, are exposed to climatic extremes, hypothermia and possibly predation. In addition, they receive little assistance to deal with difficult births, mismothering and disease, although these may occur less frequently than in other types of sheep. However, easy-care sheep may be less likely to encounter pathogens, particularly those borne by shepherds from one animal to the next, and are allowed to express greater natural behaviour patterns with less shepherd-induced problems, such as movement from the birth site, disrupted mother-young bonding, and prolonged parturition through disturbance (Fisher & Mellor 2002). This suggests that a minimal shepherding approach for animals that have been selected or bred for self-sufficiency traits may improve welfare.

Estimates of lamb mortality in New Zealand flocks range from 5 to 25% although the amount of shepherding input is rarely mentioned (Meyer & Clarke 1978; Dalton et al. 1980, McCutcheon et al. 1981; Morris et al. 2000). This figure is similar to that achieved in other countries where estimates range from 10 to 35% (Binns et al. 2002), and an average of 15% is generally accepted (see Chapter 5). Mortality in easy-care Marshall Romney flocks is reportedly 10.7% in comparison to 24.9% in control Romneys under the same conditions (Knight et al. 1988). This compares to figures of 7–10% in housed flocks in the UK (Binns et al. 2002). Thus lamb survival in easy-care flocks appears to be improved over the accepted average of 15%, although the improvement is no greater than that which can be achieved in shepherded flocks. As discussed in Chapter 1, there is no clear evidence to suggest whether shepherding enhances or decreases lamb survival (as is also discussed in Fisher & Mellor 2002), and other aspects of welfare, such as the prompt treatment of disease or injury, also need to be taken into account. It is also likely that the quality of the stockperson, and the nature of previous interactions that the sheep have had with humans, will play a large part in the detrimental effects, or otherwise, of shepherding at lambing time. In general, therefore, there is the potential that, for certain genotypes in particular environments, a low level of contact with humans at lambing time may enhance welfare over that seen with a greater level of shepherding, although this still needs to be conclusively demonstrated.

3) *Extension of the easy-care concept*: Initially, easy-care systems were concerned principally with the amount of shepherding the flock might need at lambing. However, more recently this concept has been extended to include general hardiness and fitness traits which, whilst reducing the need for labour input and shepherding, may also bring additional benefits in terms of improved sheep welfare. Although some of these breeding strategies will be discussed in more depth in Chapter 10, these include selection for resistance to footrot and lameness, selection for parasite resistance and breeding animals that are resistant to fly-strike (Scobie et al. 1999; Stear et al. 2001). As outlined above for easy-care lambing systems, if the selection procedure itself does not impact on welfare of the individual (allowing animals to suffer or die without intervention) then these breeding strategies may ultimately improve the welfare of sheep in these systems.

6.2.3 Sheep Farming in the British Isles

Large areas of the British Isles are dominated by mountains and hills upon which sheep production is the dominant land use. The climate, topography and culture have led to systems of pastoral management which differ in many respects from those seen in the rest of Europe. These pastoral systems play an important role in maintaining habitats and landscapes and maintaining social structures in some of the most depopulated areas in Europe. Scotland's land use in particular is dominated by pastoral systems for sheep production, although dairy and beef cattle systems also occur in significant numbers.

The highest annual rainfall for the British Isles is seen in some of the north-western areas (over 2000 mm per annum), and there is a marked contrast between west and east. There is considerable variation in elevation, steepness and associated poor soils. Although elevation in Britain does not exceed 1500 m above sea level (the highest point, Ben Nevis in Scotland, is only 1343 m) a combination of wet temperate climate and moderate altitude leads to a harsh environment. In addition, poor and thin soils and steep, rocky topography or wet peat areas lead to poor conditions for vegetation in many areas, and low digestibility of forage from much of the year (Hodgson & Grant 1985). The main agricultural land use systems in Scotland are heavily influenced by rainfall and topography. In the higher altitude areas, the upland areas of southern Scotland and virtually all of north-western Scotland, hill sheep with extensive beef cattle is the dominant land-use. In the very far north and fringes of the north-western mainland and Scottish islands a unique form of part-time farming, called crofting, utilises the land for hill sheep and extensive beef cattle.

The United Kingdom is the main sheep producing country in Europe (Table 6.1) and there are large contrasts between sheep systems in Scotland and continental Europe. Flock sizes are relatively larger than other European countries (the UK average at 300 sheep is approximately three times the European average), although there is considerable variation within the UK. Systems are generally based on grass with a low level of concentrate feed. Systems of sheep production do, however, differ dramatically between regions, mainly determined by topography, soils and altitude. A broad classification of British sheep systems is shown in Table 6.4.

6.2.3.1 Stratified Structure of Sheep Production in Great Britain

Some seasonal movement of stock occurs in Great Britain, with the commonplace movement of flock replacements, often over hundred of kilometres, from hill and

Table 6.4 Summary of British sheep farming systems

	System	Characteristics	Main products
Hill systems	Hill ewes producing pure-bred lambs	Bound (hefted) acclimatised stock on a high proportion of semi-natural pasture	Store lambs Pure-bred fat lambs (14–18 kg) Pure-bred fat lambs (8–12 kg) for export Draft ewes
	Hill ewes producing some crossbred lambs	Bound stock, possibly partially de-hefted from hill. Usually a higher proportion of inbye (improved) land	As above but draft ewes sales reduced and crossbred male and female lambs sold
	Crofting	Smaller scale producer but often with larger sheep 'farmers' using most land in the area	Mainly store lambs as value of draft ewes is low
Upland	Draft hill ewes producing crossbred lambs (middle tier, Fig. 6.1)	Unbound flock brought onto unit – grazing mainly sown pastures and fenced hill areas	Crossbred ewe lambs for breeding and slaughter Crossbred male lambs for slaughter
	Crossbred ewe mated to terminal sire[1] (lowest tier, Fig. 6.1)	Unbound flock brought onto unit – grazing only sown pastures and fenced hill areas	3-way cross lambs for slaughter at 17–21 kg, usually of a terminal sire breed.
Lowland	Crossbred ewe mated to terminal sire	Grassland farm Mixed farm Arable unit	3-way cross lambs for slaughter at 17–21 kg, usually of a terminal sire breed.

[1]A terminal sire is a lowland breed selected for rapid lean tissue growth and good carcase composition (predominately Suffolk, Texel or Charollais in the UK), that is used to produce improved carcase characteristics in their progeny than would be achieved with a ram of a similar genotype to the dam.

mountain regions to lowland pastures for the winter. This is a different form of migration to that found in Mediterranean regions as people do not move with the livestock, instead responsibilities for the stock are, usually, temporarily transferred to the farmer on whose farm the seasonal grazings are taken.

Although transhumance is found in many parts of Europe, stratification is common only in Great Britain and Ireland. Stratification is the development of a highly structured sheep industry based on the natural resources of different areas of the country and the sale and movement of livestock between these different areas. Typically sheep flow from the most naturally disadvantaged areas, the hills and mountains, to the more favoured areas, the lowlands. In the hills and mountains, sheep are bred to produce animals which are sold to producers in the more favourable areas for use as breeding stock or for further grazing and sale as finished livestock. The classical pattern of stratification of breeding animals and the flow of stock from hill to lowland is described in Fig. 6.1. This pattern is, however, far from a complete model and is changing. Nevertheless, it is a major difference between the sheep systems within the British Isles and those in Europe, allowing use of different categories of land by different systems in the former. The flow of genes for crossbred stock, usually with high health status (because hill farms are typically closed to female stock), from the hills and mountains is a key strength of these hill systems and of considerable strategic value to the whole sheep industry.

Socio-economic, land use and animal breeding benefits can arise from this industry structure in Britain. The National Sheep Association (NSA 1995) identified that the stratified systems allows large areas of hill and upland to support an economic enterprise. The breeding structure provides a framework that allows the

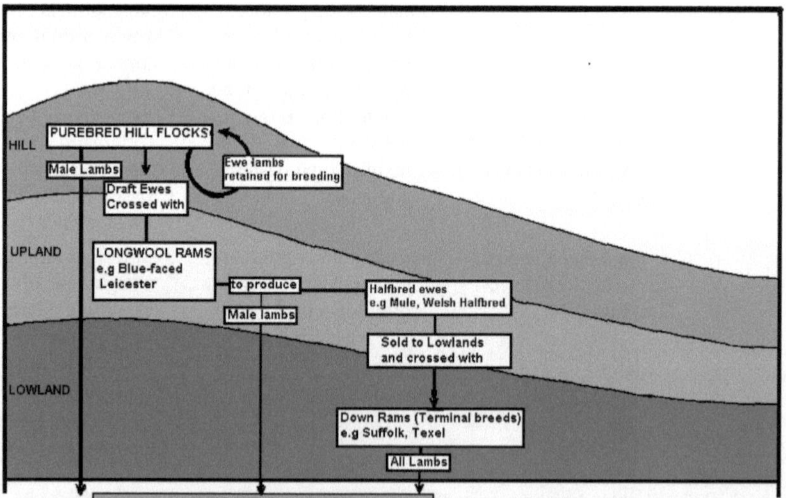

Fig. 6.1 Diagrammatic representation of the stratification of the UK sheep industry and flow of stock from the hills to lowlands (modified after Cooper & Thomas, 1991)

efficient use of breeding stock and the production of a high quality slaughter lamb. Hardy hill breeds, which are capable of withstanding periods of nutritional stress are maintained and improved through selection (both natural and supplemented by artificial selection) for maternal performance and vigour. The hill ewes contribute their valuable maternal characteristics to their progeny, while the use of Longwool rams contributes enhanced prolificacy and increased body size. Thus, the value of crossing hill ewes with Longwool rams results in a crossbred ewe of high commercial value, with the capacity to produce and raise an average of two lambs. Further crossing with a terminal sire, bred under conditions where genetic selection can be concentrated on carcass and growth characteristics, allow the production of slaughter lambs which are better able to meet the demands of the consumer.

There is a considerable variation amongst hill farms in the British Isles. For example, Welsh hill farms are small and have a much higher proportion of improved (inbye) pasture than those in England and Scotland (Cunningham & Groves 1985). Scottish hill farms are at the other end of the spectrum, large but with a low proportion of inbye pasture. Many hill farms in Scotland are many thousand hectares, with typical flock sizes over 1000 ewes, and an average of 637 ewes per labour unit. Hill farms can be further classified by the intensity of their management system (Eadie 1985; Maxwell 1994).

6.2.3.2 Hill Systems

These systems all use extensive areas of semi-natural vegetation and small areas of improved pasture. They rely on pure-bred lamb production to sustain flock replacements. The primary output is store livestock sold to lower altitude farms for breeding or for further grazing and slaughter. The breeding ewes tend to remain on the open hill and moorland throughout the year. There is little daily contact between sheep and shepherds. Hill ewes maintain themselves within an adopted home range grazing singly or in small groups (Hunter 1962; Hunter & Milner 1963; see also Chapter 2), often with unfenced boundaries between adjacent farms. Typically only a small proportion of ewes are brought onto improved pastures for mating and lambing, the rest remaining in the hills. In some regions (e.g. Angus Glens of Scotland, Welsh mountains) ewes are traditionally removed from the hill during all or part of the winter. Lambing dates are locally fixed, typically late April and May, with little flexibility due to the time of onset of spring grass growth. Levels of supplementary feed given to hill ewes have generally increased in the last two decades, but still range from zero (although emergency feeding of hay in difficult weather conditions may be carried out) to considerable inputs of concentrate feed and fodder, such as big bale silage.

It is typical for hill ewes to be mated to produce their first lamb at 2 years of age. In some regions and with some breeds (e.g. the Cheviot in Sutherland) a first lambing at 3 years old is still practised. The unmated flock replacements typically leave the hill, to be in-wintered in buildings or grazed on better pasture on the hill farm, or wintered (termed 'agistment') on a lowland farm sometimes many hundreds of kilometres away. Levels of output tend to be low, typically the number of lambs weaned per ewe is less than 0.8 (Vipond & Gunn 1985). Nevertheless, with improved health care, higher levels of supplements, improved nutrition and management at mating and lambing, increased outputs of lambs are possible on many of the better farms (McClelland et al. 1985). A proportion of ewes may be mated to rams of other breeds, either to produce lambs for slaughter (via the use of, for example, Suffolk or Texel rams) or for producing breeding females (via the use of a Longwool breed ram, such as the Blue Faced Leicester).

The sale of draft ewes, after they have produced four or five lamb crops, from hill farms to lower altitude farms is still a very important part of the British hill sheep system. These ewes may be retained for breeding for a further 1–3 years on the lower altitude farm. High health status, in relation to the absence of ovine enzootic abortion (Chlamydia) infection, is also highly valued in these animals. Many ewes, however, fail to go into this market because they are unsuitable for future breeding as a result of failures of udders, teeth or a lack of demand for particular breeds. Instead, these ewes are sold for mutton.

It is common for hill farms to be tenanted, and for hill or moorland to be grazed in common with other farmers whose farmhouses, buildings and inbye grazings border the common grazings. In this respect, some of the problems of common grazing are shared with other countries having a similar system. There is difficulty in making decisions on overall stocking rates, and difficulty in applying different management to individual flocks. Many stock management tasks require considerable co-operation and these may break down. Thus stocking rate changes may be quite large, made simultaneously by individual commoners.

There is considerable adaptation, both of stock and management practices, to the extensive hill environment. Hill breeds are hardy with physical (e.g. wool characteristics), physiological (e.g. cold tolerance) and behavioural adaptations (e.g. grazing behaviour). Maintenance of the sustainability of systems requires co-operation between neighbouring farmers and common land users to allow unfenced and unshepherded systems to be viable. Other management practices include castration of males and the purchase of replacement stock lambs from specialist breeders, and selection based on survival of ewes and 'type' of replacement females and stock rams.

6.2.3.3 Hefting or Bound Stock

A particularly important aspect of hill systems in Great Britain is the 'bound' nature of the breeding flock to the land. British hill breeds all possess a pattern of grazing and social behaviour (see Chapter 2) in which they graze a block of land as their home range, known as a 'heft'. The heft is maintained by retaining only flock replacements born on each individual block of land on the farm, and by policing and dealing with ewes that stray beyond the boundaries. As a result many hill farms have several self-contained flocks, each with their own home range, maintained partly by sheep behaviour and partly by shepherding. These self-contained hefted flocks have become known as the bound stock of the holding, and when hill farms are sold or there is a change of tenant, the bound flock is sold to the new farmer. These traditions provide both stability and lack of flexibility in making breed and stocking rate changes on hill farms.

6.2.3.4 Crofting

Crofting is a particular type of hill farming occurring only in the north western fringe of Scotland. It typically involves small areas of improved pasture land (about 50 ha) and access to common (semi-natural) hill grazings. The right to keep a small number of sheep (10–50 ewes) and cows (1–4) are called 'soumings'. This system became established during the period 1870–1905. There is a strong trend for cattle numbers to decrease and for the sheep to be kept in larger flocks by only a few crofters in any one of the villages. Problems created by the small scale of enterprise, difficult farming conditions and access to common land have all contributed to these changes.

6.2.3.5 Upland Systems

Upland sheep production can be divided into two main systems, broadly by the types of animals kept in the systems. The systems are generally based on improved pasture, typically a large proportion of sown pastures. Some semi-natural moorland or grass may be attached but this is frequently fenced or enclosed. The first system involves the purchase of draft ewes from hill farms. Each ewe then produces one to three crops of crossbred lambs before she herself is sold for mutton. The crossbred lambs may be sired by a terminal sire breed and be intended primarily for slaughter. However, a large number of crossbred female lambs will be destined to providing the breeding females in the next tier down the stratified pattern. For the upland flock producing crossbred females from draft hill ewes, the level of input is typically much lower than for crossbred ewes producing slaughter lambs. The production cycle is driven by a single date in the year: the local annual auction sale of breeding females. Surplus males are sold as store lambs or fattened on the farm.

The second system involves the purchase of crossbred ewe stock, which are then managed to produce slaughter lambs. With high levels of management it is possible for lambs from this system to be sold from June onwards, although most lambs are sold for slaughter in early autumn (August–October). Typical carcase weights are 18–21 kg. In Fig. 6.1 this upland system occupies the lowest tier or fulfils the role of a lowland farm. However, it would continue to be referred to as an upland farm in most definitions: stock are not bound (bringing in replacements from outside the unit of land) but are found within the more marginal land areas.

Because of their lower altitude, better pastures and often higher proportion of buildings, both types of upland flock systems have more flexibility in the time of lambing than hill farms. Lambs from upland flocks may be born between February and May. It is quite common for the whole flock to be housed through mid/late winter until shortly after lambing. Fodder conserved on the farm is a major component of the system, although a small proportion of upland farms use straw as their main forage. Because the ewes spend more of their time within fields and buildings, and require daily inspection and careful management through many parts of their production cycle, flock sizes and number of ewes cared for by a single shepherd are lower than for hill flocks. Furthermore the better climate, soils and availability of ploughable lands found in upland areas compared to hill areas, give the farms in this category more choice in their production system. They are more easily able to change flock numbers, to change the breed of ewe used, and the breed of lamb produced.

6.2.3.6 Lowland Systems

In many cases, other than the land and subtle differences in altitude and climate of the region, the typical lowland crossbred ewe flock and the upland crossbred ewe flock (described above) are indistinguishable in management terms. However, there is a greater variety of land and cropping systems combined with the sheep enterprise in lowland farming systems. For example: sheep on a lowland dairy or beef farm, which is mainly grassland based; sheep on mixed farms, where grass forms an important part of the arable rotation; sheep kept on predominantly arable units using small areas of grass (perhaps areas of land that are most difficult, or impossible, to cultivate), stubbles and crop residues.

6.2.3.7 Welfare Issues Specific to Scottish Sheep Farming

There are a number of welfare issues in the management of Scottish sheep farming systems that are common to other extensive systems and these will be discussed more in Section 6.2.4 below. This section will deal only with those issues that are either specific to the stratified Scottish system, or occur with great frequency there.

1) *Issues with hefted or bound stock*: These systems have the potential to provide high welfare. The system makes use of the natural behaviour patterns of the sheep, rather than attempting to adapt sheep to man's requirements. Thus, the animals are able to range in familiar social groups, making use of the naturally provided shade, forage and shelter. There is a limited ability for cultural knowledge to be passed from mother to daughter as animals remain on the same land for generations. These systems may be relatively inflexible, in terms of the ability of the farmer to use different breeds. However, as the sheep have been bred and raised in the same area they should be well adapted to surviving. These flocks are generally closed, thereby reducing the impact of imported disease conditions so they can have high health status in comparison to other flocks. The remote nature of many hefted flocks mean they are also at lower risk of contracting infection from other animals, such during as the foot and mouth (FMD) outbreak in the UK in 2001.

The main welfare concern is probably the issue of destocking and restocking hefted hill flocks. This issue came into focus following FMD and the slaughter of significant numbers of hefted sheep, particularly in the north of England. The adaptations that these animals possessed, following years of grazing the same areas, were potentially lost. In addition, there were concerns about the establishment of a new hefted flock on the now empty ground. In these cases the role of an experienced shepherd, familiar with hefted sheep and sheep behaviour, is essential to help the new flock to become familiar with the location of shelter, water and forage. Additionally, there are concerns that animals may not forage as efficiently in a new environment, at least initially, expending more energy in searching for feed whilst obtaining an intake of lower quality. This could cause problems with those animals that may already be the leanest of the flock, particularly when the availability of forage may be low or nutrient poor, resulting in higher mortality in this group. Finally, adaptation to the new environment, for example at the level of the rumen, or in terms of fleece characteristics, may also mean that sheep are initially less well able to cope in the new environment.

There has been little research into other possible welfare costs to sheep from leaving the heft. For example, as demonstrated in Chapter 4, sheep are capable of retaining memories for individuals for at least two years. Whether sheep suffer when removed from familiar animals, or indeed familiar environments, is unknown although it is likely that they will experience some stress at least in the short term.

2) *Sheep in mixed systems*: There are a number of situations where sheep in mixed systems may suffer poor welfare. Sheep are likely to be a low value part of the enterprise and, therefore, concern for their health and welfare may be a low priority, particularly if there are financial constraints. Draft ewes leave the hill environment when they are considered too old to be able to rear a lamb in the harsh climate, and are sold to farms with a more benign environment to produce at least one more lamb crop. Potential welfare issues include the possible lack of adaptation of these ewes to the new environment, having spent their lives in the hills, which may affect their health, for example exposure to new infections or disease challenge. These possible health issues may compromise their ability to survive and raise lambs in the new environment.

Sheep may be fed on root crops through the winter as a source of protein when pasture availability is low. As these areas can become very wet and muddy sheep fed on roots need a dry area on which to lie. Sheep feeding on root crops or other particularly fibrous residues may suffer mouth ulcers or lesions and this could lead to the loss of teeth. Broken-mouthed animals, or lambs growing their second teeth, will be unable to feed properly on roots.

3) *Sheep on common grazings*: The main welfare concern with sheep kept on common grazing is the health status of animals mixing from different flocks. Potentially infectious disease can pass easily from one flock to another and control of diseases such as scab rely on all commoners agreeing to implement a particular prophylactic or treatment policy.

6.2.4 Welfare Issues in Extensive Systems

6.2.4.1 Ewe Nutrition

The nutritional value of hill grazings is low and levels of supplementation are often also low, either because of the low profitability of these enterprises or because of problems with access to feed animals on remote pastures. As the nutrition of the pregnant ewe is frequently suboptimal, high levels of both ewe and lamb mortality can occur. Improved management strategies can reduce these welfare issues. In hill systems with only small areas of nutritionally improved pastures the management of this resource is important. This grass can provide better nutrition for ewes at mating if left ungrazed for a period before mating. Both conception rate and the number of lambs born per ewe lambing are increased significantly with better nutrition at conception, and there is now a good understanding of the importance of periconception nutrition for the lifelong health of the animal (Kind et al. 2006). Until relatively recently, supplementary feeding was targeted specifically at the last 6–8 weeks of gestation, when the lamb was growing most rapidly, even though it has been known for 20 years that mid-pregnancy supplementation can be beneficial in improving lamb birth weight and survival (Waterhouse & McClelland 1987). In addition, the development of ultrasonographic pregnancy scanning has had a major impact on the nutrition of the pregnant ewe. This allows farmers to provide specific better nutrition for ewes carrying twin lambs. By allowing these ewes to lamb at lower altitudes the lambs also graze with their dams on improved pastures, resulting in increased weaning weights and enabling the ewe to retain and regain body condition more effectively. A further part of the improved management at lambing involves the segregation of ewes according to risk of lambing difficulty (e.g. primiparous ewes, older ewes) such that these animals are provided with better nutrition or better supervision.

6.2.4.2 Stocking Density and Environmental Issues

Management of semi-natural hill grazing is relatively imprecise. There is a large range of plant species available and their digestibility, production and utilisation by sheep depend on many factors, and may vary from year to year. Calculation of carrying capacity and thus required stocking density on these pastures is often achieved by assessing the previous year's livestock performance and adjusting new flock replacement numbers accordingly. Although this maintains some level of balance between carrying capacity and livestock numbers there may be periods or years when livestock numbers are out of balance with the available grazing leading to undernutrition.

There may also be ecological and environmental constraints on the numbers of animals grazing in particular areas. Certain environmental pressure groups have advocated widescale reductions in sheep grazing densities for environmental reasons. These are likely to involve reduced flock sizes, lower economic returns and lower employment. These changes may also contribute to sheep welfare problems, mainly due to lower labour availability, but also as the sheep may be perceived as of low value when the main management of the land is for ecological purposes.

6.2.4.3 Housing and Handling

Although extensive sheep systems mean that sheep spend most of their time at pasture, there are times under extensive management when sheep are moved from pasture to facilities that enable their easy handling or inspection. These facilities are pens, races and sheds and are designed to move the sheep into closer and closer confinement until each sheep can be manipulated individually. However, these facilities are not necessarily designed with the welfare of the sheep in mind. As discussed in more detail in Chapter 8, these systems can have consequences for sheep welfare. These may arise as a result of human behaviour due to frustration if sheep do not move smoothly through the facilities or as a result of injury or death to sheep while they are confined in the yards, races or sheds. Although these issues are common to all sheep production systems, extensively managed sheep have little experience of human interaction or close proximity to man and dogs. Furthermore gathering of extensive animals is more likely to require the use of vehicles, dogs and horses than in other systems, and animals may be driven over large distances, often on rough ground. In addition, nearly all the encounters extensively managed sheep have with humans are aversive (involving restraint, shearing or dipping which sheep find unpleasant, or painful manipulations such as tagging, tail docking, mulesing or castration). Thus, gathering and handling are likely to be considerably more stressful for extensively managed animals than animals with more experience of human interaction.

Goddard et al. (2000) demonstrated that extensively managed lambs had higher heart rates with handling than intensively managed lambs of the same breed, suggesting that the lack of experience with human contact was more stressful in the extensively managed animals. Yarding has been shown to be moderately stressful as shown by a comparison of cortisol levels with those found in sheep on pasture (Fell & Shutt 1988). However, the presence of a dog, or dog barking, causes elevated plasma cortisol, ACTH and heart rates, above those seen on sudden exposure to humans and noise (Harlow et al. 1987; Baldock & Sibly 1990; Cook 1996; Komesaroff et al. 1998). Vigorous movement by dogs also impairs ovulation in young ewes and aggressive behaviour (biting) by the dog caused elevated plasma cortisol and is considered to be more stressful than dipping or a 90-minute truck journey (Kilgour & de Langen 1970). Thus handling, particularly with dogs, may be a potent source of stress to the extensively-managed sheep. Particular consideration to the training and temperament of dogs and stockpersons in sheep handling, the speed of movement over rough ground and in hot weather and the design of handling facilities can all help to minimise the welfare problems associated with handling sheep. Furthermore, these factors can also contribute to making the handling process more efficient and reducing stress on the operator, so leading to an improved human-animal relationship.

6.2.4.4 Labour Input and Economic issues

As mentioned above, a considerable problem with extensive sheep farming is the drive to reduce inputs and declining labour in an already low input system. This brings problems of supervision as already discussed in Chapter 1. A recent survey of extensive sheep farmer's opinions suggests that they perceive a further decrease in labour as a major constraint to their ability to maintain or improve the welfare of their flocks. The low value of draft ewes and hill lambs are drivers for the efforts to reduce inputs, and hence increase the risks of poor welfare in hill systems. The low economic worth of individuals frequently means that veterinary treatment is not provided to sick or injured animals, even if supervision is sufficient for these animals to be identified, as the monetary worth of the animal may not exceed the cost of treatment.

6.2.4.5 Expression of Natural Behaviours

The fact that most sheep in the extensive management systems are run out of doors all year round means that they have some capacity to exhibit most of their full range of natural behaviours. The major exception here is the normal socialisation that would occur in the wild (see Chapter 2), which is disrupted due to breeding considerations. These disruptions include weaning of lambs, which occurs earlier than would be the case without human interference, and the separation of the sexes.

One possible effect of early weaning could be to disrupt the bonds that form between lambs and their mothers. However, the studies in wild sheep indicate that these are not long lasting. This has also been shown to be the case where it has been studied experimentally. Such studies have shown a gradual weakening of the bond over the first six or seven months of life and a complete disappearance by twelve months of age (Arnold & Pahl 1974; Lawrence 1990) and its replacement by association with peers. Hinch et al. (1990) also showed a gradual weakening of the mother-young bond with time and its replacement with associations with peers. However, they reported continuing mother-dam associations for more than two years, although these bonds did not appear to be particularly strong. It would, therefore, appear that the extensive management of sheep has little lasting impact on the expression of their natural behaviours, and any effects that it does have has little effect on their welfare.

6.3 Intensive Management Systems

Sheep are rarely kept exclusively in the intensive management systems that are found with other farmed species. Lambs bred for meat may live in intensive finishing systems for a short period, when they will be housed in groups but do not receive the same degree of physical restriction, even then, that may be experienced by laying hens or sows. The main more intensive system that sheep may be managed under occurs in dairy sheep, although the intensive nature of sheep management in these systems may be primarily the closer supervision and contact with man that these sheep experience.

6.3.1 Dairying

More than 7 million tonnes of sheep's milk are produced in the world every year, the greater part of which is produced in southern Europe, the Near East and Middle

East (Table 6.2). This is increasing. In 2005, 8.5 million tonnes of sheep milk was produced, half of which was produced in the Mediterranean basin. In this region alone, there are about 40 million milk-producing sheep which are milked twice a day throughout a three- to six-month lactation period. However, sheep milking is also becoming more popular in the UK, USA, Central America, South Africa, Australia and New Zealand. The 'pure' dairy breeds, such as Friesian, Lacaune or Awassi, are often used in these countries, although dual purpose breeds such as the British Milksheep or Dorset Horn are also used. In the Mediterranean region there are a large number of local dairy breeds that have been used for milk production for centuries. Dairy ewes produce up to 600 l of milk per lactation and, although there is a growing market for whole milk, sheep milk is particularly suited to cheese-making and much of the production of milk is processed into cheese, yoghurt or ice-cream.

Although buildings are essential for lambing and milking, sheep can live outdoors all the year round and, where possible, live almost entirely at pasture (Mills 1989). However, there is great variation in the intensiveness of sheep dairying in different regions. In the newer sheep milking regions (e.g. UK and USA), as well as more traditional sheep dairying countries in Europe, sheep milk production is ecologically oriented and a pasture-based system. In other areas (e.g. Israel, Australia) intensive systems have been developed where the animals may be fully housed and fed silage, cereals and a high protein feed, such as whole lupins, cottonseed or soyabean meal (Gootwine & Pollott 2000; Rogan & Grant 2001). Elsewhere, dairy sheep may be kept indoors during lambing and milking period on the farm, and out of these periods on pasture (part outdoor, part indoor systems).

6.3.1.1 Reproduction and Lamb Management

As a seasonally breeding animal, in many dairy sheep systems the ewe may be productive for only 6 months of the year. Using dual purpose breeds, or crossing meat sires onto pure dairy ewes can help to make the animals more productive by increasing the value of the lamb produced. However, in more intensive systems, an accelerated lambing regime is practised to maintain economic levels of milk production (Gootwine & Pollott 2000). Using improved Awassi ewes, or Assaf flocks (a stabilised composite breed of the Awassi and East Friesian developed in Israel in the 1950s and 60s) ewes are synchronised in oestrus and multiple mating or insemination periods occurs during the year (Pollott & Gootwine 2004).

Traditionally, husbandry systems in dairy sheep include a suckling period followed by a milking-only period. In some systems there may be an intermediary period where both suckling and milking occur. The length of the period that includes suckling is 25–75 days according to the breed. In general, lambs either remain with their mothers for at least 25 days or are taken off at birth, but ewe-lamb separation is avoided between 5 and 20 days after parturition. In some countries operating more intensive systems, such as in the Czech Republic, Germany, England, Australia and Israel, the suckling period is eliminated and milking starts immediately after lambing (Astruc et al. 2002). In these systems the lambs will be either artificially reared, or multiple sucking of a late lactation or low production ewe may be practised

(Rogan & Grant 2001). As described in Chapter 3, ewes do not accept alien lambs readily and thus these ewes must be kept in a head bail for a period of days before they will readily accept the sucking attempts of their foster lambs.

Between 1970 and 1990, milk production doubled in the Lacaune breed in France, through both genetic selection and improvements in herd management and milking techniques (Barillet & Bocquier 1993). An important improvement in herd management was to increase the frequency of lambing along with a month of suckling before weaning (Marnet et al. 1998). Because of the higher milk production that resulted from these improvements, and the limited milk intake capacity of lambs, the suckling period is nowadays replaced by a mixed system of suckling and milking. The mixed system results in similar milk yields to systems with weaning occurring at birth and also results in improved lamb growth (McKusick et al. 2001). An important advantage of this system is that it may also reduce the stress of weaning (Guillouet & Barillet 1991). Such a reduction in stress is very important from both a welfare and a production point of view since numerous authors have noted milk yield losses of 30–40% after weaning (Folman et al. 1966; Labussière et al. 1979; Benyoucef & Ayachi 1991). The reduction in frequent udder emptying explains only 20–25 % of this loss (Labussière et al. 1974). Since an alien lamb was shown to be unable to obtain more milk than that collected through machine milking (Labussière et al. 1979), it appears that the quality of udder stimulation is less significant than the psychological effect of the breaking of the mother-young bond. This hypothesis is confirmed by the increased difficulty of adapting a ewe to milking in proportion to the length of contact between the ewe and her lamb (Labussière & Ricordeau 1970). All these effects can be linked to modifications in oxytocin release (Marnet et al. 1998).

One drawback of a longer period of contact between ewes and their lambs during the suckling period is that ewes adapt less readily to exclusive machine milking following weaning (Labussière et al. 1979; Labussière 1988; Gargouri et al. 1993; Marnet & Negrao 1999). In order to overcome this, Labussière et al., (1979) and Gargouri et al. (1993) have tried to habituate ewes to machine milking and to prevent some of the loss in milk yield at weaning by using a mixed management system. In this system, ewes are allowed to suckle their lambs for part of the day, and then are separated for machine milking once daily in the following morning. However, despite increases in total milk yield with the mixed system as a result of more complete udder evacuation, they still found a significant drop in milk production at the time of complete weaning (20%). These results suggested that the suckling stimulus of a ewe's own lamb was more efficient in stimulating milk ejection, and therefore milk yield, than the stimulus generated by a milking machine (Labussière et al. 1979, Marnet & Negrao 1999). They also suggest that ewes are significantly affected by the separation from their lambs.

6.3.1.2 Milk Production and Genetic Selection

In studies where data are available for the entire lactation (i.e. the intensive systems where there is no suckling period), milk production per day increases to reach a peak around 4–7 weeks after parturition and then gradually declines (Pollott & Gootwine 2004). Peak milk yield can be as high as 3.4 l per day (Gootwine & Pollott 2000), with average daily production of nearly 2 l for a lactation varying between 150 and 214 days. Ewes are dried off once milk production has dropped to around half a litre per day. Lambing intervals, in accelerated lambing systems, range between 270 and 330 days (Pollott & Gootwine 2004). Individual ewes may have as many as 10 lactations, although average lifetime is calculated at 6.4 years (El-Saied et al. 2006).

In intensive systems dairy females are mated as ewe lambs and give birth at an average age of 14 months (Gootwine & Pollott 2000), whereas in more traditional systems ewes lamb for the first time at 22 months (El-Saied et al. 2006). Primiparous ewes have lower milk yields than multiparous ewes, although there is a positive relationship between age at first lambing and total milk yield in the first lactation (Gootwine & Pollott 2000). Milk yield has also been shown to be affected by lactation number with a marked increase in milk yield in the second parity, followed by a gradual decline in milk yield, due to a combination of shorter lactations and reduced maximum secretory potential (Pollott & Gootwine 2004). Other factors influencing milk yield and composition are litter size, season, nutrition and management. Ewes with larger litters produce more milk than singleton-bearing ewes, even when the young are not suckled (Pollott & Gootwine 2004; Gootwine & Pollott 2000). The seasonal variables of heat load and day length both affect ewe daily milk yield, the effect of heat load may be due to heat stress or secondary to a reduction in feed intake at high temperatures (see Chapter 2). In addition, seasonal changes in climate, ambient temperature and herbage availability in pasture managed ewes affect milk yield and composition (Sevi et al. 2002; 2004). Pasture-managed ewes have a 30% lower milk yield than fully housed ewes (Rogan & Grant 2001), although these animals resume cycling more quickly and have a shorter lambing interval. In other studies, however, no difference has been found between housed and outdoor-managed ewes (Casamassima et al. 2001). Pasture-managed sheep have also been found to have higher concentration of long-chain unsaturated fatty acids and desirable (for human health reasons) fatty acids than sheep managed in a feed lot (Atti et al. 2006). Housing management, and factors such as poor ventilation, undernutrition, high stocking density and social stress, can have a negative impact on both milk yield and milk composition (see Section 6.3.2.1 below).

Genetic selection for increased milk production has a negative impact on reproductive function in the sheep as in dairy cattle. Greater milk production through lactation results in a longer postpartum anovulatory interval in Awassi and Assaf ewes (Gootwine & Pollott 2000; Pollott & Gootwine 2004). There is also a significant negative genetic correlation between milk yield and heat tolerance, suggesting reduced ability to deal with heat in high yielding animals (Finocchiaro et al. 2005). As the majority of milk-producing sheep are located in the hot and arid climate of the Mediterranean basin, this suggests that the welfare of high producing animals is at risk unless heat tolerance is considered in selection decisions. This reduction in heat tolerance may occur because genetic selection for milk yield increases voluntary feed intake and rumen fill capacity (Molina et al. 2001) to support the higher metabolic activity of increased milk production.

6.3.1.3 Interaction Between the Sheep and the Milking Machine

Unlike other areas of sheep management, modern dairy sheep management requires sheep to adapt to very artificial conditions, and relies upon an animal's ability to produce milk under certain technological conditions. This has the potential to cause high levels of stress, particularly in animals of high reactivity, and could result in increased fear and higher welfare risks in these animals with a temperament not suited to this management (Ivanov 1998). While the introduction of machine milking improves milking conditions, there are still many problems to be solved. These include problems inherent in the milking machine itself, human-animal interaction, times and frequencies of milking and simplification of milking methods.

Machine milking has several advantages. Firstly, the work is easier and more hygienic. Furthermore, machine milking also allows for improved control of nutrition through the capacity to feed concentrates at milking, and the quality of the milk can be much higher. Other advantages are increased hourly productivity levels of the shepherds, easy monitoring of milk yields and, as the animals are haltered in the milking parlour, prophylactic and curative treatments can be easily administered. Finally, since this is a mechanical and routine job the milking machine can be changed whenever necessary. However, there are also some disadvantages, especially those concerning sheep welfare. Firstly, although no rigorous studies have been carried out, it appears that milk production obtained from a milking machine is slightly less than that from milking by hand, suggesting milk letdown may be somewhat inhibited in machine milking. Furthermore, the vacuum pressure exerted by the milking machine can influence milk production and teat irritation (Sinapis et al. 2006), potentially affecting the incidence of mastitis.

The emotional sensitivity and temperament of the milking ewes also have important influences on their welfare, selection, and on the organisation of machine milking (Ivanov et al. 1996; Ivanov 1998; Ivanov & Djhorbineva 1999a,b;). Over 25% of the ewes in the flock are fearful during machine milking, resulting in different problems, the most important of which concerns milk let down. The milk ejection reflex is not stimulated by machine milking in a considerable proportion of ewes (16% of Sardinian ewes, 18–21% of Lacaune ewes, 45% of Ost-Friesan ewes and up to 90% of Chura ewes; Labussière 1988; Mayer et al. 1989; Marnet 1997). In particular, the first milking of primiparous ewes is associated with elevations in adrenaline and noradrenaline, and inhibited release of oxytocin, resulting in impaired milk yield (Negrao & Marnet 2003), although 75% of ewes in this study adapted to the milking procedure within 2 weeks.

With a view to establishing the emotional consistency of individual animals under different situations and over time, a method for assessing the temperament in milking animals has been established (Ivanov 1998; Ivanov & Djhorbineva 1999a,b; Ivanov et al. 1996). The method takes into account the complex factors influencing individual behaviours in the milking parlour. Moderate to high repeatability coefficients ranging from 0.50 to 0.97 have been established for such behavioural traits as 'willingness and persistency in taking place into the milking parlour', 'activity toward neighbours', 'reaction toward forage offered by the hand of a stranger' and 'reaction toward positioning of teatcups'. These measurements, made by different assessors acting independently, have also allowed computation of a Complex Score calculated over two consecutive years. Using the Complex Score values, three defined temperaments – nervous, calm and intermediate – have been distinguished.

When the productivity characteristics of ewes of the three temperaments were compared, highly significant differences between calm, intermediate and nervous ewes were found for the indicator 'Milk per lactation period'. There were no effects on wool production or live weight. The indicator 'Milk ejection latency' was also highly significantly different between the three temperaments. The conclusion from this study was that nervous ewes were difficult to milk by machine.

6.3.2 Specific Welfare Problems Associated with Dairy Sheep

There are a number of specific welfare challenges that arise through intense management, similar to those seen in other intensively managed livestock species, and some that pertain in particular to the management of dairy animals. The main risks to welfare will be discussed here.

6.3.2.1 Housing and Intensive Management

As with other livestock species, close confinement can be a source of stress and a welfare concern in housed dairy sheep. As this type of sheep management occurs less frequently than in most other species, or has only recently begun to be used for intensively managed dairy ewes, there has been considerably less research in this area than with other species. Sheep can be particularly prone to respiratory infections and good ventilation of buildings is important. Inadequate ventilation results in increased concentrations of ammonia and carbon dioxide in the air, and reduced the milk yield, protein and fat content of the milk from ewes kept under these conditions compared to ewes kept in well-ventilated buildings (Sevi et al. 2002; 2003a). Ewes in poorly ventilated housing have reduced feed intakes and increased adrenal responsiveness to adrenocorticotrophin. There is, however, no effect on immune measures of stress in the ewes. In addition to ventilation, air quality and hygiene are improved by keeping bedding clean. Ewes kept on soiled bedding had lower milk yields, higher somatic cell counts and poorer milk hygiene than those where the bedding was regularly renewed (Sevi et al. 2003b). Reduced milk yield and quality, and increased sub-clinical mastitis, is also found when the air volume per ewe is less than 7 m^3 (Sevi et al. 2001a).

Ewes kept at relatively low stocking density (2 m^2 per ewe versus 1 m^2 per ewe) had greater milk yield with a higher proportion of protein and fat than ewes at the high stocking density. These ewes also had fewer somatic cells and bacteria in their milk (Sevi et al. 1999a). This resulted in fewer cases of subclinical mastitis in the ewes at low stocking density in comparison to ewes with less space per animal. When ewes were subjected to the social stresses of being moved between pens or social groups the stressed ewes were more active and engaged in more aggressive interactions than ewes which remained in a stable group in the same pen (Sevi et al. 2001b). Socially stressed ewes had impaired immune responses, and a lower milk yield and content than the stable group. Thus housed ewes, in addition to requiring sufficient space and ventilation, have a preference for remaining within the same social group. For many of these housing parameters the interests of the lactating ewe coincide with the interests of the producer since milk yield and cheese-making quality of the milk are adversely affected by the conditions that reduce welfare of the ewe.

There have been very few studies of the preferences of sheep for different types of bedding or housing conditions. In general sheep appear to prefer to lie on straw in comparison to other types of flooring (Bøe 1990; Gorden & Cockram 1990; Faerevik et al. 2005), and spend more time lying on straw bedding. This preference is particularly expressed in shorn ewes, but less apparent in ewes with thick fleeces. In general many of the dairy breeds of ewe have thinner fleeces than meat breeds, suggesting that these ewes are more likely to require straw bedding for adequate thermoregulation, particularly during cold weather.

6.3.2.2 Undernutrition

Lactation is an energetically expensive process, and ewes kept on traditional pasture management systems run the risk of experiencing undernutrition, particularly during late pregnancy and early lactation when peak lactation occurs. Under these natural pasture systems herbage availability and quality, as discussed above for extensively managed sheep, can vary markedly both between and within years, requiring skilled management of grazing and stocking density. Since the period of undernutrition coincides with the suckling period in these systems, there may be a lower financial pressure on the producer to provide supplementary feeding to achieve good milk yields, since lamb production is not the goal of the enterprise. However, undernutrition in late pregnancy has a significant effect on udder development, which leads to a reduced milk yield even in ewes provided with adequate feed intake during lactation (Charismiadou et al. 2000; Bizelis et al. 2000). Thus it is again beneficial to the welfare of not only the sheep but also the producer (as long as financial returns from milk production exceed the costs of supplying additional nutrition) that ewe nutrition during pregnancy, as well as lactation, be carefully managed.

6.3.2.3 Early Weaning

A particular feature of all dairy systems is the early weaning of offspring so that milk production can be exploited by man. As described above, the interval between birth and early weaning can be very variable in different systems, ranging from immediate separation to weaning at 2–3 months of age, and may include a period of both milking and suckling as in the mixed systems. Generally however, if weaning does not occur within a few days after birth it is avoided until lambs are at least 3 weeks of age. As described in Chapter 3 this period is characterised by very close contact between ewe and lamb, and frequent sucking before the ewe starts to regulate the lamb's access to the udder at about 4 weeks of age.

Mother-young separation at any age before natural weaning is stressful for both partners (see Chapter 3), causing at least transient increases in stress hormone concentrations (Mears & Brown 1997; Rhind et al. 1998; Orgeur et al. 1999). Behaviourally, abrupt weaning is associated with an increase in activity, especially vocalisation, and disruption of circadian rhythms of activity (Veissier et al. 1989; Orgeur et al. 1998; 1999). Weaning may also be associated with reduced lamb growth, and an increased susceptibility to disease and parasitism through impaired immune responses, when conducted at a young age (Watson 1991; Napolitano et al. 1995). Studies attempting to reduce the stress of weaning in dairy lambs and ewes suggest that gradual separation, or allowing the lamb to remain with the ewe but preventing suckling, can be more stressful for the lamb than abrupt separation (Sevi et al. 1999b; 2003c), presumably due to frustration or repeated stresses of short-term separation. Artificial rearing of lambs can be associated with behavioural disturbances, such as redirected sucking on pen mates and other non-food substrates (termed 'pica': Stephens & Baldwin 1971; Jagusch et al. 1977), suggesting that these systems do not meet all the behavioural needs of young lambs. The system described above where lambs are fostered onto a late lactation or low producing ewe may be better for the welfare of the lamb, although there are issues with the welfare of the ewe subjected to restraint for several days to induce acceptance of the foster lambs (see Chapter 3).

6.3.2.4 Human-Animal Interaction

The close nature of the contacts between ewe and stockperson inherent in milking sheep means that the stockperson can be a source of fear for the dairy ewe, although conversely a good stockperson can bring positive welfare benefits to the flock. As described above, ewe temperament may also contribute to the degree of stress associated with milking and human contact. These temperament traits are related to innate immune responses (Dimitrov et al. 2005) such that health and welfare may be compromised in reactive ewes.

In other intensively-managed livestock species stockperson behaviour and attitudes have been shown to have an influence on the welfare of the animals in their care. Although this has not yet received much attention in sheep farming it is likely that this will also influence welfare in this species. As described in Chapter 4, the sheep has a good ability to recognise and distinguish between different handlers, and forms an emotional attachment, good and bad, to different types of handlers. Gentle handling and feeding has been shown to result in the development of a positive social bond between lambs and the stockperson (Boivin et al. 2000; Tallet et al. 2005). Thus attention to the development of a relationship with the stockperson in the early life of the dairy lamb, and positive handling during the productive life of the ewe, will help to reduce welfare issues with milking of dairy ewes.

6.3.2.5 Disease Issues

As with dairy cattle, dairy ewes are at risk of developing production-related diseases such as mastitis. The incidence of clinical intramammary infections in sheep is relatively low, at or below 5% (Bergonier et al. 2003). However the incidence of subclinical mastitis, usually identified by somatic cell count in milk, varies from 4% to more than 40% depending on country, breed and flock (Saratsis et al. 1998; Las Heras et al. 1999; Leitner et al. 2001; Albenzio et al. 2002; McDougall et al. 2002; Al-Majali & Jawabreh 2003; Batavani et al. 2003). Infections usually occur at the beginning of milking and during the first third of lactation (Bergonier et al. 2003), which may be due to the impaired responsiveness of the ewe immune system during the early transition to machine milking (Albenzio et al. 2002). Infections are fewer in first lactation ewes compared to ewes in their third or more lactation. Subclinical mastitis appears to be less with machine milking than hand milking (Las Heras et al. 1999), suggesting that attention to hygiene during milking may reduce the spread of infection since this is when most transfer of infection occurs.

Subclinical mastitis shows an unfavourable genetic relationship to milk yield (Barillet et al. 2001), suggesting that selection for milk yield alone will increase the incidence of mastitis in dairy sheep. However, somatic cell count, as an indicator of genetic resistance to mastitis, has low to moderate heritability and including this in breeding goals, alongside production traits, will help to limit or prevent an increase in mastitis incidence.

Gastrointestinal nematodes are a major health and welfare issue for small ruminants, and this may be particularly a concern with lactating animals, since the ability to maintain immunity to parasites is compromised with late pregnancy and early lactation (Houdijk et al. 2001). In sheep, unlike goats, there does not appear to be a relationship between susceptibility to nematode infection and high milk production (Hoste et al. 2006) although primiparous ewes are more at risk. This may be due to the increased stress of adapting to early weaning and milking in these animals.

The increased time spent in housing and intensification of dairy sheep production also carries the risk of increasing the spread of infection. Recent research has shown, for example, that Maedi-Visna seroprevalence is three-fold higher in intensively managed dairy flocks compared to semi-intensive flocks (Iratxe et al. 2006). Seroprevalence was 5–15 times higher in the dairy flocks in comparison to extensively managed lamb-producing flocks. Thus the potential risk of disease, and rapid spread of disease between animals, is higher for intensively managed animals than those managed more extensively.

6.4 Traditional Pastoral Management Systems

Pastoralism is the main production system of semi-arid open rangelands. These marginal areas have unpredictable climates, determined either by rainfall or elevation, and are unfavourable for agricultural cropping so allowing pastoralism to compete. Economic dependence on livestock increases as annual rainfall decreases, reaching almost total dependence in desert regions, as the most efficient means of converting the vegetation into a food product. Nomadic or migrant forms of pastoralism, by exploiting the inherent variability in these areas, allow sustainable livestock production and support more people than would be possible by other strategies. Pastoralism can be categorised by the degree of movement into three main classes (FAO 2001):

1) *Nomadic:* This is a highly mobile and flexible system of seasonal migration with no established home base. Movements are opportunistic, following pasture and water availability, so are highly dependent on the growth cycles of different plant species. Nomadic routes are not as random as they may appear however, as nomads prefer established migration routes, particularly if agreements with farmers and trade points exist along the way. However this form of livestock production is flexible and routes will diverge in the face of drought, pasture failure or disease.

2) *Transhumance:* This form of migration involves regular movement about fixed points. Transhumance can consist of vertical migrations in montane areas, which tend to be ancient routes associated with high rainfall regions (e.g. routes followed in Europe or Afghanistan). In these cases lowland winter pastures are used with permanent barns where supplementary feeds may be available from hay fields. In the summer animals are moved to high mountain grasslands, usually accompanied by herders living in tented nomadic camps. Horizontal transhumance tends to be more opportunistic, and can be altered by climate as well as economic or political change along the migration routes. Although in temperate regions migrations are largely dictated by growing seasons and forage availability, movement in the tropics is affected by rainfall cycles and insect prevalence. Transhumance has been transformed in modern Europe with the advent of vehicles so that animals can be moved without needing to be accompanied by herders. An example would be the away-wintering of hill sheep in lowland pastures as practised in UK (see above).

3) *Agropastoralism:* This pastoralism, also called ranching, is a more sedentary form of pastoralism as carried out in Australia and USA, and on a smaller scale in Europe. This differs from other forms not only by degree of movement, but also because other forms of pastoralism occur at the subsistence level, wherein the animal products maintain the family group and are not kept for commercial profit, although some trade may occur. The other main differences are a greater provision of supplementary feeding (particularly in Europe), fenced ranges and land tenure that occurs with ranching. Agropastoralism of sheep has been discussed in more detail in Section 6.2 so this section will concentrate only on nomadic and transhumance pastoralism.

Sheep are managed in nomadic or transhumance pastoral systems in many regions of the world (see Table 6.5 for main areas). In many areas herds are composed of a mix of species (e.g. sheep are kept alongside goats, cattle, camels, yaks and horses), depending on the suitability of the vegetation and climate as well as skills, preferences and labour availability. Although mixed herds can be advantageous for taking advantage of unpredictable climates, they can make pastoralism a labour intensive system where management needs to adapt to the differing requirements

Table 6.5 The main regions of the world where sheep are managed in nomadic or transhumance pastoral systems

Geographical region	Habitat types	Examples
Middle-East, Eastern Mediterranean	Hilly, semi-arid	Iran, Iraq, Afghanistan, Turkey
Central Asian steppe	Severe winters, grasslands	Mongolia, Uzbekistan, Kazakhstan
Asia and Himalayas	Arid, low temperatures, high altitude	Rajasthan, Gujarat, Ladakh, Tibetan plateau
North Africa	Arid desert	Morocco, Algeria
Sub-Saharan Africa	Dry Savannah	Sudan, Niger, Senegal, Ethiopia
Europe	Hilly, semi-arid	Spanish Basque country, Bosnia, Herzegovina, Macedonia
Americas	High altitude, semi-arid grasslands	Navajo, Andes

Source: FAO 2001

of different species. For example in East Africa, where herds may be composed of camels, cattle and sheep, the dry season requires management of the different watering regimes needed by the different species (every day for sheep, every two days for cattle and every three days for camels). For many nomadic peoples sheep are essential for the day-to-day livelihood, partly because of the large array of products that can be obtained from sheep, although other animals (such as the yak in Tibet or the horse in Mongolia) may be viewed as higher status and more desirable animals. In addition, a loss of herd size can be recovered more quickly with small ruminants as their breeding cycles occur more quickly than other species.

6.4.1 Environmental Risks and Challenges

In all areas where pastoral management of sheep occurs, the main risk to both sheep and man is the unpredictability of the climate. This has an impact on the growing season of plants and hence the vegetation available to the sheep. In arid areas seed reserves may remain dormant in the soil until suitable rainfall or climatic conditions have occurred for germination. Thus pasture availability and herbage quality may vary substantially between seasons and years. In cold-weather rangelands the problems arise not through absence or unpredictability of rainfall but because of the short growing season and severe winter (e.g. see Section 6.4.4 below).

For nomadic peoples their control of land may be temporary, thus risk of pasture availability requires negotiation and co-operation with other land users. However, this sharing of the land brings increased biosecurity risks where major outbreaks of disease can rapidly sweep through nomadic herds. In general, disease and climatic events are the major causes of catastrophic loss of livestock (and human) lives. These risks are particularly acute during mid-winter, when animals are in the poorest condition and least likely to survive any other challenge, or during the spring lambing season. Loss of neonates is perceived as especially detrimental since it affects the supply of new females entering the herd and slows recovery from catastrophe.

6.4.2 Products and Management

Nomadic herds are generally larger than what might be considered the economic optimum by an economist, and the objective of many pastoralists is to increase herd size. There are several reasons why, to a nomadic pastoralist, a large herd is desirable: (i) the herd acts as storage as there is no other facility to store animals (or capital), (ii) the main subsistence protein source is dairy products rather than meat, thus requiring a greater number of animals; (iii) a large herd provides some protection from the consequences of risk or catastrophic loss; and (iv) the wealth of a nomad is in livestock as they possess no other assets. In addition, surplus animals can be sold, traded or slaughtered as required (FAO 2001).

The importance of the sheep to the livelihood of many nomadic pastoralists is due to the large number of products that the sheep can provide. Sheep provide milk (and associated dairy products designed to lengthen the life of milk e.g. curd, cream, butter, ghee, cheese, yoghurt), mutton, hide, wool or hair (depending on temperate or tropical locations), and dung for cooking or heating. Sheep stomachs are used for butter churning and storage by Mongols, and the organs, blood and intestines for sausage (Goldstein & Beall 1990). In Tibet adult male sheep are also used as portage or pack animals (these are particularly long-legged breeds of sheep) for carrying salt or grain (up to 7.5–11.5 kg per animal; Goldstein et al. 1990), and for vertical transhumance in the Himalayas.

6.4.3 General Aspects of Sheep Management

Sheep are generally herded on foot and are moved daily from the night camps to fresh grazing grounds. At night sheep are herded back to the camps, and may be kept within stone corrals, partly for protection from predators and also to produce a readily available supply of dung for heating. In countries with significant numbers of large predators dogs accompany the herders and sheep. These are not used to herd the flocks as in the UK and Australia but are there to protect the sheep from predator attacks. In some countries (e.g. Tibet) herding is considered a low status job and is begun by children once they are 8 or 9 years old, most herders being teenagers. The adults make decisions about where the flock should forage before the herds leave in the morning, and the principal tasks of the herder are to keep the animals together, to protect them from predators, and to ensure that they graze evenly over the area. In other areas herding and sheep management may be carried out by adult men (e.g. Rajasthan) using long handled knives to cut branches with leaves and seedpods on which the sheep can forage.

In temperate regions, where sheep breed seasonally, milk is collected through the spring and summer (after the lambs are weaned), and animals are slaughtered only in winter to provide food when sheep are not lactating at the winter campsites. This is considered the optimum time for slaughter as animals are in the best condition after the summer grazing and have not started to lose weight during the winter months (Goldstein & Beall 1990). However, the selection of the animals to be slaughtered is done carefully as even males and barren ewes can be valuable, providing wool and portage. Thus animals are only slaughtered in sufficient numbers to meet the meat needs of the family group, and generally the animals least likely to survive the winter will be chosen. The meat is then dried or frozen to last through winter. In tropical areas milk can be available all year, as animals have less pronounced seasonal breeding patterns, and there is less need to find ways to preserve dairy products. Sheep milk yields in these systems are very low compared to intensive western dairying (see above) but this response is partly due to the poor nutritional intake of these animals (FAO 2001). In addition, pastoralists need to balance the milk yield of their sheep with the consequences of early weaning for offspring survival, as the mortality of young lambs threatens the future of their herds.

Specific sheep management practices in temperate and tropical environments will be considered in the following sections for sheep production in Asia (Tibet and India) and Africa (Sudan and Ethiopia).

6.4.4 Nomadic Pastoralism in Tibet

The Tibetan plateau is the harshest and highest pastoral area in the world located at 4000 m above sea level (Miller & Craig 1997; Miller 1999). Daily temperatures average around 0°C in summer and vary between −28°C and −40°C in the winter (Goldstein et al. 1990). With only a 3.5 month growing season livestock have to survive for the majority of the year on dried dead forage. In addition, access to forage may be a problem in the winter where snowfall can block routes to pasture. Estimates suggest that, in Tibet, 25 adult sheep are required per person for a pastoral family group to be out of poverty (Miller 1999), emphasising the importance of the sheep to pastoral livelihoods. Herd size in different regions range from between 60 and 250 sheep. On average, herds are made up of 30% adult males (to provide income from wool and portage), and 44% lactating females (to provide milk and new offspring to increase the size of the herd). About 20–30 animals each year are sold in trade or slaughtered for consumption by the family group.

Livestock exist by grazing all year round in Tibet, even though grass growth may be for only 3.5 months of the year. New spring foliage is seen in late April or early May, and begins to form a significant part of sheep diet selection about a month later. In summer animals are moved between grazing areas so that summer pastures can regenerate. Milking of ewes begins mid June when the lambs are weaned. Growth of plants ceases in late September, therefore, sheep need to eat enough vegetation in the summer and autumn months to lay down sufficient fat stores to last through the winter when they subsist on dried vegetation. This is achieved by rotation around grazing pastures: sheep remain on the home pastures until autumn when they migrate to different grazing grounds. As all the land available to the nomads is at high altitude and grass is not growing elsewhere these migrations may only be relatively short (10 – 40 miles) (Goldstein et al. 1990; Goldstein & Beall 1990). Sheep and herders remain away at these camps until winter when they return to the home pastures to eat the plants that have been left by the autumn migration.

Additional satellite camps may be used for males and non-lactating females during the summer and autumn, although the lactating females need to be near the home camp for milking.

Although herding is often considered an important but menial task, at lambing time the work is critical and is taken over by more experienced adults. Special 'birthing' pastures may be used for pregnant females, and herders live with the flocks to keep track of the progress of births. During the day herders keep parturient ewes from straying too far from the flock and help to dry the neonates to protect them from hypothermia in the very cold temperatures. The newly born lamb and its mother have to keep up with the rest of the flock during the day so herders may carry the newborn lamb in a sling. At night herders sleep in the corral with the pregnant ewes, and lambs may be kept together in specially constructed clay enclosures covered with blankets to protect them from wind chill. After weaning, in summer the lactating ewes are milked twice daily by hand. Ewes are caught in pairs and tied head to head in a long line to allow the women access to milk them.

6.4.5 Sheep Production in Rajasthan

Sheep (and goat) production is the main form of pastoralism in the arid region of Rajasthan and 20–30% of India's sheep population are kept in this state. Male lambs are sold for meat, with wool and dung the other sources of income from sheep. Milk is not sold but kept for home consumption and the production of products such as ghee. Sheep herds vary in size from 20 to 200 animals and herd size influences their management (Geerlings 2004). Larger herds are migratory, spending up to 10 months travelling, whilst smaller herds are sedentary and taken out to graze daily. During drought years, however, all herds may become migratory (Singh & Köhler-Rollefson 2005). Sheep graze on common lands in the dry season, and in the forest in the wet. Feed obtained by grazing may also be supplemented by seeds, grain, straw, ghee and vegetable oil (Geerlings 2004). Men are traditionally responsible for moving animals to grazing and participating in sheep husbandry decisions determined at village level, whereas women are responsible for milking ewes, caring for newborn lambs and preparing supplementary feeds and milk products. Recent developments in local politics and the development of irrigation practices have restricted access to grazing lands, and reduced the amount of grazing available (Singh & Köhler-Rollefson 2005), making lack of grazing the most significant issue in sheep production.

In tropical regions lambing may occur twice a year, depending on rainfall, although there is regional variation in when lambing occurs and how frequently (Geerlings 2001). For example, some herders will manage breeding such that lambs are born during or just after the rainy season when there is better forage availability. Flocks are frequently made up of a mixture of breeds. The *Boti* (Marwari) breed is most common and is considered to be very drought and disease resistant, and adapted for long migration. However this breed grows slowly and has low milk yield and thin wool (Singh & Köhler-Rollefson 2005). The *Bhakli* and *Dumi* breeds (also known as Sonadi and Patanwadi respectively) have better milk production and growth but lower reproduction, need better quality feed and water access and may not fare well on migrations. Breeding males are selected in the first week after birth on the basis of their weight, condition and appearance, although drought resistance and long legs for walking are also preferred (Geerlings 2004). Unlike other lambs, they are allowed to suckle unrestricted from their mother and may also be supplemented by milk of other ewes. In addition they may be fed food supplements, such as ghee and turmeric. All animals are named from birth and have a close relationship with their human carers, responding to the herder's calls.

In addition to their role as wealth and food providers, sheep also play an important role in cultural and religious rituals (Geerlings 2004). Before shearing (which is considered to be a sacred communal function) some of the best animals are selected to participate in ceremonies for Laxmi (the goddess of money) to ask for a good wool harvest and price. The selected sheep are washed, decorated with paint (*tika*) and silver jewellery, and fed jaggery (unrefined sugar from sugar cane and date palm) and coconut. Particular lambs, born on days with important religious significance, become devotional animals and neither they nor their offspring are sold or slaughtered but confer status and respect on their owners.

6.4.6 Traditional Sheep Management in Ethiopia and Sudan

Ethiopia and Sudan have some of the largest populations of sheep in Africa (Table 6.1). While there is some modernisation of animal production systems in these two countries, the majority of animals are raised under traditional systems. These systems pose difficulties for the collection of accurate information on factors that affect animal welfare, and the following is an attempt to draw together information that comes from many sources and is not always easily accessible.

6.4.6.1 Ethiopia

Ethiopia is a land-locked country situated in the region known as 'The Horn of Africa'. Elevation ranges from below sea level in the north-east of the Great Rift Valley to more than 4000 m above sea level. The main geographical feature is the complex of mountains running down the centre of the country from north to south. This complex is dissected into the northern and southern highlands by the Great Rift Valley, which runs through the country from south-west to north-east. The highlands then descend into lowlands, steppes or semi-desert. Many rivers arise in the highlands, the most notable being the Blue Nile which runs through north-western Ethiopia into Sudan. Several rivers also flow south out of Ethiopia and into Somalia.

The elevation and geography produce three climatic zones, the cool zone where elevation is greater than 2400 m, the hot zone where elevation is below 1500 m and the temperate zone in between. The temperate zone has a bimodal rainfall pattern with a short rainy season running from March to April and a longer wet season with heavy rain running from June to October (Mukasa-Mugerawa & Tekelye 1988). Average annual rainfall varies from 1000 to 1200 mm (Tolera 1998). Approximately 75% of Ethiopia's sheep are kept on small-scale mixed farms in the cool highland regions where the annual rainfall exceeds 700 mm while the rest are found in the lowlands and in the Great Rift Valley.

As with other data in this section, much of the information concerning sheep management in Ethiopia has come from surveys of sheep owners. The following information is derived from recent surveys of Ethiopian households where the number of households ranged from 80 to more than 400. In the Highlands, the principal breeds are the Horro and the Menz, whereas in the pastoral lowlands, the principal breed is the Black Headed Ogaden. All three are fat-tailed breeds. In the Highlands, the majority of farms keep sheep, although this number decreases with decreasing altitude as the emphasis on cropping increases. Alemayahu & Fletcher (1991) surveyed farms in the northern highlands and showed that of farms at altitudes of 2200 m and above, 90% kept sheep, whereas of farms at altitudes between 1300 m and 2200 m in the lowlands only 47% kept sheep. At the higher altitudes, the average sheep flock was 4.6 compared to 2.3 at the lower altitudes. In the very north of Ethiopia, the average farm size is 2.2 ha, the average household contains 5.7 people and the average number of sheep per farm is 5.3 (Holden & Shiferaw 2000). The main purposes of sheep rearing are, in order of declining importance, the sale of live sheep (Mekoya et al. 1999), particularly at festive times of the year, the production of cash income, wool for household use, meat for household consumption and animals for sacrifice.

In the Northern Highlands, all farmers surveyed had privately owned land (Mekoya et al. 1999). There is some communal grazing land but this is limited with areas of 1–2 ha being shared by five to ten households. These communal lands are generally used during the rainy season when sheep are excluded from the crops and pasture on the farmer's own lands. During the dry season, sheep usually graze on wheat or barley stubble on the farms. Sheep generally graze for 10–11 h a day. Sixty-seven percent of the flocks are taken to water once per day while the remainder are taken every second day. Similar findings were reported by Agyemang et al. (1985) in a survey of flocks in the Northern Highlands. In this survey, the main feed source was native pasture on fallow land or communal grazing areas or grazing of crop stubble when available. Grazing time was 10–11 h per day with watering every day at around noon. The source of water was usually a river one to two kilometres from the homestead.

All households housed their sheep (Mekoya et al. 1999). Almost all (95%) housed them in buildings separate from the main dwelling, whereas the remainder housed them in the main dwelling, usually in underground floors. The majority of farmers in the Highlands kept newborn lambs indoors for the first one to three days after birth (Agyemang et al. 1985, Mekoya et al. 1999). In the rainy season, this could extend to the first 15 days (Mekoya et al. 1999) or up to 60 days (Agyemang et al. 1985). Shearing is carried out twice per year in October-November and again in April to June. The main reason is harvest of wool for use in the household as well as the removal of ticks and mud.

In the survey of 80 households in the south, Tolera (1998) reported an average landholding per household of 3.5 ha. About 90% of the respondents had no access to communal grazing and had to use roadside grazing or grazing of natural pastures and crop residues on their private holdings. In this study, water was reportedly not a problem, with rivers, springs and ponds being the main sources. About 80% of respondents took their sheep to water once per day, 10% twice a day and 3% once every two days. The rest did not bother as they considered their animals able to look after themselves. However, in the drier parts of the country, pastoralists have to migrate with their flocks in order to find pasture in the dry months from October to March. These migrations range from 50 to 200 km (Adugna 1992).

Similar reasons for keeping sheep were found in the South as in the Highlands: sale for cash, wool and meat for household use and sacrifice (Tolera (1998). Two advantages of small ruminants over the keeping of larger livestock are low capital cost and ease of management by women and children. The mean number of sheep per owner was 15.1, but ownership was quite complex in that 43% of flocks were owned by one person with the rest owned by two, three or four people. Ninety percent of respondents housed their sheep at night; of these 85% did so in a shed separate from the main dwelling and 15% in the main dwelling. The sheep were normally enclosed in pens at night and lambs were usually kept at home for the first month of their lives while the ewes went out to graze (Adugna 1992).

Little information is available on sheep kept in urban areas. However, in study of animals kept in backyards in the city of Awassa the most common animals kept were sheep (Taye & Abebe 2000). Of these households, 78% kept sheep for food and/or income generation and the mean flock size was 4.6. About 60% practised herding using roadsides, crop residues and communal grazing areas. These were readily available since about 60% of the city is vacant land. Herding was mainly the responsibility of the boys before and after school but some households employed shepherds who are shared among households. Average grazing time was seven hours per day and all respondents housed their sheep at night, 89% providing housing separate to the main dwelling, the rest inside the main dwelling.

6.4.6.2 Sudan

Sudan lies adjacent to the north-west of Ethiopia. In contrast to Ethiopia, Sudan is relatively flat, the major geographical features being plateaus and plains (Fig. 6.4). Two major rivers run through Sudan, the Blue Nile and the White Nile, which join just north of Khartoum. In the north of the country, most of which is desert, rainfall is rare. However, in the south, sufficient rain falls to give rise to rainy seasons of six to nine months.

The country can be divided into six agro-ecological zones (Fadlalla & Ahmed 1997). These are desert, with less than 75 mm of rainfall per year, semi-desert where yearly rainfall ranges from 75 to 300 mm, low rainfall (301–500 mm per year) savannah, high rainfall savannah (0.01–1000 mm per year), mountains and the flood zone of the Nile river. This results in a gradual change in agricultural pursuits from pastoralism in the north to crop-based mixed farming in the south.

Traditional Sudanese agriculture has been classified as pastoral where more than 50% of household income is from livestock, agropastoral where from 10 to 50% of household income is from livestock with the remainder coming from crop production, agricultural with less than 10% of household income from livestock and urban/peri-urban where livestock are raised in backyards (Wilson 1986). The traditional classification can be further divided into nomadic and sedentary. The general movement of the nomads is from the northern semi-desert where they spend the wet season to the savannah for the dry season (Fadlalla & Ahmed 1997). Nomads comprise about 11% of the human population but own more than 90% of the animal wealth and produce 90% of the country's meat (Harbi 1992). On the flood plan of the Nile there is also a practice of migratory agropastoralism where sheep are raised in rainfed, settled villages but livestock graze the plain of the Nile and are moved away in times of flood and back again as the floodwaters recede (Fadlalla & Ahmed 1997).

Recent increases in the local and overseas demand for desert sheep has resulted in an increase in their numbers. The most recent data indicate a total population of sheep of 47 million (Table 6.1), a marked rise from the estimate of 16.2 million in the animal census of 1975/1977 (Sulieman 1986). In a study of sedentary pastoralism in the Kordofan region of central Sudan, holdings ranged in size from 3.5 to 14.5 ha (Anon 2000).

6.4.7 Welfare Issues in Traditional Management Systems

6.4.7.1 Environmental Impacts

As with all forms of pastoralism the animals are subject to environmental stressors. As these systems for livestock rearing take place in areas with harsh and unpredictable climates these challenges may be frequent and sustained. Breeds used in these areas have adapted for years to the arid or cold climates but even these may suffer during harsh winters, where in Tibet mortality may be 30%, particularly of lambs (Miller 1999). Introduction of stock selected for production traits rather than survival characteristics can cause high levels of suffering as these animals are not suited to the environment. In addition they have little impact on the genetic make up of the nomadic herds as mortality of these animals is extremely high.

In general the main environmental disasters affecting livestock are drought and blizzards (FAO 2001). Under blizzard conditions the stock are cut off from feed and large numbers may die simultaneously from hypothermia or suffocation regardless of their body condition. The effects of drought are progressive and cumulative. Thus deaths will be slower than under blizzard conditions and weaker animals die first. Nomads respond to drought by long distance opportunistic movement of stock but national borders or cultivation may prevent or curtail these movements. Cultivation may further affect water availability by altering watercourses for irrigation. Long distance migrations also have a negative impact on livestock health leading to an increase in mortalities (Johnson 1975). Provision of irrigation points can partially offset some of these problems but too many watering points can also create problems (FAO 2001). In good years the numbers of animals in herds increase and become more sedentary. However in drought years there is not enough water and the land can become exhausted from the large numbers of animals feeding there. At that point, as annual migrations have not occurred, herds can be too far from water to walk there resulting in large numbers dying of thirst and malnutrition.

While water is not considered a problem in some areas of Ethiopia, it is one of the main limiting factors in the arid pastoral areas, so much so that intervals between waterings may be as long as ten days (Adugna 1992). Similar watering intervals of 4–5 days in the summer and 10 days in the winter have been reported for Sudanese desert sheep (Osman & Fadlalla 1974). Often, the aim of such long intervals is to conserve water. In an experiment where watering frequencies varied from ad libitum to every seven days, in Blackhead Ogaden sheep, there was no impact on growth of watering as infrequently as every three days (Sisay 1999). However, this experiment did show that, at watering intervals of two to seven days, sheep appeared very thirsty and, when offered water, drank their fill in within three minutes. This suggests that, although the sheep could survive this interval without effects on growth there was a need for water that was not being met by watering at intervals of greater than two days.

In common with other pastoral systems sheep flocks are vulnerable to predation from wolves, snow leopards, lynx and other large carnivores. The use of dogs during the day in some communities, and bringing the animals back to corrals at night, can reduce the risks of predation. However effective protection of sheep relies on the herder's ability to keep the flocks together during the day to prevent straying when there is increased risk that isolated animals will be attacked.

6.4.7.2 Diseases

The effects of disease outbreaks in pastoral livestock systems can be catastrophic with major epizootics spreading rapidly and potentially wiping out whole herds (FAO 2001). The sharing of grazing between herds can exacerbate the impact of infectious disease, which can spread rapidly from one herd to the next. Disease does, however, function to keep herds small enough to be managed on the available pastures. Pastoralists are often very knowledgeable about the behaviour and physiology of their livestock, and this ethnoveterinary knowledge of animal health has been used to avoid areas of disease or insects in certain times of year. However, modern veterinary medicines allow these areas to be exploited, as long as vaccination and medication programmes are adhered to. In the long-term, however, this may increase herd size beyond that which can be sustained by the land and reduce the nutrient quality of these areas, particularly in years of drought.

For many pastoralists in tropical countries, disease can be one of the biggest barriers to production and good welfare. In Ethiopia farmers list liver fluke, coenurus (sheep turning in circles) diarrhoea and anthrax as the most common diseases in that order (Agyemang et al. 1985). Other studies report similar findings and include lungworm, bloat, parasitic gastroenteritis, dermatophilus infection, sheep pox and pneumonia as significant disease challenges (Adugna 1992; Tolera 1998; Mekoya et al. 1999). The principal causes for mortality are pneumonia, fascioliasis, gastrointestinal parasites and various mechanical causes such as foreign objects in the rumen and injury (Bekele et al. 1992). In India similar disease conditions of diarrhoea, pneumonia, liver fluke, foot and mouth disease and abortion are cited (Singh & Köhler-Rollefson 2005). Inadequate feed in the dry season can contribute to mortalities. In many cases the decision to treat sick animals depends on the availability of cash to the farmer. However, in Rajasthan, indiscriminate use of broad-spectrum antibiotics to treat any situation where the animal is not considered to be performing optimally has been reported (Singh & Köhler-Rollefson 2005). Antibiotics were injected, often at the site of the problem rather than intramuscularly, applied to wounds and given orally. This situation is exacerbated by a lack of veterinarians with specific knowledge about sheep diseases.

6.4.7.3 Management Interventions

In common with all sheep farming systems there are some management interventions that may be detrimental to sheep welfare. For subsistence pastoral systems, however, in many cases these may not be as severe as in other systems. Firstly, as the health and survival of livestock is more closely related to the livelihood and survival of the managers of subsistence systems, there is a greater vested interest for the manager to ensure the welfare of their stock. Animal welfare and profitability will be more closely aligned in these systems. Secondly, the close relationship between herder and livestock in these systems means that any loss of welfare may be detected more quickly than in systems where the animals are viewed less frequently. Thirdly, this close affiliation may reduce the fear aspects of some handling practices in comparison to that seen in other systems. Because sheep are taken to pasture and often to water during the day and herded at night either in or close to the main dwelling, they are frequently in close contact with human beings. This should mean that the contact between humans and sheep is, from the animals' point of view, relatively benign. Several studies suggest that sheep are known by name and respond to herders' calls (Adugna 1992; Geerlings 2001) indicating a close relationship between human and animal. The small flock size would also suggest that the individual behavioural characteristics of each animal are well known to the farmer and that management can be adjusted accordingly.

Despite these benefits of a close human-animal relationship, there are some management practices that can impact on welfare. Castration of male lambs is common in these systems, which causes pain and suffering as in other systems (see Chapter 8). The main reason for this is to obtain a higher price for the animals at sale due to fatness. In Africa, rams are generally kept entire for some period of their lives to allow them to be used for breeding and then castrated at a later age. This is done at an age between the eruption of the second pair of incisors and the eruption of the fourth pair although no age is given (Agyemang et al. 1985) or at an average age of 2.4 years (Mekoya et al. 1999). The normal method is crushing the vas deferens using stones or by beating it between two sticks especially designed for the purpose (Adugna 1992). However, because of the difficulty of identifying the best breeding stock at a young age other management practices are also used to prevent indiscriminate breeding. For example, genital covers are used in parts of Western Asia, and in Africa and India a rope is tied around scrotum and penis to prevent erection although the ram is still able to urinate (Geerlings 2001). Males may also be kept separate from females and lambs at night. Management solutions to prevent random mating may be more humane than castration.

Sheep need to be marked as they are often grazed in co-operation with other herds. In Tibet, ear cuts are made when lambs are a few months old so that they can be identified. In Ethiopia and Sudan, marking sheep for identification is done by cutting pieces out of the ear with a knife or by marking with fire-heated metal on the ears, face, back of the head or forelegs (Adugna 1992). These practices, in common with the ear-tagging and notching carried out in other systems, are done without anaesthesia and cause pain. In addition to marks, as mentioned above, in many systems each animal will be named, regardless of the size of the flock.

Other sources of stress to sheep, similar to other systems, are the interactions involved in milking, shearing (which may occur as frequently as three times a year in the tropics) and slaughter. To some extent the stressfulness of these procedures may be mitigated by the good human animal relationship as discussed above. Not all tropical sheep are shorn, and little in the way of detail is available for those that are. One method of shearing involves restraint of the sheep by one person while a second person removes the wool using hand shears (Johannes Alemseged, personal communication). The sheep are shorn in the shed in which they are housed during the night. The fleece is removed from only the wool bearing portion of the body, so that sheep do not undergo the removal of the whole fleece as happens in other systems.

A particular form of stress to sheep on the Tibetan plateau is their use to carry salt. Male sheep are herded to the salt flats and then harnessed to saddlebags, which are loaded with salt. The sheep then carry this load continuously for up to a month before arriving back at the home pastures. Goldstein & Beall (1990) describe the struggle to harness the animals, suggesting that this is a stressful procedure. In addition, there are considerably impacts on the health and survival of animals used for portage in this way such that wealthy families tend not to risk their animals on the trip.

6.4.7.4 Lamb and Adult Mortality

Neonate and offspring survival is very important for the future of the herd and particular care is taken with very young animals, as illustrated above. However there is also a need to wean lambs early to take the milk for human consumption, which may impact on lamb welfare and survival.

In Ethiopia, perinatal lamb mortality in the Adal, Horro and Menz breeds appears to be relatively low, ranging from 4 to 20% (Wilson et al. 1985; Awgichew & Hailu 1986; Mukasa-Mugerawa et al. 1994; 2000), and can be improved by maternal nutritional supplementation (Mukasa-Mugerawa et al. 1994). However, pre-weaning mortality rates are considerably higher ranging from 17 to 32% (Mukasa-Mugerawa et al. 2002) and mortality by one year of age as high as 26–60% (Mukasa-Mugerawa et al. 2000; 2002). Similarly in the Sudan, data from El-Huda Sheep Research Station in central Sudan indicate mortality rates of 5.9% on the first day of life, 15.2% at 30 days, 28.5% at 120 days and 45.1% at one year. Other estimates of mortality rates are 47 and 41% in Shugor and Dubasi lambs (Sulieman et al. 1978) and of 41, 43 and 30% to 120 days in Shugor, Dubasi and Watish lambs respectively (Sulieman & Wilson 1989). In addition, high abortion rates are reported in Sudan and Rajasthan. Other figures suggest pre-weaning death rates are 35.3% under sedentary pastoralism and 25.1% under migratory pastoralism (Wilson et al. 1985). In Sudan adult mortality rates are 10.4% per year in females and 12.8% in males (Sulieman et al. 1990). Mortality rates in lactating adult ewes under maintenance feeding are 11%, but increase to 23% during the dry season (Cook & Fadlalla 1987). Much of this increase is attributed to severe phosphorus deficiency caused by low phosphorus content of the pasture and exacerbated by low pasture availability. Thus it appears that, although lambs may have good perinatal survival, in comparison to other systems e.g. temperate extensive farming, they are considerably more susceptible to morbidity and mortality from disease before reaching adulthood. The apparent improved survival in migratory pastoralism may be attributable to better pasture and water availability, or to reduced disease risks in comparison to sedentary flocks.

6.4.7.5 Sea transport

In the lowland plains of south-east Ethiopia, there is a growing trade in live sheep to both Saudi Arabia and Yemen, and live transport of sheep also occurs from India to the Middle East. In 1996, the estimated number of sheep exported from Ethiopia was 1.5 million (Anon 1998) and 48,000 sheep were exported from Sudan in 2003. Sheep are generally taken to assembly points and trucked to sea ports in neighbouring countries such as Somalia (Anon 1998). Occasionally, Saudi Arabia rejects entire shipments of animals citing poor health inspection in Ethiopia as the reason (Anon 1998). The Middle East is a relatively short sailing distance from the sea ports in neighbouring countries, so the sheep are not on board for protracted periods as is the case with transport from Australia (see Section 6.2.1.1), although sheep will often be transported in smaller boats or dhows which may not be designed for sheep transport. There is little information about mortalities, and other welfare issues with sheep sea transport from these countries, but many of the same issues discussed above in Section 6.2.1.1. are relevant.

6.5 Conclusions

Sheep are kept in a vast number of different systems, and for a variety of different reasons, and we have been able to discuss only a few of these systems in this Chapter. For the most part sheep are maintained outdoors for large parts of their lives and fed primarily on pasture, whether this is in large, managed but semi-wild flocks or shepherded and in close contact with humans on a daily basis. Although there are welfare issues common to all sheep systems, for example: threats from disease, management practices such as castration and early weaning, there are also welfare challenges that are peculiar to particular sheep systems. These include such diverse practices as salt portage in Tibet, mulesing of Merino animals in Australia and interactions with the milking machine in dairy ewes.

All the systems mentioned here have both advantages and disadvantages from the perspective of sheep welfare. On the whole, however, it appears that, properly managed, pasture-based systems can be generally good for sheep welfare allowing the expression of many aspects of the natural behaviour of the sheep. Potentially the biggest threat to sheep welfare in these systems comes from interactions with humans where the animals have little experience of positive interactions with man, and contacts are generally aversive and stressful. The exceptions to this are traditional systems, where animals have considerable social contact with their human carers, and dairy systems, where lambs have the opportunity to form social bonds with humans early in life. The increasing intensification of dairy production in sheep will potentially increase welfare problems, as has been the case for other livestock species. However, outdoor production systems have the potential to be as profitable as indoor sheep dairy systems, and the close association between sheep welfare and milk quality and yield suggest that management can be optimised to benefit both the sheep and the producer. In general, the benefits in reducing stress of the close human contact between sheep and man in more traditional systems could be developed in the extensive systems to improve sheep welfare. Likewise, these more traditional systems could learn from the better health care practices of the extensive sheep systems described here.

A feature of sheep systems is the great diversity of sheep breeds managed worldwide. These breeds often have specific adaptations to local conditions that allow the effective exploitation of diverse habitats. Welfare problems can arise if an animal that is highly productive in one environment is transported to another area where they may be poorly adapted, or have little resistance to local disease conditions. Genetic selection can exploit the adaptations that sheep possess to improve welfare in a particular area or under a particular style of management, as has been the case with the development of 'easycare' breeds in New Zealand, or by selecting stock with long legs in migratory management systems. Extension of this concept could be applied to the developing areas of sheep management such as the selection of dairy animals with calm or relaxed temperaments for ease of milking. By working with the physical adaptations and behaviour patterns of the sheep all the sheep management systems described here have the capacity to provide good welfare for the animals kept within them, provided that adequate resources (e.g. supplementary feed, labour) can be given when required.

Acknowledgments We acknowledge the expertise and persistence of Vicki Drew and Lisa Howarth (both from NSW Department of Primary Industries, Trangie, Australia) in obtaining the literature relating to sheep production in Ethiopia and Sudan, some of which was quite difficult to obtain.

References

Adugna, G. (1992). Black Head Ogaden sheep under traditional management practices in south-eastern Ethiopia. Proceedings of the First Biennial Conference of the African Small Ruminant Research Network, International Laboratory for Research on Animal Diseases (ILRAD), Nairobi, Kenya, 10–14 December 1990. B. Rey, S.H.B. Lebbie & L. Reynolds (Eds.), pp. 133–143.

Agyemang, K., Akalework, N., Voorthuizen, A. & Anderson, F.M. (1985). A rapid survey of sheep production in the traditional sector of Debre Berhan, Ethiopian highlands. In R.T. Wilson and D. Bourzat (Eds.). *Small Ruminants in African Agriculture.* ILCA, Addis Ababa, Ethiopia, pp. 174–185.

Albenzio, M., Santillo, A., Caroprese, M., Marino, R., Centoducati, P. & Sevi, A. (2002) Effect of different ventilation regimens on ewes' milk and Canestrato Pugliese cheese quality in summer. *Journal of Dairy Research* 72: 447–455.

Alemayahu, Z. & Fletcher, I. (1991). Small ruminant productivity in the central Ethiopian mixed farming systems. Proceedings of the fourth National Livestock Improvement Conference, 13–15th November 1991, Addis Ababa, Ethiopia, pp. 141–147.

Al-Majali, A. M. & Jawabreh, S. (2003) Period prevalence and etiology of subclinical mastitis in Awassi sheep in southern Jordan. *Small Ruminant Research* 47: 243–248.

Anon (1977). Annual Report, Animal Health Group, Department of Agriculture, Victoria, p. 45.

Anon (1998). Livelihoods and food security in Ethiopia's Somali region. Famine Early Warning System Special Report 98-1. www.fews.net (accessed August 2007).

Anon. (2000). Sudan: increasing sheep production. Quarterly Newsletter of the Association for Strengthening Agricultural Research in Eastern and Central Africa. www.asareca.org/agriforum/articles17 (accessed 7 April, 2004).

Arnold, G.W. & Pahl, P.J. (1974). Some aspects of social behaviour in sheep. *Animal Behaviour* 22: 592–600.

Astruc, J.M., F. Barillet, M. Fioretti, M., Gabina, D., Gootwine, E., Mavrogenis, A.P., Romberg, F.J, Sanna, S.R. & Stefanake, E. (2002). Report of the working group on milk recording of sheep. International Congress on Animal Reproduction, Hungary, 2002.

Atti, N., Rouissi, H. & Othman, M.H. (2006) Milk production, milk fatty acid composition and conjugated linoleic acid (CLA) content in dairy ewes raised on a feedlot or grazing pasture. *Livestock Science* 104: 121–127.

Awgichew, K. & Hailu, G. (1986). Review of small ruminant research and development activities in Ethiopia. Proceedings of the Workshop on the Improvement of Small Ruminants in Eastern and Southern Africa, 18th–22nd August 1986, Nairobi, Kenya, Organization of African Unity, pp. 269–274.

Baldock, N.M. & Sibly, R.M. (1990) Effects of handling and transportation on the heart rate and behaviour of sheep. *Applied Animal Behaviour Science* 28: 15–39.

Barillet, F. & Bocquier, F. (1993). Le contexte de production des ovins laitières en France: principaux objectifs de recherche – développement et conditions de leur mise en oeuvre. *Productions Animales* 1: 17–24.

Barillet, F., Rupp, R., Mignon-Grateau, S., Astruc, J.M. & Jacquin, M. (2001) Genetic analysis for mastitis resistance and milk somatic cell score in French Lacaune dairy sheep. *Genetics Selection Evolution* 33: 397–415.

Batavani, R.A., Mortaz, E., Falahian, K. & Dawoodi, M.A. (2003) Study on frequency, etiology and some enzymatic activities of subclinical ovine mastitis in Urmia, Iran. *Small Ruminant Research* 50: 45–50.

Bergonier, D., De Cremoux, R., Rupp, R., Lagriffoul, G. & Berthelot, X. (2003) Mastitis of dairy small ruminants. *Veterinary Research* 34: 689–716.

Bekele, T., Woldeab, T., Lahlou-Kassi, A. & Sherington, J. (1992). Factors affecting morbidity and mortality on-farm and on-station in the Ethiopian highland sheep. *Acta Tropica* 52: 99–109.

Benyoucef, M.T. & Ayachi, A. (1991). Mesure de la production laitière de brebis Hamra durant les phases d'allaitement et de traite. *Annales de Zootechnie* 40: 1–7.

Binns, S.H., Cox, I.J., Rizvi, S. & Green, L.E. (2002) Risk factors for lamb mortality on UK sheep farms. *Preventative Veterinary Medicine* 52: 287–303.

Bizelis, J.A., Charismiadou, M.A. & Rogdakis, E. (2000) Metabolic changes during the perinatal period in dairy sheep in relation to level of nutrition and breed. *Journal of Animal Physiology and Animal Nutrition* 84: 73–84.

Bøe, K. (1990) Thermoregulatory behaviour of sheep housed in insulated and uninsulated buildings. *Applied Animal Behaviour Science* 27: 243–253.

Boivin, X., Tournadre, H. & Le Neindre, P. (2000) Hand-feeding and gentling influence early-weaned lambs' attachment responses to their stockperson. *Journal of Animal Science* 78: 879–884.

Bradford, G.E. & Meyer, H.H. (1986) Economic evaluation of breeding objectives for sheep and goats: practical considerations and examples. Proceedings of the 3rd World Congress on Genetics applied to Livestock Production. IX. Breeding programs for dairy and beef cattle, water buffalo, sheep and goats, Lincoln, Nebraska, USA. pp. 479–492.

Buckman, P.G. (1983). Losses of Sheep After Shearing due to Adverse Weather. Proceedings of a seminar by the Australian Society of Animal Production (Western Australian Branch). Eds S.K. Baker, H.M. Chapman and I.H. Williams pp. 11–19.

Bush, R.D., Toribio, J.A. & Windsor, P.A. (2006) The impact of malnutrition and other causes of losses of adult sheep in 12 flocks during drought. *Australian Veterinary Journal* 84: 254–260.

Casamassima, D., Sevi, A., Palazzo, M., Ramacciato, R., Colella, G.E. & Bellitti, A. (2001) Effects of two different housing systems on behaviour, physiology and milk yield of Comisana ewes. *Small Ruminant Research* 41: 151–161.

Charismiadou, M.A., Bizelis, J.A. & Rogdakis, E. (2000) Metabolic changes during the perinatal period in dairy sheep in relation to level of nutrition and breed. I. Late pregnancy. *Journal of Animal Physiology and Animal Nutrition* 84: 61–72

Cook, C.J. (1996) Basal and stress response cortisol levels and stress avoidance learning in sheep (*Ovis ovis*). *New Zealand Veterinary Journal* 44: 162–163.

Cook, R.H. & Fadlalla, B. (1987) Seasonal variations in plasma phosphorus levels of transhumant sheep in Kordofan, Sudan. *Tropical Animal Health and Production* 19: 57–62.

Cooper, M.McG. & Thomas, R.J. (1991) *Profitable Sheep Farming.* Farming Press Books, UK.

Cunningham, J.M.M. & Groves, C.R. (1985) The hills and uplands of the UK. In *Hill and Upland Livestock Production.* T.J. Maxwell & R.G. Gunn (Eds.). BSAP Occasional Publication No 10, pp. 1–8.

Dalton, D.C., Knight, T.W. & Johnson, D.L. (1980) Lamb survival in sheep breeds on New Zealand hill country. *New Zealand Journal of Agricultural Research* 23: 167–173.

Dimitrov, I., Djorbineva, M., Sotirov, L. & Tanchev, S. (2005) Influence of fearfulness on lysozyme and complement concentrations in dairy sheep. *Revue de Medicine Veterinaire* 156: 445–448.

Eadie, J. (1985) The future contribution of the hills and uplands to agricultural output. In *Hill and Upland Livestock Production.* T.J. Maxwell & R.G. Gunn (Eds.). British Society for Animal Production Occasional Publication 10, pp. 123–130.

El-Saied, U.M., de la Fuente, L.F. & San Primitivo, F. (2006) Phenotypic study on total and partial lifetime traits for dairy ewes. *Journal of Dairy Research* 73: 101–108.

Evans, I. & Joshua, E. (2003). Mulesing: accredited contractors June 2003. Agnote DAI-262, NSW Agriculture.

Fadlalla, B. & Ahmed, F.A. (1997). Sudan country paper. In *'Global Agenda for Livestock Research'*. (eds. E.F. Thomson, R. von Kaufamann, H. LiPun, T. Treacher & H. van Houten) Proceedings of a Consultation on Setting Livestock Research Priorities in West Asia and North Africa (WANA) Region, Aleppo, Syria, 12–16th November 1997, pp. 77–83.

Faerevik, G., Andersen, I.L. & Bøe, K. (2005) Preferences of sheep for different types of pen flooring. *Applied Animal Behaviour Science* 90: 265–276.

FAO (2001) Pastoralism in the new millennium. FAO Animal Production and Health paper 150. http://www.fao.org/docrep/005/Y2647E/y2647e00.htm

Fell, L.R. & Shutt, D.A. (1988). Salivary cortisol and behavioural indicators of stress in sheep. *Proceedings of the Australian Society of Animal Production* 17: 186–189.

Fell, L.R. & Shutt, D.A. (1989). Behavioural and cortisol responses to acute surgical stress in sheep. *Applied Animal Behaviour Science* 22: 283–294.

Finocchiaro, R., van Kaam, J.B.C.H.M., Portolano, B. & Misztal, I. (2005) Effect of heat stress on production of Mediterranean dairy sheep. *Journal of Dairy Science* 88: 1855–1864.

Fisher, M. (2003) New Zealand farmer narratives of the benefits of reduced human interventions during lambing in extensive farming systems. *Journal of Agricultural and Environmental Ethics* 16: 77–90.

Fisher, M. & Mellor, D.J. (2002) The welfare implications of shepherding during lambing in extensive New Zealand farming systems. *Animal Welfare* 11: 157–170.

Folman, Y., Volcani, R. & Eyal, E. (1966). Mother-offspring relationships in Awassi sheep. I. The effect of different suckling regimes and time of weaning on the lactation curve and milk yield in dairy flocks. *Journal of Agricultural Science, Cambridge* 67: 359–368.

Gargouri, A., Caja, G., Such, X., Ferret, A., Casals, R. & Peris, S. (1993) Evaluation of a mixed system of milking and suckling in Manchega dairy ewes. Proceedings of the 5th International Symposium on Machine-Milking of Small ruminants. *Hungarian Journal of Animal Production* (Suppl. 1): 484–499.

Geerlings, E. (2001) Raika sheep husbandry and ethnoveterinary knowledge. MSc thesis, Wageningen University, Netherlands.

Geerlings, E. (2004) The black sheep of Rajasthan. *Seedling* October 2004, pp. 11–16.

Goddard, P.J., Fawcett, A.R., Macdonald, A.J. & Reid, H.W. (2000) The behavioural, physiological and immunological responses of lambs from two rearing systems and two genotypes to exposure to humans. *Applied Animal Behaviour Science* 66: 305–321.

Goldstein, M.C. & Beall, C.M. (1990) *Nomads of Western Tibet: The Survival of a Way of Life*. London, Serindia.

Goldstein, M.C., Beall, C.M. & Cincotta, R.P. (1990) Traditional nomadic pastoralism and ecologial conservation on Tibet's northern plateau. *National Geographic Research* 6: 139–156.

Gootwine, E. & Pollott, G.E. (2000) Factors affecting milk production in improved Awassi dairy ewes. *Animal Science* 71: 607–615

Gorden, G.D.H. & Cockram, M.S. (1990) A comparison of wood slats and straw bedding on the behaviour of sheep. *Animal Welfare* 4: 131–134.

Granger, W. (1944). Selection of breeding ewes. Dependence of practicable degree of selection of young ewes upon vital ststistics. *Australian Veterinary Journal* 20: 253–260.

Guillouet, P. & Barillet, F. (1991). La oveja Lacaune y la produccion de leche en la region de Roquefort. *Ovis, Tratado de Patologia y Produccion Ovina* 14: 29–49.

Harbi, M.S.M.A. (1992). Role of livestock production in the rural economy in the Sudan. In 'Livestock production in rural development: Development of livestock policies'. Proceedings of the international workshop. Eds C.P. Gootjes, G. den Hertog, R. Jong and A.J. Nell, A.J. International Agricultural Centre, Wageningen (The Netherlands) pp. 256–264.

Harlow, H.J., Thorne, E.T., Williams, E.S., Belden, E.L. & Gern, W.A. (1987) Adrenal responsiveness in domestic sheep (*Ovis aries*) to acute and chronic stressors as predicted by remote monitoring of cardiac frequency. *Canadian Journal of Zoology* 65: 2021–2027.

Harris, D.J. & Nowara, G. (1995). The characteristics and causes of sheep losses in the Victorian Mallee. *Australian Veterinary Journal* 72: 331–340.

Higgs, A.R.B., Norris, R.T. & Richards, R.B. (1991) Season, age and adiposity influence death rates in sheep exported live by sea. *Australian Journal of Agricultural Research* 42: 205–214.

Higgs, A.R.B. Norris, R.T., Love, R.A. & Norman, G.J. (1999) Mortality of sheep exported by sea: Evidence of similarity by farm group and of regional differences. *Australian Veterinary Journal* 77: 729–733.

Hight, G.K. (1978/9) Agricultural research in the New Zealand Ministry of Agriculture and Fisheries. Annual Report of the Research Division, pp. 134–136.
Hinch, G.N., Lynch, J.J., Elwin, R.L. & Green, G.C. (1990). Long-term associations between Merino ewes and their offspring. *Applied Animal Behaviour Science* 27: 93–103.
Hodgson, J. & Grant, S.A. (1985) The grazing ecology of hill and upland swards. In *Hill and Upland Livestock Production.* T.J. Maxwell & R.G. Gunn (Ed.). BSAP Occasional Publication No10, pp 77–84.
Holden, S. & Shiferaw, B. (2000). Development paths and policies for sustainable land management in Andit Tid, North Shewa: an exploration. In *Policies for Sustainable Land Management in the Highlands of Ethiopia.* M.A. Jabbar, J. Pender & S.K. Ehui (Eds.). Socio-Economics and Policy Research Working Paper Number 30. pp. 45–48. International Livestock Research Institute, Nairobi, Kenya.
Holm Glass, M. & Jacob, R.H, (1992). Losses of sheep following adverse weather after shearing. *Australian Veterinary Journal* 69: 142–143.
Hoste, H., Rulie, A.C., Prevot, F., Bergeaud, J.P., Grisez, C., De La Farge, F., Jacquiet, P. & Dorchies, P. (2006) Differences in receptivity to gastrointestinal infections with nematodes in dairy ewes: Influence of age and of the level of milk production. *Small Ruminant Research 63*: 150–155.
Houdijk, J.G.M., Jessop, N.S. & Kyriazakis, I. (2001) Nutrient partitioning between reproductive and immune functions in animals. *Proceedings of the Nutrition Society* 60: 515–525.
Hunter, R.F. (1962) Hill sheep and their pasture: A study of sheep grazing in south-east Scotland. *Journal of Ecology* 50: 651–680.
Hunter, R.F. & Milner, C. (1963) The behaviour of individual, related and groups of South Country Cheviot hill sheep. *Animal Behaviour* 11: 507–513.
Hutchinson, J.C.D. (1968) Deaths of sheep after shearing. *Australian Journal of Experimental Agriculture and Animal Husbandry* 8: 393–400.
Iratxe, L., Ramon, A.J., Barandika, J., Amorena, B., De Andres, D., Lujan, L., Badiola, J. & Berriatua, E. (2006) Extensive rearing hinders Maedi-Visna (MV) infection in sheep. *Veterinary Research* 37: 767–778.
Ivanov, I.D. (1998). Behavioural responses and assessment of behavioural type in milking ewes. PhD Thesis, Inst. of Cattle and Sheep Husbandry – Stara Zagora, Bulgaria.
Ivanov, I.D. & Djhorbineva, M.K. (1999a). Behavioural reactions and assessment of temperament of dairy sheep. Proceedings of the 6th Intern. Symp. on the milking of Small Ruminants, Athens, Greece, Sept. 1998; European Association of Animal Production Publication N 95, 1999, 309–312.
Ivanov, I.D. & Djhorbineva, M.K. (1999b). Assessment of fear reactions in dairy sheep. Proceedings of the 6th Intern. Symp. on the milking of Small Ruminants, Athens, Greece, Sept. 1998; European Association of Animal Production Publication N 95, 1999, 312–315.
Ivanov, I.D., Djhorbineva, M.K. & Varlyakov, I.S. (1996). Behavioural reactions of ewes during machine milking. *Bulgarian Journal of Agricultural Science* 2: 399–403.
Jagusch, K.T., Gumbrell, R.C., Mobley, M.C. & Jay, N.P. (1977) Effect on growth of early weaning lambs on to pure stands of lucerne. *New Zealand Journal of Experimental Agriculture* 5: 15–18.
Johnson, D.L. (1975) The state of pastoral nomadism in the Sahelian Zone. In *MAB (Man and the Biosphere) Technical Notes. The Sahel: Ecological Approaches to Land Use,* Paris, UNESCO, pp. 75–87.
Jordan, D.J. & LeFeuvre, A.S. (1989). The extent and cause of perinatal lamb mortality in three flocks of Merino sheep. *Australian Veterinary Journal* 66: 198–201.
Kilgour, R.J. (1992). Lambing potential and mortality in Merino sheep as ascertained by ultrasonography. *Australian Journal of Experimental Agriculture* 32: 311–313.
Kilgour, R. & de Langen, H. (1970). Stress in sheep resulting from management practices. *Proceedings of the New Zealand Society of Animal Production* 30: 65–76.
Kilgour, R. & de Langen, H. (1980) Neonatal behaviour in 'easycare' sheep. In *Reviews in Rural Science: Behaviour in Relation to Reproduction, Management and Welfare of Farm Animals.*

M. Wodzicka-Tomaszewska, T.N. Edey & J.J. Lynch (Eds.).Vol. 4, pp. 117–118. University of New England Press, Armidale, Australia.

Kind, K.L., Moore, D.V.M. & Davies, M.J. (2006) Diet around conception and during pregnancy – effects on fetal and neonatal outcomes. *Reproductive Biomedicine Online,* 12: 532–541.

Kleemann, D.O., Walker, S.K., Grosser, T.I., Grimson, R.J. & Smith, D.H. (1990). Reproductive loss in in South Australian commercial Merino flocks. *Proceedings of the Australian Society of Animal Production* 18: 504 (Abstract).

Knight, T.W., Oldham, C.M., Smith, J.F. & Lindsay, D.R. (1975). Studies in ovine infertility in agricultural regions in Western Australia. *Australian Journal of Experimental Agriculture and Animal Husbandry* 15: 183–188

Knight, T.W., Lynch, P.R., Hall, D.R.H. & Hockey, H.U.P. (1988) Identification of factors contributing to the improved lamb survival in Marshall Romney sheep. *New Zealand Journal of Agricultural Research* 31: 259–271.

Knight, T.W., McMillan, W.H., Hall, D.R.H., Kilgour, R. & Hockey, H.U.P. (1989) Effect of slope of lambing site on lambs slipping and lamb mortality. *New Zealand Journal of Agricultural Research* 32: 199–206.

Komesaroff, P.A., Esler, M., Clarke, I.J., Fullerton, M.J. & Funder, J.W. (1998) Effects of estrogen and estrous cycle on glucocorticoid and catecholamine responses to stress in sheep. *American Journal of Physiology* 275: E671–E678.

Labussière, J. (1988) Review of the physiological and anatomical factors influencing the milking ability of ewes and the organisation of milking. *Livestock Production Science* 18: 253–274.

Labussière, J., Combaud, J.F. & Douaire, G. (1979). Effect of weaning conditions on the milk production and behaviour at milking of Préalpes du Sud ewes. Proceedings of the Deuxième Symposium sur la Traite Méchanique des Petits Ruminants, Alghero, Italy, 22nd–27th May 1978 pp. 146–163.

Labussière, J., Combaud, J.F. & Petrequin, P. (1974). Influence de la fréquence des traites et des tétées sur la production laitière des brebis Préalpes du Sud. *Annales de Zootechnie* 23: 445–457.

Labussière, J. & Ricordeau, G. (1970). Aptitude à la traite méchanique des brebis de race Préalpes du Sud Croisées Frisons x Préalpes. Etude aux différents stades de lactation. *Annales de Zootechnie* 19: 159–190.

Las Heras, A., Dominguez, L. & Fernandez-Garayzabal, J.F. (1999) Prevalence and aetiology of subclinical mastitis in dairy ewes of the Madrid region. *Small Ruminant Research* 32: 21–29.

Lawrence, A.B. (1990) Mother-daughter and peer relationships of Scottish hill sheep. *Animal Behaviour* 39: 481–486.

Lee, C. & Fisher, A.D. (2007) Welfare consequences of mulesing of sheep. *Australian Veterinary Journal* 85: 89–93.

Leitner, G., Chaffer, M., Zamir, S., Mor, T., Glickman, A., Winkler, M., Weisblit, L. & Saran, A. (2001) Udder disease etiology, milk somatic cell count and NAGase activity in Israeli Assaf sheep throughout lactation. *Small Ruminant Research* 39: 107–112.

Luff, A.F. (1980). A service for all seasons. Final report, Australian Wool Corporation Priject K/1/1051.

Mackereth, V. (1979) How 'Marshall' Romneys began. *New Zealand Farmer* 11 January, 1979, pp. 24–26.

Marnet, P.G. (1997). Ewe management for improved milk yield and quality. Proceedings of the 3rd Great Lakes Dairy Sheep Symposium, 4th April, 1997, USA, 5–11.

Marnet P.G. & Negrao, J.A. (1999). The effect of a mixed-management weaning system on the release of oxytocin, prolactin and cortisol in ewes during suckling and machine milking. Proceedings of the 6th International Symposium on the milking of Small Ruminants, Athens, Greece, Sept. 1998. European Association of Animal Production Publication N 95, 1999.

Marnet, P.G., Negrao, J.A. & Labussière, J. (1998). Oxytocin release and milk ejection parameters during milkin of dairy ewes in and out of natural season of lactation. *Small Ruminant Research* 28: 183–191.

Maxwell T.J. (1994). Future of animal production n hill and upland areas. In *Livestock Production and Land Use in Hills and Uplands*. T.L.J. Lawrence, D.S. Porter & P. Rowlinson (Eds.). British Society of Animal Production Occasional Publication No. 18, pp. 75–84.

Mayer, H., Weber, F. & Segessemann, F. (1989) Oxytocin release and milking characteristics of Ostfreisian and Lacaune dairy sheep. Proceedings of the 4th International Symposium on Machine-Miling of Small Ruminants, Tel-Aviv, Israel.

McCutcheon, S.N., Holmes, C.W. & McDonald, M.F. (1981) The starvation-exposure syndrome and neonatal lamb mortality: A review. *Proceedings of the New Zealand Society of Animal Production* 41: 209–217.

McClelland T.H., Armstrong R.H., Thompson J.R. & Powell T.L. (1985) Sheep Production Systems in the Uplands. In *Hill and Upland Livestock Production.* T.J. Maxwell & R.G. Gunn (Eds.). BSAP Occasional Publication No. 10, pp. 85–94.

McKusick, B.C., Thomas, D.L. & Berger, Y.M. (2001) Effects of weaning system on commercial milk production and lamb growth of East Freisian dairy sheep. *Journal of Dairy Science* 84: 1660–1668.

McDougall, S., Pankley, W., Delaney, C., Barlow, J., Murdough, P.A. & Scruton, D. (2002) Prevalence and incidence of subclinical mastitis in goats and dairy ewes in Vermont, USA. *Small Ruminant Research* 46: 115–121.

Mears, G.J. & Brown, F.A. (1997) Cortisol and β-endorphin responses to physical and psychological stressors in lambs. *Canadian Journal of Animal Science* 77: 689–694.

Mekoya, A., Yami, A. & Mariam, M.H. (1999). Management of traditional sheep production in Lallomamma Mider Woreda, North Shoa, Amhara region. Proceedings of the 7th Annual Conference of the Ethiopian Society of Animal Production (ESAP), Addis Ababa, Ethiopia, 26–27 May 1999.

Meyer, H.H. & Clarke, J.N. (1978) Genetic and environmental effects on incidences and causes of lamb mortality. *Proceedings of the New Zealand Society of Animal Production* 38: 181–184.

Miller, D. J. (1999) Nomads of the Tibetan plateau rangelands in Western China, Part two: Pastoral production practices. *Rangelands* 21: 16–19.

Miller, D.J. & Craig, S.R. (1997). *Rangelands and pastoral development in the Hindu Kush-Himalayas*. Proceedings of a Regional Experts Meeting, International Centre for Integrated Mountain Development (ICIMOD) Publications, Kathmandu, Nepal.

Mills, O. (1989). *Practical Sheep Dairying*. Thorsons Publishers Ltd., Wellingborough, Northamptonshire, England

Molina, E., Ferret, A., Caja, G., Calsamiglia, S., Such, X. & Gasa, J. (2001) Comparison of voluntary food intake, apparent digestibility, digesta kinetics and digestive tract content in Manchega and Lacaune dairy sheep in late pregnancy and early and mid lactation. *Animal Science* 72: 209–221.

Morris, C.A., Hickey, S.M. & Clarke, J.N. (2000) Genetic and environmental factors affecting lamb survival at birth and through to weaning. *New Zealand Journal of Agricultural Research* 43: 515–524.

Mukasa-Mugerawa, E. & Tekelye, B. (1988). The reproductive performance of Ethiopian highland sheep. *Animal Reproduction Science* 17: 95–102.

Mukasa-Mugerawa, E., Said, A.N., Lahlou-Kassi, A., Sherington, J. & Mutiga, E.R. (1994). Body weight as a risk factor for perinatal lamb mortality, and the effects of pregnant ewe supplementation and gestation weight gain in Ethiopian Menz sheep. *Preventive Veterinary Medicine* 19: 45–56.

Mukasa-Mugerawa, E., Lahlou-Kassi, A., Anindo, D., Rege, J.E.O., Tembley, S., Tibbo, M. & Baker, R.L. (2000). Between and within breed variation in lamb survival and the risk factors involved with major causes of mortality in indigenous Horro and Menz sheep in Ethiopia. *Small Ruminant Research* 37: 1–12.

Mukasa-Mugerawa, E., Anindo, D., Sovani, S., Lahlou-Kassi, A., Tembley, S., Rege, J.E.O., & Baker, R.L. (2002). Reproductive performance and productivity of Menz and Horro sheep lambing in the wet and dry seasons in the highlands of Ethiopia. *Small Ruminant Research* 45: 261–271.

Napolitano, F., Marino, V., de Rosa, G., Capparelli, R. & Bordi, A. (1995) Influence of artificial rearing on behavioural and immune response of lambs. *Applied Animal Behaviour Science* 45: 245–253.

National *Sheep Association (1995) Sheep UK, Our Natural Hidden Asset.* NSA, Malvern

Negrao, J.A. & Marnet, P.G. (2003) Cortisol, adrenalin, noradrenalin and oxytocin release and milk yield during first milkings in primiparous ewes. *Small Ruminant Research* 47: 69–75.

Norris, R.T. (2005) Transport of animals by sea. *Scientific and Technical Review of the OIE* 24: 673–681.

Norris, R.T., Richards, R.B. & Dunlop, R.H. (1989) An epidemiological study of sheep deaths before and during export by sea from Western Australia. *Australian Veterinary Journal* 66: 276–269.

Norris, R.T. & Richards, R.B. (1989) Deaths in sheep exported by sea from Western Australia – analysis of ship Master's reports. *Australian Veterinary Journal* 66: 97–102.

Orgeur, P., Bernard, S., Naciri, M., Nowak, R., Schaal, B. & Lévy, F. (1999) Psychobiological consequences of two different weaning methods in sheep. *Reproduction, Nutrition, Development* 39: 231–244.

Orgeur, P., Mavric, N., Yvore, P., Bernard, S., Nowak, R., Schaal, B. & Levy, F. (1998) Artificial weaning in sheep: consequences on behavioural, hormonal and immuno-pathological indicators of welfare. *Applied Animal Behaviour Science* 58: 87–103.

Osman, H.E. & Fadlalla, B. (1974). Effect of level of water intake on some aspects of digestion and nitrogen metabolism of the 'desert sheep' of the Sudan. *Journal of Agricultural Science, (Cambridge)* 82: 61–69.

Peel, L.J. (1986). History of the Australian Pastoral Industries. In *The Pastoral Industries of Australia.* Revised edition. G. Alexander & O.B. Williams (Eds.). Sydney University Press, Sydney, Australia pp. 41–57.

Pollott, G.E. & Gootwine, E. (2004) Reproductive performance and milk production of Assaf sheep in an intensive management system. *Journal of Dairy Science* 87: 3690–3703.

Rhind, S.M., Reid, H.W., McMillen, S.R. & Palmarini, G. (1998) The role of cortisol and β-endorphin in the response of the immune system to weaning in lambs. *Animal Science* 66: 397–402.

Richards, R.B., Norris, R.T., Dunlop, R.H. & McQuade, N.C. (1989) Causes of death of sheep exported live by sea. *Australian Veterinary Journal* 66: 33–38.

Richards, R.B., Hyder, M.H., Fry, J.M., Costa, N.D., Norris, R.T. & Higgs, A.R.B. (1991) Seasonal metabolic factors may be responsible for deaths in sheep exported by sea. *Australian Journal of Agricultural Research* 42: 215–226.

Rogan, I. & Grant, T. (2001) Dairy Sheep: I. Commercial scale management and breeding procedures, A report for Rural Industries Research and Development Corporation (RIRDC). RIRDC Publication No. 01/29, Barton, Australia.

Saratsis, P., Leontides, L., Tzora, A., Alexopoulos, C. & Fthenakis, G.C. (1998) Incidence risk and aetiology of mammary abnormalities in dry ewes in 10 flocks in Southern Greece. *Preventive Veterinary Medicine* 37: 173–183.

Scobie, D.R., Bray, A.R. & O'Connell, D. (1999) A breeding goal to improve the welfare of sheep. *Animal Welfare* 8: 391–406.

Scobie, D.R., O'Connell, D., Morris, C.A. & Hickey, S.M. (2007) A preliminary genetic analysis of breech and tail traits with the aim of improving the welfare of sheep. *Australian Journal of Agricultural Research* 58: 161–167.

Sevi, A., Massa, S., Annicchiaro, G., Dell'Aquila, S. & Muscio, A. (1999a) Effect of stocking density on ewes' milk yield, udder health and microenvironment. *Journal of Dairy Research* 66: 489–499.

Sevi, A., Napolitano, F., Casamassima, D., Annicchiaro, G., Quarantelli, T. & De Paola, R. (1999b) Effect of gradual transition from maternal to reconstituted milk on behavioural, endocrine and immune responses of lambs. *Applied Animal Behaviour Science* 64: 249–259.

Sevi, A., Taibi, L., Albenzio, M., Annicchiaro, G. & Muscio, A. (2001a) Airspace effects on the yield and quality of ewe milk. *Journal of Dairy Science* 84: 2632–2640.

Sevi, A., Taibi, L., Albenzio, M., Muscio, A., Dell'Aquila, S. & Napolitano, F. (2001b) Behavioral, adrenal, immune and productive responses of lactating ewes to regrouping and relocation. *Journal of Animal Science* 79: 1457–1465.

Sevi, A., Albenzio, M., Annicchiaro, G., Caroprese, M., Marino, R. & Taibi, L. (2002) Effects of ventilation regimen on the welfare and performance of lactation ewes in summer. *Journal of Animal Science* 80: 2349–2361.

Sevi, A., Taibi, L., Albenzio, M., Caroprese, M., Marino, R. & Muscio, A. (2003a) Ventilation effects on air quality and on the yield and quality of ewe milk in winter. *Journal of Dairy Science* 86: 3881–3890.

Sevi, A., Albenzio, M., Muscio, A., Casamassima, D. & Centoducati, P. (2003b) Effects of litter management on airborne particles in sheep houses and on the yield and quality of ewe milk. *Livestock Production Science* 81: 1–9.

Sevi, A., Caroprese, M., Annicchiarico, G., Albenzio, M., Taibi, L. & Muscio, A. (2003c) The effect of a gradual separation from the mother on later behavioural, immune and endocrine alterations in artificially reared lambs. *Applied Animal Behaviour Science* 83: 41–53.

Sevi, A., Albenzio, M., Marino, R., Santillo, A. & Muscio, A. (2004) Effects of lambing season and stage of lactation on ewe milk quality. *Small Ruminant Research* 51: 251–259.

Shafron, W., Martin, P. & Ashton, D. (2002). Profile of Australian wool producers 1997–98 to 2001. Australian Bureau of Agriculture and Resource Economics (ABARE) Research Report 02.7, Canberra, Australia.

Shutt, D.A., Fell, L.R., Connell, R., Bell, A.K., Wallace, C.A. & Smith, A. I. (1987). Stress-induced changes in plasma concentration of immunoreactive β-endorphin and cortisol in response to routine surgical procedures in lambs. *Australian Journal of Biological Sciences* 40: 97–103.

Shutt, D.A., Fell, L.R., Connell, R. & Bell A.K. (1988a). Stress responses in lambs docked and castrated surgically or by the application of rubber rings. *Australian Veterinary Journal* 65: 5–7.

Shutt, D.A., Smith, A.I., Wallace, C.A., Connell, R. & Fell, L.R. (1988b). Effect of myiaisis and acute restraint stress on plasma levels of immunoreactive ß-endorphin, adrenocorticotrophin (ACTH) and cortisol in sheep. *Australian Journal of Biological Sciences* 41: 297–301.

Sinapis, E., Diamantopoulos, K., Abas, Z. & Vlachos, I. (2006) Effect of vacuum level on milking efficiency, somatic cell counts (SCC) and teat end wall thickness in ewes of Greek mountain Boutsiko breed. *Livestock Science* 104: 128–134.

Singh, C. & Köhler-Rollefson, I. (2005) Sheep pastoralism in Rajasthan: still a viable livelihood option? Report on a State Level Workshop, Sadri, Rajasthan, India. The LIFE Initiative (Local Livestock for Empowerment of Rural People).

Sisay, Z. (1999). Water economy and its implication on pastoral livestock production. In 'Challenges and Opportunities for Research and Development in Regional and Agro-Pastoral Ethiopia'. Proceedings of the National Workshop, 16–18 December 1998, Mekelle University College, Mekelle Ethiopia, pp. 77–88.

Smith, I.D. (1962). Reproductive wastage in a Merino flock in central western Queensland. *Australian Veterinary Journal* 38: 500–507.

Stear, M.J., Bishop, S.C., Mallard, B.A. & Raadsma, H. (2001) The sustainability, feasibility and desirability of breeding livestock for disease resistance. *Research in Veterinary Science* 71, 1–7.

Stephens, D.B. & Baldwin, B.A. (1971) Observations of the behaviour of groups of artificially reared lambs. *Research in Veterinary Science* 12: 219–224.

Sulieman, A.H (1986). The small ruminants of the Sudan. Proceedings of the Workshop on the Improvement of Small Ruminants in Eastern and Southern Africa, 18th–22nd August 1986, Nairobi, Kenya, pp. 333–342. Organization of African Unity, Addis Ababa Ethiopia.

Sulieman, A.H., El-Amin, F.M. & Osman, A.H. (1978). Reproductive performance of Sudan indigenous sheep under irrigated Gezira conditions. *World Review of Animal Production* 14: 71–79.

Sulieman, A.H., Sayers, A.R. & Wilson, R.T. (1990). Evaluation of Shugor, Dubasi, and Watish subtypes of Sudan desert sheep at the El-Huda National Sheep Research Station, Gezira Province, Sudan. ILCA Research Report No. 18, International Livestock Centre for Africa, Addis Ababa, Ethiopia.

Sulieman, A.H. & Wilson, R.T. (1989) Productivity of three sub-types of Sudan desert sheep at El Huda research station. In *African Small Ruminant Research and Development*. R.T. Wilson & M. Azeb (Eds.). International Livestock Centre for Africa, Addis Ababa, Ethiopia.

Tallet, C., Veissier, I. & Boivin, X. (2005) Human contact and feeding as rewards for the lamb's affinity to their stockperson. *Applied Animal Behaviour Science* 94: 59–73.

Taye, D. & Abebe, G. (2000). Socioeconomic aspect and husbandry practices of sheep kept in Awassa city. In 'The opportunities and Challenges of Enhancing Goat Production in East Africa.' Proceedings of a conference held at Debub University, Awassa, Ethiopia, 10–12th November 2000. Awassa College of Agriculture, Awassa, Ethiopia.

Tolera, A. (1998). Production situation and some productivity and physical characters of traditionally managed sheep and goats in Kochore district, southern Ethiopia. *Journal of Applied Animal Research* 13: 49–59.

Turner, H.N., Dolling, C.H.S. & Sheaffe, P.H.G. (1959). Vital statistics for an experimental flock of Merino sheep. 1. Death rates in adult sheep, in relation to method of selection, age, and sex. *Australian Journal of Agricultural Research* 10: 581–590.

Veissier, I., Le Neindre, P. & Trillat, G. (1989) The use of circadian behaviour to measure adaptation of calves to changes in their environment. *Applied Animal Behaviour Science* 22: 1–12.

Vipond J.E. & Gunn R.G. (1985) The management and performance of sheep in the hills and uplands. In *Hill and Upland Livestock Production.* T.J. Maxwell & R.G. Gunn (Eds.). British Society of Animal Production Occasional Publication No 10, pp. 49–54.

Waterhouse, A. & McClelland, T.H. (1987). Supplementation of Scottish Blackface gimmers during mating and mid-pregnancy. *Animal Production,* 44: 471.

Watson, D.L. (1991) Effect of weaning on antibody responses and nematode parasitism in Merino lambs. *Research in Veterinary Science,* 51: 128–132.

Wilson, R.T. (1986). Mixed species stocking and the vital statistics and demography of domestic animals in the arid and semi-arid zones of northern tropical Africa. In 'Rangelands: a resource under siege.' Proceedings of the second international rangeland congress. Eds P.J. Joss, P.W. Lynch, P.W. and O.B. Williams, O.B. pp 375–378, Australian Academy of Science, Canberra.

Wilson, R.T., Traoré, A., Peacock, C.P., Mack, S. & Agyemang, K. (1985). Early mortality of lambs in African traditional livestock production systems. *Veterinary Research Communications* 4: 295–301.

Chapter 7
Nutrition and the Welfare of Sheep

J.P. Hogan, C.J.C. Phillips, and S. Agenäs

Abstract Sheep are often kept on marginal lands and inadequate nutrition is therefore a common problem that affects their welfare. Their digestive system, which utilises a complex microflora in a large fermentation chamber to extract nutrients from low quality forages, is robust against short-term deprivation of energy and protein requirements. The ability of sheep to withstand short term food and water deprivation, such as occurs when they are transported long distances to slaughter, largely depends on their condition at the start. Animals with low levels of reserves or that have been stressed during handling and gathering pre-transport will be most susceptible to the stress of a long journey. Long term feed deficiencies are unlikely to cause major changes in stress-related behaviour because sheep evolved as prey animals, and effects of feed deficiency on welfare are usually assessed by the effects on body condition, or by changes in hormones and metabolites associated with body tissue reserves, such as leptin and creatinine. As well as undernutrition, sheep are also regularly exposed to situations where toxic substances may compromise their welfare. This includes toxic plants, but many are specific to local conditions and this chapter only describes some examples. Of growing concern is the exposure of sheep to toxic pollutants from industrial emissions, which have been mainly of interest because of their potential to contribute to human accumulation of these elements, but which are now known also to compromise the welfare of the sheep themselves.

Keywords Undernutrition · Food and water deprivation · Rumen · Transport · Toxic · Plants · Heavy metals · Welfare

7.1 Introduction

Sheep have evolved to utilise low quality roughage for food by fermenting it in their rumen, with the assistance of a complex and comprehensive microflora. Being adapted to utilise low quality feed, sheep are commonly kept in low input systems on

J.P. Hogan
Centre for Animal Welfare and Ethics, University of Queensland, Australia

marginal or climatically-challenged land, which means that they are often subjected to periods of low food availability, which characterise these fragile environments. Undernutrition is therefore a more common problem, or malnutrition when the diet becomes unbalanced, than for farm animals kept in intensive management systems. Undernutrition is covered in detail in this chapter, although the specificity of malnutrition to the region that sheep are kept in necessitates only brief coverage of this field, with more detail provided in the excellent book by Garland and Barr (1998). For details of specific mineral disorders caused by malnutrition, the reader is referred to Underwood and Suttle (1999).

Although it is delicately balanced, the enlarged foregut with its fermentation system gives sheep a certain resistance to short term food and water deprivation (FWD) caused by climatic variation. In addition, sheep that are kept in remote marginal lands are usually subjected to several days with restricted food and water access during transport to slaughter. This practice is of concern not just for the welfare issues, but also because of the risk of the rumen becoming colonised by zoonotic micro-organisms that could be transferred to humans post mortem.

In addition to the risk of insufficient or unbalanced food available for sheep on marginal land, they are also at risk of consuming plants containing or polluted by toxic agents. As well as numerous toxic plants, a problem that has developed with the recent industrialisation processes is the pollution of forages with heavy metals, particularly in the vicinity of smelters. The welfare concerns are often unclear because the focus so far has been on the potential to contribute to the accumulation of toxic element accumulation in humans (e.g. Prankel et al., 2005). However, it is increasingly clear that there are serious welfare consequences for the sheep themselves (Phillips and Prankel, 2007), and this is also considered in this chapter.

Overnutrition, leading to obesity, is a rare condition in sheep, confined to some highly intensive production systems and systems where dominant animals can take more than their share of high energy and protein feeds. It is therefore not considered further in this chapter.

7.2 The Effect of Short-Term Food and Water Deprivation in Sheep

7.2.1 Introduction

Sheep are regularly transported, sometimes for long distances from breeding properties to sale yards, or to farms to permit further growth or to abattoirs. The animals undergo feed and water deprivation (FWD) for some hours before loading onto transport, and frequently during the journey, to reduce the fouling of other animals and of transport vehicles with faeces and urine and to permit buyers to assess carcase weight more accurately. The effect of FWD on sheep is further complicated by the stressors associated with movement from familiar surroundings, loading into unfamiliar vehicles, travel, unloading and being mixed with unfamiliar and often aggressive companions. These stressors result in the release of cortisol from the adrenal cortex, and of adrenalin from the adrenal medulla, to provide the animal with additional energy, produced through the catabolism of muscle glycogen and of fat. Depleted levels of glycogen in muscle at the time of slaughter are associated with high ultimate pH levels because the lack of substrate prevents the production of lactic acid post-mortem and the meat is undesirably tough. The aim of this part of the Chapter is to examine the effects on sheep of FWD and factors that influence recovery when FWD is ended.

7.2.2 *Effects of FWD on Body Weight*

The most obvious effect of FWD is weight loss which occurs most rapidly in the first 24 hours but continues at a slower rate for another 36 or more hours. About 80% of weight loss is water (Cole, 1995), the remainder being faecal organic matter, minerals in urine and faeces and respired gases, carbon dioxide and methane. The respired gases represent an appreciable proportion of daily weight loss but 48 hours FWD reduces carbon dioxide output from (g/d) 709 to 267 and methane output from 18 to 1 (Blaxter and Graham, 1955). Apart from overall weight losses, weight changes occur in the reticulo-rumen, again most rapidly in the first 24 hours, with the loss of both water and particulate material but where subsequently the removal of weight as feed particles is masked by the inflow of saliva.

7.2.3 *Effects of FWD on Metabolism and Physiology*

The consequences of FWD affect the rumen and its microbes, (including those responsible for controlling infection by enteropathogenic bacteria), tissue metabolism, the maintenance of homeostasis, and the systems involved in muscle to produce tender meat.

7.2.3.1 Effect of FWD on the Rumen and its Microbes

The effect of FWD on the rumen and its microbes will depend on the weight and composition of digesta present in the rumen when fasting commences. Weston (1985) expressed the weight of rumen digesta (RD) relative to the weight of the animal free of rumen digesta (RFW), that is the RD/RFW ratio. He showed that RD/RFW (in g/kg) ranged from 100 with highly digestible feeds to 300 with very mature feeds. For example, the weight of rumen contents in a sheep with RFW of 60 kg could thus vary between 6 and 18 kg with forages of advancing maturity. There is a corresponding wide range of retention times of plant material in the rumen calculated as the ratio of dry matter in the rumen divided by dry matter intake (Thornton and Minson, 1973). Hence, in comparison with animals fed mature roughages, animals grazing lush pastures would have appreciably less feed residue in the rumen at the start of FWD and would probably remove that feed residue more rapidly. The difference could be even greater if animals begin a period of FWD before commencing the day's grazing (Hogan, 1964), when the weight of rumen digesta is probably minimal. For a particular roughage, RD/RFW is likely to increase as feed intake increases, for instance during lactation (Weston, 1989), and presumably in response to mild cold stress, during compensatory growth and following the correction of a mineral deficiency.

The abrupt cessation of the arrival of nutrients that occurs with FWD is likely to have its most immediate effect on those bacteria that ferment low molecular weight carbohydrates because such substrates are most readily exhausted. For instance, Coop (1949) used rumen liquor from a lucerne (*Medicago sativa*)-fed sheep undergoing FWD as inoculum in an *in vitro* system studying glucose fermentation. The time required for glucose concentration to fall by one half increased for inoculum removed at times of fasting of 0, 24, 48 and 72 hours from (h) 8 to 20, 38 and 50. Cellulolytic bacteria probably survive for longer periods in the rumen, as they adhere to or are closely associated with feed particles (Miron et al., 2001). This may explain why Fluharty et al. (1996) observed no differences in the rate of disappearance of feed from artificial fibre bags before and after fasting.

Rumen microbes represent the first line of defence not only against plant toxins but also against enterotoxic bacteria such as *Clostridium welchii (perfringens)* (Bullen et al., 1953) and *Salmonella Spp* and *Escherichia coli* (*E coli*) (Brownlie and Grau, 1967). The latter authors also showed that this protection was lost when feed intake was reduced or interrupted by fasting. It is possible that a decline in the production of volatile fatty acids is involved (Wallace et al., 1989). Currently, the emphasis has shifted to the Shiga toxin producing *E coli* serotype 0157: H7, the numbers of which also increase following the withdrawal of feed for 6–48 hours (Callaway et al., 2002: Vanselow et al., 2005).With cattle, the development of *E coli* seems to be most severe with animals fasted on removal from pasture (Gregory et al., 2000). There appear to be benefits to feeding high quality hay for a few days before fasting (Gregory et al., 2000; Jacobson et al., 2002) but the quantity involved and the mechanism through which the benefit is delivered have not been fully investigated.

When feed is again made available intake may not return to pre-fast levels for several days. Many factors apart from rumen microbial activity could be involved (Weston et al., 1989; Weston, 1995; Forbes, 2003). Certainly Cole (1991) found no improvement in feed intake following FWD when rumen contents from fasted sheep were replaced with those from fed sheep.

7.2.3.2 Homeostasis During FWD

Homeostasis, the preservation of such features of the physiological environment as pH, osmotic pressure (OP) and acid-base balance, is affected by changes in the concentration of ions, notably sodium, potassium, bicarbonate and phosphate through intra-and extra-cellular fluids (ICF and ECF, respectively). The most important of these is the sodium ion (Michell, 1995) which is largely confined to the ECF. However the concentration of sodium in the rumen is generally about 110 m eq./l

(Payne et al., 1970b) compared with 145 m eq./l in plasma and it has been suggested (Hogan et al., 2007) that sodium in the rumen represents about one quarter of the total in the animal. It is not clear to what extent the water and sodium in the rumen can be regarded as readily accessible components of ECF. Sodium transport from the rumen may be as little as 14 m eq./h (Warner and Stacy, 1965) which would not suggest a clear connection of the rumen with ECF. Nevertheless, the calculations of Coghlan et al. (1977) that only 15–20% of sodium drained from sheep, via fistula of the parotid salivary duct, was withdrawn from ECF suggest that sodium in the rumen is readily accessible. Certainly Parker et al. (2003) observed that steers maintained acid-base balance following 48 hours transport with FWD.

During FWD, plasma OP may increase and a rise of 1–2%, that is 3–6 m. osm/kg, is sufficient to induce thirst (Denton et al., 1999) acting through osmoreceptors in the hypothalamus via angiotensin II. Transported sheep undergoing FWD can exhibit such changes in plasma OP but may show little interest in drinking until after eating (Knowles et al., 1994). The reason is not clear but may reflect an inhibition by cortisol of the renin-angiotensin-aldosterone axis during and immediately after transport. Cortisol levels presumably fall in sheep that are sufficiently relaxed to begin eating. In addition, during eating, sheep may transfer to the rumen, via saliva, a volume of water equivalent to 10–20% of the water in extracellular fluid and plasma (Scott, 1975), which would be sufficient to induce thirst (Denton et al., 1999). Eating too causes the almost complete cessation of urine formation (Stacy and Brook, 1964) in response to the release of arginine vasopressin. There is possibly involvement also of the rennin-angiotensis-aldosterone system. Saliva production during eating results in the transfer to the rumen of sodium, normally present in mixed saliva at a concentration of approximately 105 m eq./l in cattle (Bailey and Balch, 1961). If the concentration in sheep saliva is similar, saliva production of 2 L/h would extract in one hour about one tenth of the sodium in ECF. If aldosterone is released during eating in FWD sheep, the level of salivary sodium could be reduced (McSweeney and Cross, 1992; Richter et al., 1998). There is clearly need for further information on the movement of sodium in the animal during and immediately after FWD.

7.2.3.3 The Effect of FWD on Dehydration, Stress and Meat Quality

This subject has recently been reviewed by Harper (1999) and Ferguson et al. (2001). In brief there are two major effects of FWD, the consequences of dehydration and of exposure to stressors arising from animal management associated with transport. Dehydration particularly affects lambs, especially those lambs who are still milk-fed at the time of transport (Jacob et al., 2006). Presumably the severity of dehydration depends on the extent to which the lamb has developed rumen fermentation, the amount and composition of feed residues in the rumen and the experience that the lamb has had in drinking water from a trough. Dehydration, which results in weight loss and hence in financial loss to the owner, can be largely reversed within a few hours of the intake of water (Wythes et al., 1980). Meat quality, generally judged by tenderness (Harper, 1999), is affected by the inability of muscle, post-mortem, to reduce pH to 5.7 or less (Shorthose, 1977). The change in pH comes, as indicated earlier, from the production of lactic acid from glycogen, but muscle glycogen is diminished during FWD plus transport by stressors, acting through the release of cortisol and adrenalin from the adrenal glands. The same situation causes a loss of nitrogen from the animal (Cole et al., 1986), but the extent to which muscle protein is depleted is not clear. Glycogen depletion and presumably any protein loss from muscle can be remedied by feeding and rest for animals for several days before slaughter (Ferguson et al., 2001).

7.2.4 Conclusion on FWD

The consequences of FWD plus transport are clearly affected by the physiological state and temperament of the sheep and lambs involved. FWD depletes the sheep's reserves of minerals and of muscle glycogen, and reduces the fermentative capacity of the rumen microbes. The extent of these effects and of the rate of recovery post-FWD depends on the nature of the diet and hence on the amount and composition of digesta in the rumen when FWD commences. Despite mild dehydration, the transported animals may not be thirsty until after they have eaten, because cortisol, released in response to stressors from transport suppresses signals of thirst arising from the hypothalamus. Further research is needed on the management practices pre-fasting that will minimize the effects of FWD and transport on sheep and lambs.

7.3 Long Term Energy and Protein Undernutrition

7.3.1 Introduction

In a situation where the daily feed intake fails to provide sufficient energy or protein over a longer period of time than the FWD described above, the animal shows two principal responses. One is to continue the nutrient mobilisation from body tissues initiated during the first days of food deprivation. This is achieved via catabolic endocrine signals that stimulate mobilisation of glucose from glycogen stores, fatty acids from adipose tissue and amino acids from muscle tissue. The result is weight loss and loss in body condition. The other principal response is to decrease energy and protein output by adjusting production levels to the nutrient availability, leading to decreased growth and decreased or ceased milk production. Sexually mature animals can respond by reduced sexual function, primarily in the form of absent oestrus and aborted or resorbed foetuses in females (reviewed by Abecia et al., 2006). Sheep can also show impaired wool production when nutrient intake is insufficient (Reis and Sahlu, 1994). Within reasonable limits the low nutrient intake does not necessarily compromise animal welfare. In fact, ruminants are generally well prepared for seasonal variation in nutrient availability and, as long as normal catabolic metabolism can adjust for the low nutrient intake, the health and welfare of the animal may not be severely impaired. However, it is very difficult to tell at what point the low energy and protein intake causes severe discomfort in the animals. Long term undernutrition occurs in many different types of situations and added effects of nutrient shortage and other factors, such as unsatisfactory access to drinking water, restricted dry matter access and concomitant space restrictions (as in the case of transported animals) will exaggerate any stress experienced. The effects of undernutrition are severe when the changes in metabolism are irreversible, not allowing the animal to return to an anabolic state when nutrient availability increases again. The effects are dependent on the stage of growth of the body. For example, in lambs weaned onto cereal grain and hay, which contain levels of calcium less than required to balance the amount of phosphorus available, the development of permanent teeth may be impaired (Franklin, 1950). In a similar situation, calves may develop lameness known as 'pegleg'. Neither condition responds to the subsequent provision of adequate amounts of calcium and phosphorus, and the reduction in chewing ability and mobility imposes reduced feed intake and hence a lifetime loss in productivity.

One of the potential contributors to long term undernutrition is pregnancy, especially where the ewe is carrying twins or triplets (Kenyon et al., 2007). This is considered in more detail in Chapters 3 and 6. Lambs are also at more risk than mature adults, because of the requirements for growth. Allden (1970) determined how heavy a Merino lamb should be when the pasture dies off in South Australia about mid-October if it was to survive the following summer. He also showed that lambs that survived the first summer, even if they were at less than the critical weight, showed no impaired wool growth in later years.

7.3.2 Feed-Seeking and Nutrition in Sheep Experiencing Long Term Undernutrition

The conversion of feed to nutrients is elegant in ruminants as they can convert cellulose-rich plants with low quality protein to energy and high-quality animal protein. Sheep, as with many other ruminant species, evolved to survive in conditions with restricted access to feed of high nutrient concentration, and their ability to survive in poor conditions is well known. However, the success of sheep depends on selective feed intake, which requires access to large areas for feed seeking. The normal behaviour of sheep includes around ten hours of grazing per day, divided into four to seven periods, and rumination for approximately eight hours per day. For sheep that are kept outdoors the time for feed seeking may be greater than this when feed quality or access to feed decreases and, if they are restricted to feed of low digestibility, the time for rumination may increase. If the environment offers plenty of dry matter, but with low nutrient concentration, sheep will select the most digestible parts of the plants or feed. If they are left with no other choice they will eat straw or yellow grass but will still select the most digestible parts. The rougher parts will not be eaten and this leads to a build up of sometimes large amounts of feed refusals, which may wrongly lead to the assumption by the animal manager that there is an abundance of feed when the animals are actually struggling to support themselves with daily nutrition. Also, damage to the mouth and pain or obstruction in the oesophagus may also limit feed intake even if feed is available. Other contributing factors include broken-mouth in ewes (where the ewe's teeth are maloccluded, crowded, crooked, out of alignment or broken and she is unable to masticate her food) and parasitism, which restricts nutrients available to the sheep. Undernutrition may also be caused by unsatisfactory feeding of sheep kept indoors or in small fields, in which case the possibilities for the sheep to exert natural feed-seeking behaviour are limited and the welfare is then affected, both by the undernutrition and the stress of behaviour restriction.

7.3.3 Metabolism During Long Term Undernutrition

The metabolic response to low nutrient intake is determined by the condition of the animal at the beginning of the undernutrition and by the severity and duration of the nutrient restriction. Therefore it is complicated to give an absolute description of the metabolism of sheep in undernutrition. It is very important that we learn to distinguish animals in a functional catabolic state, coping with reasonable limitations in nutrient availability, from those suffering from severe long-term undernutrition that causes both acute stress and irreversible damage. Effects of short term restrictions in nutrient intake are discussed above. This section considers effects over longer periods of time, where nutrient intake is unsatisfactory for weeks or months.

When a healthy ruminant shifts from anabolic metabolism to a catabolic state the mobilisation of energy from adipose tissue causes an increase in the concentration of non-esterified fatty acids (NEFA) and β-hydroxybutyrate (BHB) (Reist et al., 2002). Furthermore, reduced concentrations of albumin, total protein and urea in plasma indicate short-term negative protein balance (Payne et al., 1970b). Changes in these metabolites can easily be detected with a simple blood sample and are often used to assess the metabolic state of high-producing dairy animals. The experience with these animals is sometimes wrongly transferred to sheep and cattle in long-term undernutrition (Agenäs et al., 2006). Concentrations in plasma of NEFA and other metabolites of tissue catabolism are only raised when there is tissue to mobilise. Over a longer time of limited nutrient intake the tissue reserves will be limited, as they have already been mobilised. Adipose tissue releases an endocrine peptide called leptin. The release has a daily rhythm in sheep (Bertolucci et al., 2005), and when adipose tissue volume decreases plasma leptin is also reduced. Decrease in plasma leptin has been shown within 50 h of fasting in sheep (Bertolucci et al., 2005). The changes in plasma leptin seem to mediate effects of low body condition on reproductive function, for example in male (Blache et al., 2006) and female (Forcada and Abecia, 2006) sheep. However, leptin is also reduced by factors other than reduced adipose tissue and is, for example, low in lactating animals regardless of body condition (Chilliard et al., 2005). The suppression of leptin in lactating animals may be caused by lipolysis but this has not yet been established experimentally. The main reason for the suppression in early lactation is probably to support a high feed intake (leptin inhibits appetite), but the mechanism for suppressing leptin in lactating animals is not known yet.

Creatinine is always released from muscle tissue and reduced concentrations of creatinine in plasma indicate reduced muscle bulk (Istasse et al., 1990). When tissue reserves of glycogen, fat and muscle protein are depleted the animal will even mobilise bone marrow. This process is irreversible and is a sign that animals are starving to death.

7.3.4 Indicators of Long Term Undernutrition

The most common method for monitoring the effects of nutrition on animals is visual inspection of the animals and feed residuals. Attempts have been made to standardise the subjective visual assessment of the animal, resulting in different body condition scoring systems. These are based on visual inspection and palpation of the lumbar and sacral areas. People experienced in condition scoring may be capable of achieving a high degree of repeatability between different occasions, but the risk of inter-person variation remains (Calavas et al., 1998; Veerkamp et al., 2002). However, there has been little attempt to introduce allowances for individual and breed differences in body size, conformation and tissue composition, and there remains an element of subjectivity in the assessment. Body condition scoring is often useful in animal management, in monitoring changes in animals, but the method is unsatisfactory in determining the nutritional state of animals or herds at a single occasion since the body condition score does not say if the animals are gaining or losing body condition. Also, body condition scoring is often performed on a 5 or 10 point graded scale and unless these points are subdivided this can be a bit blunt. Variation in rumen contents, typically between 6 and 18 kg as previously mentioned (see Section 7.2.3), are likely to distort the subjective score given to the animal for the degree of muscle and fat cover over the vertebrae.

In dairy cow management, metabolic profiling, including variables like BHB, urea and glucose, have been used for decades, in order to get more exact information about the nutritional state of the animals (Payne et al., 1970a). However, the interest in metabolic profiles has mainly been focused on high producing cows (Rowlands et al., 1977) and data for ruminants in long-term undernutrition is very limited. Some veterinarians and researchers believe that metabolic profiling is inadequate, particularly in cases of prolonged or extreme undernutrition (Payne & Payne, 1987).

7.4 Toxicity of Dietary Components for Grazing Sheep

7.4.1 Introduction

Substances ingested by the grazing sheep may differ appreciably from the pasture grasses, legumes and supplements of energy, protein and minerals provided by farmers. As Reid (1973) indicated, additional substances selected by the animal may include parts of trees and shrubs, water plants, bedding, wool, hair, rubbish, paint and animal carcases. In addition there may be the inadvertent intake of soil, sand, fungi and fungal-generated toxins, other micro-organisms, chemicals, fungicides, herbicides and insecticides together with pollutants and contaminants of the water supply or soil. Many substances ingested in plants are potentially toxic to animals, even in production systems involving close management of pastures and animals. Such substances pose an even greater threat to animals maintained under more extensive management systems where the forages on offer are frequently those that have survived uncontrolled grazing. Plant survival is often achieved by the production of secondary, protective chemicals that are toxic to micro-organisms, such as bacteria and fungi, and hence potentially toxic to the microbes responsible for the fermentation of ingested plant material in sheep. Alternatively, plants may contain chemicals produced by two types of fungus, endophytes that develop in the plant after infecting the seed and saprophytes that attack dead plant material. Such chemicals often reduce the palatability of feed. Selective grazing may remove all the more desirable plant species leading to the development of pastures dominated by species containing undesirable chemicals. The grazing animal then has no option but to ingest potentially toxic plant components. Intake of toxic substances may also occur in regions marked by annual drought, during which grazing sheep begin to browse trees and shrubs or resort to eating the leaves that fall from such plants (Lowry et al., 1993).

7.4.2 Factors Affecting the Toxicity of Plants for Sheep

The severity of toxic effects depends on the amount consumed, the form in which it reaches the rumen, the rate of its release in the rumen, the extent to which rumen microbes can metabolise it, the rate at which it is absorbed from the digestive tract and the rate of its detoxication in the liver or kidney. These processes are well illustrated by the capacity of the animal to make use of urea, which, ingested in moderation, provides a source of ammonia to rumen bacteria but in excess can cause death from ammonia poisoning. When Coombe et al. (1960) fed sheep on hay and molasses, with 100 g urea added to the hay, which was consumed gradually over the day, rumen pH remained at 6.5–7 and levels of ammonia did not exceed 750 mg/l. By contrast, when 25 g urea in solution was dosed into the rumen, ammonia levels increased to 1140 mg/l and pH reached 7.9. Both the higher ammonia concentration and higher rumen pH favour increased rates of absorption of ammonia across the rumen wall (Hogan, 1961). At the same time, the alkaline rumen pH will retard the rate of absorption of volatile fatty acids, thus reducing the energy available to a liver already being challenged to detoxicate the ammonia reaching it. Although, with their large nutrient intake, the sheep involved in that experiment were able to detoxify the ammonia produced, data quoted by Coombe et al. (1960) indicated that as little as 10 g urea dosed into the rumen in other circumstances may prove fatal for sheep.

7.4.3 *Toxic Substances in Forages*

Reviewers of this subject include James et al. (1992), Cheeke (1995), Garland and Barr (1998), D'Mello (2000) and Acamovic et al. (2003). A few examples will be mentioned here to illustrate the factors affecting the toxicity to sheep of cyanide, fluoroacetate, nitrate, tannins and phenolics, alkaloids, and fungal metabolites.

7.4.3.1 Cyanide

Cyanide exists in plants as cyanogenetic glycosides which must be hydrolysed in the rumen before cyanide is released. Cyanide, present in some Brassica species (Duncan and Milne, 1993), is detoxified by conversion to thiocyanate, the reaction requiring the presence of adequate sulphide frequently provided by dosage of thiosulphate (Rose, 1941). The amount of cyanide escaping detoxication depends on the rates of intake and hydrolysis of the glycoside relative to the rate of conversion of cyanide to thiocyanate. If sufficient surplus cyanide is absorbed into the portal blood supply, conversion of haemoglobin into cyanohaemoglobin may be fatal (Coop and Blakley, 1950). A complication in the protection of sheep against cyanide poisoning is the threat posed by thiocyanate to the foetal thyroid.

7.4.3.2 Fluoroacetate

Fluoroacetate was first identified as a toxin in the South African 'Gifblaar' (*Dichapetalum cymosum*) and also occurs in a variety of plants including *Gastrolobium* and *Oxylobium Spp* and some acacias such as *A georginae*. After absorption from the rumen, fluoroacetate in the tissues acts to inhibit the tricarboxylic acid cycle involved in glucose metabolism by converting citrate to fluorocitrate. Sheep dosed slowly into the rumen with 2 mg fluoroacetate daily lost appetite in 14–18 days when fed chopped wheaten hay but not when the hay was supplemented with 100 g gluten or replaced with lucerne hay (Jarrett and Packham, 1956). However as lucerne–fed sheep succumbed when dosed directly into the abomasum, the beneficial effects of the more nutritious diets are presumably exerted through the rumen microbes rather than through tissue metabolism. Increased microbial protection against the toxin might be anticipated in the future from a genetically-engineered rumen bacterium (Gregg et al., 1998), though anecdotal information on the rapidity of action of fluoroacetate suggests that microbial detoxication might not prevent fatalities. The action of fluoroacetate in the tissues becomes more severe if muscle metabolism is increased and the management of livestock exposed to fluoroacetate-containing plants includes the avoidance of undue exercise.

7.4.3.3 Nitrate

Nitrate is probably present at low concentration in all green plants but increases in response to nitrogen fertiliser and accumulates in greater concentration under conditions of cold, dull weather that are not conducive to complete protein synthesis

(Deinum, 1966). Such climatic conditions also depress the amount of soluble carbohydrate in plant cells (Wright and Davidson, 1964) and hence reduce the amount of readily available energy required by rumen bacteria to metabolise nitrogen in nitrate into microbial protein. Nitrate is reduced in the rumen to nitrite and subsequently to ammonia. Nitrite is readily absorbed into the portal venous system and converts haemoglobin to methaemoglobin (Wright and Davidson, 1964). Many outbreaks of toxicity in sheep have been ascribed to nitrate (Hurst, 1942). However, the absence of methaemoglobin in some sheep dead after grazing *Phalaris aquatica* led to the discovery of several other substances now thought to be the causal agents (Bourke and Corrigan, 1992). The extent to which nitrate, and perhaps ammonia are involved in outbreaks of toxicity to animals is not now clear.

7.4.3.4 Tannins

Tannins, phenolic molecules of varying molecular size, are classified as hydrolysable and condensed. The toxicity of hydrolysable tannins found in plants such as *Terminalia oblongata* and the Indonesian harendong *Clidemia hirta* (Murdiati et al., 1991) is consistent with that of simple phenolics liberated in the rumen. Those authors suggest that the action of hydrolysable tannins on protein digestion probably occurs post-ruminally. Phenols, including those liberated from lignin, a component of plant fibre, are detoxicated in the liver or kidney by combination with the amino acid glycine to form hippuric acid, which is excreted in the urine (Lowry et al., 1993). In animals fed low protein diets, the provision of glycine for phenol detoxication can thus impose a significant burden on the nitrogen economy of the animal. Condensed tannins occur in a wide range of plants including oak, acacia and many pasture plants. Their particular action in the rumen involves binding with feed proteins which are thus rendered inaccessible to attack by rumen microbes. Condensed tannins, too, have been observed to depress the numbers of proteolytic bacteria in rumen contents (Min et al., 2002), although the amount of microbial protein leaving the stomach is not affected. The binding of tannins to dietary protein may be nutritionally beneficial or detrimental to the animal depending on whether protein binding is reversed in the acid conditions of the abomasum to permit the release of amino acids in the small intestine. Useful attributes ascribed to condensed tannins include a reduction in methane emissions in goats (Purchala et al., 2005) and a reduction in intestinal helminths in sheep (Molan et al., 2002). The effects of condensed tannins can be reduced by providing an alternative binding site in the form of polyethylene glycol of molecular weight about 4000 (Pritchard et al., 1988). Although deer have adapted to diets rich in condensed tannins by producing a proline-rich protein in saliva that binds condensed tannins, sheep and cattle have not done so (Austin et al., 1989).

7.4.3.5 Isoflavones

Two types of isoflavones with oestrogenic properties are found in clover plants. For example, the 5-hydroxyisoflavones found in *Trifolium subterraneum, cultivar Clare*

(Shutt et al., 1970) include biochanin A and genistein, whereas red clover, *Trifolium pratense*, contains the 5-deoxy isoflavone formononetin. Genistein is oestrogenic to guinea pigs whereas formononetin is not (Batterham et al., 1965). However in sheep, that situation is reversed. Biochanin A is demethylated in the rumen to genistein and after conversion to p-ethyl phenol ceases to be oestrogenic, whereas formononetin, after demethylation to daidzein is finally converted to equol, a potent oestrogen (Batterham et al., 1965). Rumen microbes may thus protect their host against toxicity in one situation, but cause toxicity in another. However the capacity of rumen microbes to detoxify genistein was not instantaneous but has to be developed over 4–5 days, during which time the clover remains oestrogenic (Shutt et al., 1970). This is an example of what is probably a common situation in which the animal presumably has rumen microbes capable of metabolising a toxin but needs some days to build the population of those microbes to a size capable of detoxifying the amount ingested each day.

7.4.3.6 Alkaloids

Alkaloids appear to make their primary attack on the liver, but the symptoms are often observed in other locations in the animal. Some, such as the pyrrolizidines, are secondary plant chemicals. These are found in a variety of plants including Heliotrope *Heliotropium europaeum* (Bull et al., 1956) and *Echium plantagineum* (Culvenor et al., 1984), known in different parts of Australia as 'Paterson's Curse' or 'Salvation Jane'. More recently the role of fungal alkaloids has been recognised. These include alkaloids that cause tremorgenic (staggering) effects in animals, produced following the infection of the seed of ryegrass (*Lolium Sp)* by the endophyte *Neotyphodium lolii* and of Tall Fescue (*Festuca arundinacea*) by *Neotyphhodium coenophialum*.

A further example is locoweed poisoning, which occurs in North and South America and China. The causal agent in endophyte-infected locoweeds Astragalus Spp and Oxytropis Spp (Braun et al., 2003) is the alkaloid Swainsonine, which produces abnormalities in lysosomal catabolism and glycoprotein metabolism (Garland and Barr, 1998). The poisoning has an insidious onset, with no symptoms until the grazers have eaten locoweed for a considerable time. Clinical symptoms include those suggestive of a derangement of the nervous system, with staggering gait, muscle tremors, ataxia and nervousness, especially when roused.

Fungi developing on dead plant material can also produce alkaloids toxic to sheep. Examples are phomopsins produced by *Phomopsis leptostromiformis* on lupin stubble and Sporidesmin, a product of the pasture saprophyte *Pithomyces chartarum*. Damage to liver function causes a variety of secondary problems but two well-recognised ones are photosensitization and chronic copper poisoning. In the normal animal, chlorophyll is converted to phylloerythrin and, after removal from the digestive tract, is returned to the small intestine via the gall bladder. Following liver damage phylloerythrin and bile pass into the peripheral blood system and the animal, thus jaundiced, becomes photosensitive as shown by the eruption of scabby sores, referred to as Facial Eczema, in skin areas exposed to sunlight. Chronic copper poisoning may be observed following the intake of Heliotrope (Dick, 1953) and *Echium* (Seaman and Dixon, 1989). When the sheep liver loses its capacity to form copper thiomolybdate, the molecule that permits the removal of surplus copper via the bile, copper storage can reach toxic levels in the liver and kidneys. Diagnosis of the cause of chronic copper poisoning has been difficult because damage to the liver from the ingestion of toxic quantities of forage might have preceded the appearance of clinical signs by six or twelve months.

7.4.4 Control of Toxicity

Although sheep in some situations do not possess the appropriate bacteria to metabolise toxins, as Dodson (1959) found with oxalate and Jones and Megarrity (1986) with mimosine (the toxic amino acid in *Leucaena glauca*), ruminants will generally develop bacteria capable of metabolising plant toxins if given the time to do so. Culvenor et al. (1984) for example, found that the feeding of relatively small amounts of alkaloids in *Echium* induced a high rate of breakdown of the alkaloid in the rumen. However, with rapidly acting toxins, such as nitrate and cyanide, this may not be an option. Similarly it is not an option with plants containing 5-deoxy isoflavones, which bacteria convert into the oestrogen equol. However it may be possible to adopt animal management practices to give animals limited but gradually increasing access to dangerous plants or to ensure that toxic material is diluted by more nutritious forage. Many reports of toxicity refer to animals transported to a new environment following a dry season or drought and encountering forages with which they are not familiar (Hurst, 1942).

7.4.5 Managing Animals and Toxic Plants

The accumulation of knowledge on the causes of outbreaks of toxicity has permitted the development of management strategies both of pastures and of animals grazing them to minimize consequences. Progress has been made to replace existing pasture plants with low toxicity cultivars or seed stocks free from endophyte. Management strategies with animals may involve replacing sheep with cattle or perhaps deer, or replacing more susceptible breeding females and lambs with less susceptible mature castrate males. Other options include reducing and gradually increasing grazing time (Provenza et al., 1992) on pastures suspected of toxicity, and removing animals when climatic conditions favour the development of saprophytic fungi. In some situations animals may have to be totally excluded from areas. This might apply with Gifblaar, for instance, which sends out green shoots from an extensive root system following rain at the end of a dry season and for some time is the only green feed on offer. In extreme situations attempts can be made to control undesirable plants mechanically or chemically or, as is currently happening with *Echium plantagineum* in Australia, by the release of herbivorous insects. Changes to pasture structure are likely to be slow as newly planted seed has to compete with what is often a very large amount of seed deposited in the soil from existing plants. The management of plant and animal resources to minimize stock losses within an economically viable production enterprise will thus continue to present challenges to the farmer into the future.

7.4.6 Toxic Metals

The contamination of sheep pastures with toxic elements can occur on soils that are naturally rich in metals (e.g. Martin and Coughtrey, 1982), following accidental or anthropogenic events, such as fall-out of radiocaesium on grassland after the accident at the Chernobyl nuclear reactor (Howard and Beresford, 1994) or following the prolonged use of sewage sludge or contaminated fertilizers on pastureland (Wilkinson et al., 2001, 2003). Two of the metallic elements that cause most concern are discussed here, that is cadmium and lead.

7.4.6.1 Cadmium

Cadmium is one of the very few metals for which there is no known function in animals or plants, but it is a natural contaminant of some zinc ores, sewage sludge and phosphate fertilizers (Bruwaene et al., 1984; FAO, 1991), and when the latter are spread on grassland as fertilisers cadmium can accumulate to potentially toxic levels. The limit for cadmium concentration in pasture in the European Union is one part per million but this can be easily exceeded on regularly fertilised pasture. However, the source of cadmium is being controlled by regulatory limits on cadmium concentration in fertilisers. There is considerable variation in cadmium concentration in the different sources of phosphate fertiliser, those in Senegal and Togo being high, and those from Kola in the former Soviet Union being naturally low. Decadmiation is possible by calcination of phosphate rock or a range of other physico-chemical processes.

A serious concern in relation to cadmium contamination of animal pastures is the accumulation of cadmium through the human food chain, including sheep offal (Prankel et al., 2004, 2005), and its eventual distribution on the land in the form of sewage sludge. Another important source of cadmium is industrial effluent. Cadmium is used for electroplating, for the production of batteries, and this can contaminate rivers and streams with a toxic sediment leading to pollution of both the river banks, which sheep may graze, and micro-organisms at the start of the food chain. This form of cadmium pollution has been controlled in some industrial countries by regulatory limits on cadmium effluent concentrations. Livestock, such as sheep, may be exposed to levels of 5 mg/kg herbage near localised sources of contamination, particularly ore processing sites (Washington, 1985). Sheep concentrate feeds contain little cadmium (McDowell, 1992).

Cadmium acts as an intracellular toxin, affecting the mitochondria, which are the cell's respiratory organelles. Hence the sites of the body where cadmium does the most damage is where there is most rapid blood flow, for example, the testis and the kidney. In the mitochondria cadmium is sequestered by the lysosome system for detoxification, but when the capacity of this process is exceeded cadmium damages the respiratory capacity of the cells. The main place of entry of cadmium into the animal body is via the gastrointestinal tract, because cadmium contaminates the food. However, the processes affecting absorption are not well understood, but organic binding will increase its availability (Phillips et al., 2005). The impact of sequestration of the cadmium by micro-organisms is unknown, and this could have a serious effect in ruminants.

When it enters the bloodstream cadmium will be rapidly sequestered by metallothioneins, which are produced in the liver, the small intestine and a number of other sites. It may remain in these organs until their capacity to absorb the cadmium is exceeded. The half life of cadmium in the mammalian body is very long, perhaps 30 years, thus most sheep will excrete only a very small proportion of their intake. Hence long-term perturbations in cadmium intake are most relevant in determining the animal's ability to cope with the toxicity. The main receptor organ for cadmium removed from the bloodstream is the liver, and when the liver's capacity to absorb this cadmium is exceeded it is returned to the bloodstream and filtered out by the proximal tubules of the kidney. Here it attacks the mitochondria and, when the capacity of the lysosome to detoxify it is exceeded, it reduces the respiratory capacity of the cells.

The testis is particularly sensitive to cadmium toxicity (Parizek, 1957, 1960), because the high rate of blood flow facilitates circulatory failure in the testicular vasculature (Mason et al., 1964; Waites and Setchell, 1966). Specific damage to the spermiogenic epithelium is also believed to occur with cadmium toxicity (Parizek, 1957, 1960) and, in the long term, testosterone production is probably reduced (Nordberg, 1975). Spermiation failure has been observed within 24 hours after an intraperitoneal injection of 1 mg $CdCl_2$/kg bodyweight in rats (Hew et al., 1993) and a subcutaneous injection of 0.5 mg $CdCl_2$ in frogs (Kasinathan et al., 1987).

There have been few studies to determine the effects of chronic cadmium on grazing ruminants. There is evidence that they can detect and partially avoid the element on pasture, since Phillips et al. (2005) found that sheep ate less grass when it had a high cadmium concentration. In relation to the effects on reproduction, Doyle et al. (1974) found no effects of a relatively high level of cadmium (60 mg/kg) fed to young rams, but the sperm concentration was very low, being collected by electroejaculation. Hogue et al. (1984) found an increase in testis size and a tendency for cadmium to accumulate in the tissues of rams grazing on sludge-amended soil, but there were no effects on sperm quality, which have been observed in other animals (Kasinathan et al., 1987; Hew et al., 1993). Berry et al. (1999) found no effect of a daily dose of 7.5 mg Cd, estimated to be the likely intake in a polluted region, on spermatogenesis. However, the cadmium increased grazing time, which the authors attributed to a possible interference with nutrients, such as zinc, in the herbage. Cadmium also reduced rumination, a possible corollary of increased grazing time, and the number of Flehman responses, which may derive from reduced testosterone. Other authors have found reduced testosterone levels, and reduced sexual and aggressive behaviour with cadmium fed to sheep (Parizek, 1957; Neathery et al., 1973). This may be due to a zinc deficiency, which often accompanies cadmium toxicity and has been observed to result in a decline in androgen production (Underwood and Somers, 1969; Martin and White, 1992).

7.4.6.2 Lead

Lead is a common source of poisoning on farms, although cattle are more likely than sheep to become intoxicated, because of their inquisitive nature. Fortunately, little of the soil lead that may accumulate beside roads is taken up by plants, and little ingested lead is absorbed (only about 1–2%, Blaxter, 1950). However, sheep grazing beside a contaminated roadside do develop elevated blood lead over a period of several months, as a result of ingestion of contaminated soil and inhalation of exhaust fumes (Ward et al., 1978). Other sources of potential toxicity for sheep include land previously contaminated by munitions works, often when recently ploughed, since lead is quite immobile in the soil. Absorbed lead accumulates in the brain, kidneys, spleen and bones of sheep. In the kidneys it causes nephrotoxicity. Sheep display lethargy, anorexia, abdominal pain, and usually diarrhoea during acute lead toxicity. Anaemia is common during chronic lead ingestion (National Academy of Science, 1972).

7.4.7 Setting Appropriate Toxicity Limits for Animal Welfare

Given that toxic elements have adverse physiological, behavioural and metabolic consequences for the animal, it is appropriate to consider how critical limits of exposure should be established. In human exposure, the precautionary approach is to establish a No Observable Adverse Effect Level (NOAEL). This is an exposure level at which there are no statistically or biologically significant increases in the frequency or severity of adverse effects between the exposed population and its appropriate control; some effects may be produced at this level, but they are not considered as adverse, or as precursors to adverse effects. Protection of every individual in the population in the establishment of a NOAEL involves building in cumulative uncertainty factors, such as differences between individuals (x 10), subchronic rather than chronic exposure (x 3), knowledge of the lowest observed adverse effect level (LOAEL), rather than NOAEL (x 10). These can be summated to give a cumulative uncertainty factor.

For animals it is necessary to consider the types of response that indicate an adverse effect. Physiological responses are likely to cause adverse welfare consequences, but may only be transient. Behavioural responses may indicate the animal's attempt to rectify a deficiency, for example by increasing the feeding range. In human exposure to odours, for example, limits are set in response to reported irritation or unpleasant sensory stimulation (Paustenbach and Gaffney, 2006), but even if this can be recorded for sheep in the form of coughs, lacrimation etc, it would be difficult to achieve for nutritional discomfort. Choice tests can be used to indicate avoidance, and these have successfully demonstrated, for example, that ruminants will avoid feeds contaminated with lead (Phillips and Strojan, 2002). An important consideration is the proportion that should be protected from unpleasant experiences, if it can be proven that they are such. Paustenbach and Gaffney (2006) have considered this question in relation to human exposure to odours and suggest that the majority (which they consider might be 80–95%) should be protected from irritation or unpleasant sensory stimulation. A further consideration is the length of acceptable exposure, which will depend on the time that it takes for the effects to develop. Absence of good exposure data for different time periods will usually lead to the establishment of a NOAEL which is applicable for any time period. However, if there is reliable time dependent exposure effects data, it may be possible to set higher limits where exposure is only for a brief period of time.

7.5 Conclusion

Nutrition has the potential to have very major effects on the welfare of sheep. Of localised significance is undernutrition, which is most likely to occur when feed supply is depleted in exceptional weather conditions, usually during a drought. The low value of the output from sheep often means that supplements are uneconomic to supply in drought condions, but new Duty of Care legislation in many countries should encourage the worst welfare cases to be averted during drought conditions. Routine reduction of feed supply from sheep for periods of the year when grass growth is restricted is common, but they are able to survive this by depleting body tissues. The extent to which this affects welfare needs studying. Also feed and water supply are usually restricted when sheep are transported long distances, and this also needs further research to evaluate the welfare impact.

Routine undernutrition is rare, because the output of the sheep is reduced, but sheep are also subjected to malnutrition, when their feed is imbalanced nutritionally. This may be common in sheep that act as scavengers or when their feed is not carefully controlled. Sheep may also be potentially intoxicated when they eat poisonous plants or plants contaminated with toxic elements, such as the heavy metals cadmium or lead. Further research is needed to examine the growing contamination of pastures with toxic elements and the impact that this has on sheep welfare.

References

Abecia, J.A., Sosa, C., Forcada, F. & Meikle, A. (2006) The effect of undernutrition on the establishment of pregnancy in the ewe. *Reproduction, Nutrition and Development* **46**: 367–378.

Acamovic, T., Stewart, C.S. & Pennycott, T. W. (eds.) (2003) *Poisonous Plants and Related Toxins.* p. 608. CAB International, Wallingford, UK.

Agenäs, S., Heath M.F., Nixon, R.M., Wilkinson J.M. & Phillips, C.J.C. (2006) Indicators of undernutrition in cattle. *Animal Welfare* **15**: 149–160.

Allden, W.G. (1970) Body composition and herbage utilization of grazing Merino and crossbred lambs during periods of growth and summer undernutrition. *Australian Journal of Agricultural Research* **21**: 261.

Austin, P.J., Suchar, L.J., Robbins, C.T. & Hagerman, A.E. (1989) Tannin-binding proteins in the saliva of deer and their absence in the saliva of sheep and cattle. *Journal of Chemical Ecology* **15**: 1335–1339.

Bailey, C.B. & Balch, C.C. (1961) Saliva secretion and its relation to feeding in cattle 2. The composition and rate of secretion of mixed saliva in the cow during rest. *British Journal of Nutrition* **15**: 383–402.

Batterham, T.J., Hart, N.K., Lamberton, J.A. & Braden, A.W.H. (1965) Metabolism of oestrogenic isoflavones in the sheep. *Nature, London* **206**: 509.

Berry, N.R., Axford, R.F.E., Ap Dewi, I., Chiy, P.C. & Phillips, C.J.C. (1999) The effect of a low dose of cadmium on spermatogenesis in rams. *Small Ruminant Research* **32**: 97–102.

Bertolucci, C., Caola, G., Foa, A. & Piccione, G. (2005) Daily rhythms of serum leptin in ewes, effects of feeding, pregnancy and lactation. *Chronobiology International* **22**: 817–27.

Blache, D., Zhang, S. & Martin, G. B. 2006. Dynamic and integrative aspects of the regulation of reproduction by metabolic status in male sheep. *Reproduction, Nutrition and Development* **46**: 379–90.

Bourke, C.A. & Corrigan, M.J. (1992) Mechanisms underlying Phalaris aquatica 'Sudden death' syndrome in sheep. *Australian Veterinary Journal* **69**: 165–167.

Blaxter, K.L. (1950) Lead as a nutritional hazard to farm livestock.III. Factors influencing the distribution of lead in tissues. *Journal of Comparative Pathology and Therapeutics* **60**: 177–189.

Blaxter, K.L. & Graham, N. McC. (1955) Plane of nutrition and starch equivalents. *Journal of Agricultural Science* **46**: 292–306.

Braun, K., Romero, J., Liddell, C. & Creamer, R. (2003) Production of Swainsonine by fungal endophytes of locoweed. *Microbiological Research* **107**: 980–988.

Brownlie, L.E. & Grau, F.H. (1967) Effect of feed intake on growth and survival of salmonellas and *Escherichia coli* in the bovine rumen. *Journal of General Microbiology* **46**: 125–134.

Bruwaene, R.V., Kirchmann, R. & Impens, R. (1984) Cadmium contamination in agriculture and zootechnology. *Experentia* **40**: 43–52.

Bull, L.B., Albiston, H.E., Edgar, G. & Dick, A.T. (1956) Toxaemic Jaundice of sheep, phytogenous chronic copper poisoning, heliotrope poisoning and hepatogenous chronic copper poisoning. *Australian Veterinary Journal* **32**: 229–236.

Bullen, J.J., Scarisbrick, R. & Maddock, A. (1953) Enterotoxaemia of sheep. The fate of *Clostridium welchii* Type D introduced into the rumen of normal sheep. *Journal of Pathology and Bacteriology* **65**: 209–219.

Calavas, D., Sulpice, P., Lepetitcolin, E. & Bugnard, F. (1998) [Assessing the accuracy of the practice of a method of scoring the body condition of ewes within a professional framework]. *The Veterinary Research* **29**: 129–138.

Callaway, T.R., Elder, R.O., Keen, J.E., Anderson, R.E. & Nisbet, D.J. (2002) Forage feeding to reduce pre-harvest *Escherichia coli* population in cattle. *Journal of Dairy Science* **86**: 852–860.

Cheeke, P.R. (1995) Endogenous toxins and myco-toxins in forage grasses and their effects on livestock. *Journal of Animal Science* **73**: 3909–3918.

Chilliard, Y., De Lavaud, C. & Bonnet, M. (2005) Leptin expression in ruminants, nutritional and physiological relation with energy metabolism. *Domestic Animal Endocrinology* **29**: 3–22.

Coghlan, J.P., Fan, J.S.K., Scoggins, B.A. & Shulkes, A.A. (1977) Measurement of extracellular fluid volume and blood volume in sheep. *Australian Journal of Biological Science* **30**: 71–84.

Cole, N.A. (1991) Effects of animal- to- animal exchange of ruminal contents on the feed intakes and ruminal characteristics of fed and fasted lambs. *Journal of Animal Science* **69**: 1795–1803.

Cole, N.A. (1995) Influence of a three day feed and water deprivation period on gut fill, tissue weights and tissue composition in mature wethers. *Journal of Animal Science* **73**: 2548–2557.

Cole, N.A., Phillips, W.A. & Hutcheson, D.P. (1986) The effect of pre-fast diet and transport on nitrogen metabolism in calves. *Journal of Animal Science* **62**: 1719–1731.

Coombe, J.B., Tribe, D.E. & Morrison, J.W.C. (1960) Some experimental observations on the toxicity of urea to sheep. *Australian Journal of Agricultural Research* **11**: 247–256.

Coop, I. (1949) The effect of starvation and of feeding after starvation on metabolic activity in the rumen. *New Zealand Journal of Science and Technology* **31A**: 1–12.

Coop, I.E. & Blakley, R.L. (1950) The metabolism and toxicity of cyanide and cyanogenetic glycosides in sheep. III. Toxicity. *New Zealand Journal of Science and Technology* **31A**: 44–58.

Culvenor, C.C.J., Jago, M.V., Peterson, J.E., Smith, L.W., Payne, A.L., Campbell, D.G., Edgar, J.A. & Frahn, J.L. (1984) Toxicity of *Echium plantagineum* Paterson's Curse. 1. Marginal toxic effects in Merino wethers from long term feeding. *Australian Journal of Agricultural Research* **35**: 293–304.

D'Mello, J.P.F. (2000) Antinutritional factors and mycotoxins. In: *Farm Animal Metabolism and Nutrition* (ed. J.P.F. D'Mello), pp. 383–403, CAB International, Wallingford, UK.

Deinum, B. (1966) Climate nitrogen and grass, Research into the influence of light intensity, temperature, water supply and nitrogen on the production and chemical composition of grass. *Mededelingen Landbouwhogeschool Wageningen* **64**: 91.

Denton, D., Shade, R., Zamarippa, F., Egan, G., Blair-West, J., McKinley, M. & Fox, P. (1999) Correlation of regional blood flow and change of plasma sodium concentration during genesis and satiation of thirst. *Proceedings of the National Academy of Science USA* **96**: 2532–2537.

Dick, A.T. (1953) Influence of inorganic sulphate on the copper-molybdate interrelationship in sheep. *Nature, London* **172**: 637–638.

Dodson, M.E. (1959) Oxalate studies in sheep. *Australian Veterinary Journal* **35**: 225–233.

Doyle, J.J., Pfander, W.H., Grebing, S.E. & Pierce, J.O. (1974) Effect of dietary cadmium on growth, cadmium absorption and cadmium tissue levels in growing lambs. *Journal of Nutrition* **104**: 160–166.

Duncan, A.J. & Milne, J.A. (1993) Effects of oral administration of Brassica secondary metabolites, allyl cyanide, allyl isothiocyanate and dimethyl disulphide on the voluntary food intake and metabolism of sheep. *British Journal of Nutrition* **70**: 631–645.

FAO (1991) Status of Cadmium, Lead, Cobalt and Selenium in soils and plants of thirty countries. *FAO Soils Bulletin*.

Ferguson, D.M., Bruce, H.L., Thompson, J.L., Egan, A.F., Perry, D. & Shorthose, W.R. (2001) Factors affecting beef quality – farm gate to chilled carcass. *Australian Journal of Experimental Agriculture* **41**: 879–891.

Fluharty F.L., Loerch, S.L. & Dehority, B.A. (1996) Effects of feed and water deprivation on ruminal characteristics and microbial population of newly weaned and feedlot adapted calves. *Journal of Animal Science* **72**: 2969–2979.

Forbes, J.M. (2003) The multifactorial nature of food intake control. *Journal of Animal Science* **81**: E139–E144.

Forcada, F. & Abecia, J.A. (2006) The effect of nutrition on the seasonality of reproduction in ewes. *Reproduction, Nutrition and Development* **46**: 355–365.

Franklin, M.C. (1950) *Diet and dental development in sheep*. Bulletin 252 Commonwealth Scientific and Industrial Research Organization, Melbourne Australia.

Garland, T. & Barr, A.C. (1998) *Toxic Plants and Other Natural Toxicants*. CAB International, Wallingford, United Kingdom.

Gregg, K., Hamdorf, B., Henderson, K., Kopeckny, K. & Wong, C. (1998) Genetically modified rumen bacteria protect sheep from fluoroacetate poisoning. *Applied and Environmental Microbiology* **64**: 3486–3498.

Gregory, N.G., Jacobson, L.H., Nagle, T.A., Muirhead, R.W. & Leroux, G.T. (2000) Effect of pre-slaughter feeding system on weight loss, gut bacteria and the physicochemical properties of digesta in cattle. *New Zealand Journal of Agricultural Research* **43**: 351–361.

Harper, G.S. (1999) Trends in skeletal muscle biology and the understanding of toughness in beef. *Australian Journal of Agricultural Research* **50**: 1105–1129.

Hew, K.W., Ericson, W.A. & Welsh, M.J. (1993) A single low cadmium dose causes failure of spermiation in the rat. *Toxicology and Applied Pharmacology* **121**: 15–21.

Hogan, J.P. (1961) The absorption of ammonia through the rumen of sheep. *Australian Journal of Biological Sciences* **14**: 448–460.

Hogan, J.P. (1964) The digestion of food by the grazing sheep. II The production of ammonia in the rumen. *Australian Journal of Agricultural Research* **15**: 397–407.

Hogan, J.P., Petherick, J.C. & Phillips, C.J.C. (2007) The nutritional impact on sheep and cattle of feed and water deprivation prior to and during transport. *Nutrition Research Reviews* **20**: 17–28.

Hogue, D.E., Parrish, J.J., Foote, R.H., Stouffer, J.R., Anderson, J.L., Stoews, G.S., Telford, J., Bache, C.A.; Gutenmann, W.H. & Lisk, D.J. (1984) Toxicologic studies with male sheep grazing on municipal sludge-amended soil. *Journal of Toxicology and Environmental Health* **14**: 153–161.

Howard, B.J. & Beresford, N.A. (1994) Radiocaesium contamination of sheep in the United Kingdom after the Chernobyl accident. In: *Pollution in Livestock Production Systems*, (eds. I. Ap Dewi, R.F.E. Axford, I. Fayez Marai & H.M. Omed), pp. 97–118, CAB International, Wallingford, UK.

Hurst, E. (1942) *The poison plants of New South Wales*. New South Wales Poison Plants Committee, Sydney, Australia.

Istasse, L. Vaneenaeme, C, Evrard, P, Gabriel, A., Baldwin, P, Maghuinrogister, G, & Bienfait, J.M. (1990) Animal performance, plasma hormones and metabolites in Holstein and Belgian Blue growing fattening cattle. *Journal of Animal Science* **68**: 2666–2673.

Jacob, R.H., Pethick, D.W., Ponnapalam, E., Speijers, J. & Hopkins, D.M. (2006) The hydration status of lambs after lairage at two Australian abattoirs. *Australian Journal of Experimental Agriculture* **46**: 909–912.

Jacobson, L.H., Nagle, T.A., Gregory, N.T., Bell, R.G., Leroux, G. & Haines, J.M. (2002) Effect of feeding pasture-finished cattle different conserved forages on *Escherichia coli* in the rumen and faeces. *Meat Science* **62**: 93–106.

James, L.F., Panter, K.E., Nielsen, D.B. & Molyneux, R.J. (1992) The effect of natural toxins on reproduction in livestock. *Journal of Animal Science* **70**: 1573–1579.

Jarrett, I.G. & Packham, A. (1956) Response of the sheep to sub-lethal doses of fluoroacetate. *Nature, London* **177**: 580–581.

Jones, R.J. & Megarrity, R.G. (1986) Successful transfer of DHP-degrading bacteria from Hawaiian goats to Australian ruminants to overcome the toxicity of *Leucaena. Australian Veterinary Journal* **63**: 259–262.

Kasinathan, S., Veeraraghavan, K. & Ramakrishnan, S. (1987) Effect of cadmium on the spermatogenesis of Rana hexadactyla Lesson. *Acta Morphologica Hungaria* **35**: 183–187.

Kenyon, P.R., Stafford, K.J., Jenkinson, C.M.C., Morris, S.T., West, D.M. (2007) The body composition and metabolic status of twin- and triplet-bearing ewes and their fetuses in late pregnancy. *Livestock Science* **107**: 103–112.

Knowles, T.G., Warriss, P.D., Brown, S.N. & Kestin, S.C. (1994) Long distance transport of export lambs. *The Veterinary Record* **134**: 107–110.

Lowry, J.B., Sumpter, E.A., McSweeney, C.S., Schlink, A.C. & Bowden, B. (1993) Phenolic acids in the fibre of some tropical grasses, effect on feed quality and their metabolism in sheep. *Australian Journal of Agricultural Research* **64**: 41–49.

Martin, G.B. & White, C.L. (1992) Effects of dietary zinc deficiency on gonadotrophin secretion and testicular growth in young male sheep. *Journal of Reproduction and Fertility* **96**: 497–507.

Martin, M.H. & Coughtrey, P.J. (1982) *Biological Monitoring of Heavy Metal Pollution*. Applied Science Publishers, London, UK.

Mason, K.E., Brown, J.A., Young, J.O. & Nesbit, R.R. (1964) Cadmium-induced injury of the rat testis. *Anatomical Record* **149**: 135–148.

McDowell, L. (1992) *Minerals in Animal and Human Nutrition*. Academic Press, San Diego, California, USA.

McSweeney, C.S. & Cross, R.B. (1992) Effect of sodium intake on sodium conservation, digestion and mineral metabolism in growing ruminants fed *Stylosanthes hamata cv verano. Small Ruminant Research* **7**: 299–313.

Michell, A.R. (1995) Physiological roles for sodium in mammals. In *Sodium in Agriculture* (eds. C.J.C. Phillips & P.C. Chiy), pp. 91–106, Chalcome Publications, Canterbury.

Min, B.R., Attwood, G.T., Reilly, K., Sun, W., Peters, J.S., Barry, T.N. & McNabb, W.C. (2002) *Lotus corniculatus* condensed tannins decrease in vivo populations of proteolytic bacteria and affect nitrogen metabolism in the rumen of sheep. *Canadian Journal of Microbiology* **48**: 911–921.

Miron, J., Ben Ghadalia, D. & Morrison, M. (2001) Adhesion mechanisms of rumen cellulolytic bacteria. *Journal of Dairy Science* **84**: 1294–1309.

Molan, A.L., Waghorn, G.C. & McNabb, W.C. (2002) Effect of condensed tannins on egg hatching and larval development of *Trichostrongylus colubriformis* in vitro. *The Veterinary Record* **150**: 65–69.

Murdiati, T.R., McSweeney, C.S. & Lowry, J.B. (1991) Complexing of toxic hydrolysable tannins of yellow-wood *Terminalia oblongata.* and harendong *Clidemia hirta.* with reactive substances, an approach to preventing toxicity. *Journal of Applied Toxicology* **11**: 333–338.

National Academy of Science, 1972. Lead. Airborne lead in perspective. *Committee on Biological Effects of Atmospheric Pollutants.* ISBN 0309019419 9780309019415. Division of Medical Sciences, Washington, National Academy of Sciences.

Neathery, M.W., Miller, W.J., Blackmon, D.M., Pate, F.M. & Gentry, R.P. (1973) Effects of long term zinc deficiency on feed utilisation, reproductive characteristics and hair growth in the sexually mature male goat. *Journal of Dairy Science* **56**: 98–104.

Nordberg, G.F. (1975) Effect of long term cadmium exposure on the seminal vesicles of mice. *Journal of Reproduction and Fertility* **45**: 165–167.

Parizek, J. (1957) The destructive effect of cadmium ions on testicular tissue and its prevention by zinc. *Journal of Endocrinology* **15**: 56–63.

Parizek, J. (1960) Sterilisation of the male by cadmium salts. *Journal of Reproduction and Fertility* **1**: 294–309.

Parker, A.J., Hamlin, G.P., Coleman, C.J. & Fitzpatrick, L.A. (2003) Quantitative analysis of acid-base balance in *Bos indicus* cattle subjected to transportation of long duration. *Journal of Animal Science* **81**: 1434–1439.

Paustenbach, D.J. & Gaffney, S.H. (2006) The role of odor and irritation in risk perception and the setting of occupational exposure limits. *International Archives of Occupational and Environmental Health* **79**: 339–342.

Payne, J.M. & Payne, S. (1987) *The Metabolic Profile Test*. Oxford University Press, Oxford.

Payne, J.M., Dew, S.M., Manston, R. & Faulks, M. (1970a) The use of a metabolic profile test in dairy herds. *The Veterinary Record* **87**:150–158.

Payne, J.M., Dew, S.M., Manston, R. & Vagg, M.A. (1970b) Metabolic disorders of the ruminant, hypocalcaemia and hypomagnesaemia. In: *Physiology of digestion and metabolism in the ruminant* (ed. A. T. Phillipson), pp. 584–598. Oriel Press, Newcastle-upon-Tyne.

Phillips, C.J.C. & Strojan, S.T. (2002) The detection and avoidance of lead on herbage by dairy cows. *Journal of Dairy Science* **85**: 3045–3053.

Phillips, C.J.C., Chiy, P.C. & Zachou, E. (2005) The effects of cadmium in herbage on the apparent absorption of elements by sheep, in comparison with inorganic cadmium added to their diet. *Environmental Research* **99**: 224–234.

Phillips, C.J.C. & Prankel, S.H. (2007) The effects of cadmium on the welfare of animals. In: *Proceedings of a Workshop on the Effects of Heavy Metals on the Wellbeing of Humans and Animals*, (eds. L. Tudoreanu & C.J.C. Phillips), Bucharest, 2007. In Special Issue of Environmental Geochemistry and Health (in press).

Prankel, S.H., Nixon, R.M. & Phillips, C.J.C. (2004) Meta-analysis of experiments investigating cadmium accumulation in the liver and kidney of sheep. *Environmental Research* **94**: 171–183.

Prankel, S.H., Nixon, R.M. & Phillips, C.J.C. (2005) Implications for the human food chain of models of cadmium accumulation in sheep. *Environmental Research* **97**: 348–358.

Pritchard, D.A., Stocks, D.C., O'Sullivan, B.M., Martin, B.R., Hurwood, I.S. & O'Rourke, P.K. (1988) The effect of polyethylene glycol (PEG) on wool growth and live weight of sheep consuming Mulga *Acacia aneura.* diet. *Proceedings of the Australian Society of Animal Production* **17**: 290–293.

Provenza, F.D., Pfister, J.A. & Cheney, C.D. (1992) Mechanisms of learning in diet selection with reference to phytotoxicoses in herbivores. *Journal of Range Management* **45**: 36–45.

Purchala, R., Min, B.R., Goetsell, A.L. & Sahlu, T. (2005) The effect of a condensed tannin containing forage on methane emission in goats. *Journal of Animal Science* **83**: 182–186.

Reid, C.S.W. (1973) Limitations to the productivity of the herbage-fed ruminant that arise from the diet. In: *Chemistry and Biochemistry of Herbage* (eds. G. W. Butler & R. W. Bailey), Vol. 3, pp. 215–262, Academic Press, New York.

Reis, P.J. & Sahlu, T. (1994) The nutritional control of the growth and properties of mohair and wool fibers, a comparative review. *Journal of Animal Science* **72**: 18.

Reist, M., Erdin, D., von Euw, D., Tschuemperlin, K., Leuenberger, H., Chilliard, Y., Hammon, H.M., Morel, C., Philipona, C., Zbinden, Y., Kuenzi, N. & Blum, J.W. (2002) Estimation of energy balance at the individual and herd level using blood and milk traits in high-yielding dairy cows. *Journal of Dairy Science* **85**: 3314–3327.

Richter, P., Hinton, J.W. & Reinhold, S. (1998) Effectiveness in learning complex problem solving and salivary ion indices of psychological stress and activation. *International Journal of Psychophysiology* **30**: 329–337.

Rose, A.L. (1941) A clinical survey of intoxications of cattle by Sudan Grass Sorghum sudanense. *Australian Veterinary Journal* **17**: 211–219.

Rowlands, G.J., Little, W. & Ketchenham, B.A. (1977) Relationship between body composition and fertility in dairy cows – a field study. *Journal of Dairy Research* **44**: 1–7.

Scott, D. (1975) Changes in water, mineral and acid-base balance associated with feeding and diet. In: *Digestion and Metabolism in the Ruminant* (eds I. W. McDonald & A. C. I. Warner), pp. 205–215. University of New England Publishing Unit, Armidale.

Seaman, J.T. & Dixon, R.J. (1989) Investigations into the toxicity of *Echium plantagineum* in sheep. 2. Pen feeding studies. *Australian Veterinary Journal* **66**: 286–292.

Shorthose, W.R. (1977) The effect of resting sheep after a long journey on concentrations of plasma constituents, post mortem changes in muscles and meat properties. *Australian Journal of Agricultural Research* **28**: 509–520.

Shutt, D.A., Weston, R.H. & Hogan, J.P. (1970) Quantitative aspects of phyto-oestrogen metabolism in sheep fed on subterranean clover *Trifolium subterraneum* Cultivar Clare and Red Clover *T pratense*. *Australian Journal of Agricultural Research* **17**: 713–722.

Stacy, B.D. & Brook, A.H. (1964) The renal response of sheep to feeding. *Australian Journal of Agricultural Research* **15**: 289–298.

Thornton, R.F. & Minson, D.J. (1973) The relationship between apparent retention time in the rumen, voluntary intake and apparent digestibility of legume and grass diets by sheep. *Australian Journal of Agricultural Research* **24**: 889–898.

Underwood, E.J. & Suttle, N.F. (1999) *The Mineral Nutrition of Livestock, 3rd Edition*. CABI International, Wallingford, UK.

Underwood, E.J. & Somers, M. (1969) Studies of zinc nutrition in sheep. 1. The relation of zinc to growth, testicular development and spermatogenesis in young rams. *Australian Journal of Agricultural Research* **20**: 889–897.

Vanselow, B.A., Krause, D.O. & McSweeney, C.S. (2005) The Shiga-toxin producing *Escherichia coli*, their ruminant hosts and potential on-farm interventions, a review. *Australian Journal of Agricultural Research* **56**: 219–244.

Veerkamp, R.F., Gerritsen, C.L., Koenen, E.P., Hamoen, A. & De Jong, G. (2002) Evaluation of classifiers that score linear type traits and body condition score using common sires. *Journal of Dairy Science* **85**: 976–983.

Waites, G.M.H. & Setchell, B.P. (1966) Changes in blood flow and vascular permeability of the testis, epididymis and accessory organs of the rat after the administration of cadmium chloride. *Journal of Endocrinology* **34**: 329–342.

Wallace, R.J., Falconer, M.L. & Bhargava, P.K. (1989) Toxicity of volatile fatty acids prevents enrichment of *Escherichia coli* by rumen contents. *Current Microbiology* **19**: 277–281.

Ward, N.I., Brooks, R.R. & Roberts, E. (1978) Blood lead levels in sheep exposed to automotive emissions. *Bulletin of Environmental Contamination and Toxicology* **20**: 44–51.

Warner, A.C.I. & Stacy, B.D. (1965) Solutes in the rumen of the sheep. *Quarterly Journal of Experimental Physiology* **50**: 169–184.

Washington, J.W. (1985) *Metallic contamination surrounding a zinc smelter*. MSc Thesis, Pennsylvania State University, University Park.

Weston, R.H. (1985) The regulation of feed intake in roughage-fed ruminants. *Proceedings of the Nutrition Society of Australia* **10**: 55–62.

Weston, R.H. (1989) Factors limiting the intake of feed by sheep 12. Digesta load and chewing activities in relation to lactation and its attendant voluntary roughage consumption. *Australian Journal of Agricultural Research* **39**: 671–677.

Weston, R.H. (1995) Some aspects of constraint to forage consumption by ruminants. *Australian Journal of Agricultural Research* **47**: 175–197.

Weston, R.H., Lindsay, J.R., Peter, D.W. & Buscall, D.J. (1989) Factors limiting the intake of feed by sheep. 14. Comparisons of voluntary feed consumption and various transactions in the alimentary tract between lambs and sheep fed roughage diets. *Australian Journal of Agricultural Research* **40**: 625–642.

Wilkinson, J.M., Hill, J. & Livesey, C.T. (2001) Accumulation of potentially toxic elements by sheep grazed on grassland given repeated applications of sewage sludge. *Animal Science* **72**: 179–190.

Wilkinson, J.M. Hill, J. & Phillips, C.J.C. (2003) The accumulation of potentially toxic metals by grazing ruminants. *Proceedings of the Nutrition Society* **62**: 267–277.

Wright, M.J. & Davidson, K.L. (1964) Nitrate accumulation in crops and nitrite poisoning in cattle. *Advances in Agronomy* **16**: 197–247.

Wythes, J.R., Shorthose, W.R., Schmidt, P.J. & Davis, C.B. (1980) Effect of various rehydration procedures after a long journey on liveweight, carcasses and muscle properties of cattle. *Australian Journal of Agricultural Research* **31**: 849–855.

Chapter 8
The Management of Sheep

P.J. Goddard

Abstract The processes used to manage sheep in the broadest sense and the way these are perceived by the sheep will have a critical impact on the welfare of individual animals. While these impacts may be difficult to assess and more difficult to quantify, it is clear that the skilled stockperson has the ability to reduce the impact of some of the negative experiences that potentially aversive procedures – individual handling or transportation for example – may engender. There are additional challenges both to evaluate the perception of animals of long-term or chronic challenges in comparison with short-term, acute events and to reconcile the welfare of individuals with that of the flock as a whole. The importance of the human-animal relationship as a defined component of stockmanship is being more widely recognised. Selection and training of stockpersons will become increasingly important, yet identification and assessment of the complete range of desirable characteristics may not be easy. Mechanisms to reward those responsible for livestock – drivers of livestock vehicles for example – on the basis of animal welfare outcomes rather than work efficiency goals should be encouraged. Different management systems may expose sheep to different amounts of human contact. Where such contact is limited the reactivity of sheep to individual management procedures may be greater. Selection of breeds, or individuals within breeds, which are more tolerant of a reduced level of handling may be one way to deliver welfare benefits in the future, since it is hard to envisage an alternative situation where adequate adaptive experiences which modulate the sheep's reactivity could be provided under practical farming conditions. As the needs of sheep become better understood there is greater opportunity to fit the system to the animals. There are specific management situations where considerable research effort has delivered the potential to improve welfare – transport for example – and where legislation has been enacted to enforce higher standards. There are also a number of situations (castration for example) where a review of what is considered a routine activity, based on a cost: benefit assessment, may be valuable. Current moves towards more extensive or ranched management systems reinforce the need to ensure that well-adapted breeds are selected which are appropriate to the environment and that stockpersons have the skills needed to

P.J. Goddard
Macaulay Institute, Craigiebuckler, Aberdeen, AB15 8QH, UK

C.M. Dwyer (ed.), *The Welfare of Sheep*,

work with the sheep under these conditions and keep welfare considerations in the forefront of their mind.

Keywords Welfare · Management · Cost-benefit · Stockmanship

8.1 Introduction

There is a range of reasons why a flock of sheep may need to be managed but the way the animals are cared for in any particular system is key to ensuring a high standard of animal welfare. Classically the purpose of flock management is to optimise production (for example of meat or wool output). This may be achieved variously through efficient use of feed resources, control of breeding, and prevention of straying, predation and of theft. Individual flocks may be focused on a sub-set of objectives, for example fat lamb production or the breeding of terminal sires. How objectives are best achieved will depend on a range of resource factors including climate, terrain, natural feed availability, labour skills and cost and seasonal variation. These have been discussed elsewhere in this book. It is likely that in many cases objectives and resources will be well-matched through tradition in a particular area. More specific details of particular farming systems are provided in Chapter 6. Such traditional production goals and the methods used to achieve them are increasingly being influenced by public concern for animal welfare and it is the purpose of this chapter to consider a number of common management methods and actions from a welfare perspective. Additionally sheep can play a role as environmental "managers" whereby their grazing activity is required to deliver a particular visual or biodiversity outcome. Even in these situations, economic costs can be minimised by adopting appropriate management systems and the same range of concerns applies.

In some countries livestock, including sheep, are viewed as the "bank" to be cashed in through sale or slaughter when needs arise. In areas of the world where subsistence livestock farming is practised, effort is crucially directed at keeping animals alive. When this is the management target it may be useful to employ the five freedoms described in Chapter 1 as an aid to achieving this objective, rather than acting as a more specific framework for welfare assessment in their own right.

In any system where man manages animals there should be an assumed level of responsibility for their well-being. It could be argued that as the "amount" of management increases, so the level of responsibility increases. Issues about the ability to perform natural or instinctive behaviours have been addressed in Chapter 1. Any management action has the ability to alter the welfare state of an animal – in both positive and negative ways. Some actions have the potential to unequivocally improve the welfare state of the animal (for example the provision of an optimal level of nutrition). Others clearly reduce the welfare state, for example long distance transport from which the individual usually derives no benefit. The majority of actions fall somewhere between these extremes. Generally the financial constraints of the system require that some management actions impinge on the sheep but may not be in their direct interests. Animal welfare advocates in a number of countries have focused on the need to reduce the impact of these actions – for example pressing for the adoption of more humane methods of castration and questioning the conditions of transport of live animals. Indeed the need to perform some techniques at all is being questioned (for example, export for slaughter). Thus if we accept that management requires the performance of some techniques which the sheep would not themselves "choose" in exchange for a viable or profitable system, it is crucial that any negative impact on the sheep be minimised. In general this can be addressed through skilled stockmanship and the development of systems that focus more on the needs of the sheep rather than the needs of the enterprise. As readers will understand from other chapters in this book, our knowledge of responses of sheep and their needs is rapidly increasing to the point where it is possible to promote management systems devised with the sheep's perspective in mind. In any event, a system which works *with* the animals has many advantages.

Some actions will benefit individuals, others may be neutral or benefit the flock overall but is it right to consider the welfare of a flock as a valid entity? The concept of welfare is generally recognised to be something applying to an individual but when considering management activities applied to large groups/flocks there may be some merit in looking for a potential overall utility. Indeed it is common to refer to "flock welfare" though this is seldom defined. If this is accepted as an appropriate stance can something similar to the approach of Broom and Johnson (1993) be adopted? They suggested that a measure of how poor welfare is be multiplied by the duration of poor welfare to give a useful measure of the overall magnitude of a problem. Can we in a similar way simply derive a mathematical average a number of welfare "scores" in relation to individual impacts? A pictorial expression of this argument is given in Fig. 8.1. Perhaps an example might help. Consider a hill sheep flock in which an individual has suffered an injury which would be recognised as having a high impact on the individual. To gather the flock to enable the individual to be treated would impose a small but measurable welfare cost on a large number of individuals. Where does the balance lie? (Of course in many countries, codes of welfare practice would require action to be taken in any case).

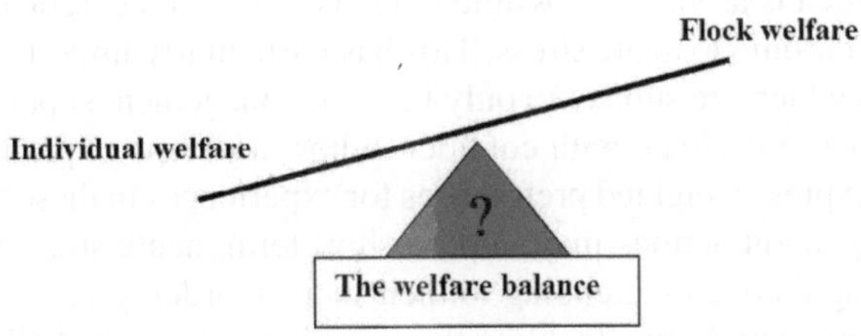

Fig. 8.1 The notional balance between individual and flock welfare

Addressing this dichotomy is important when partitioning the welfare benefit or disbenefit which may follow from some of the more specific management activities to be considered in subsequent sections. It is also helpful when making a welfare audit of a production system: if the welfare of the majority of the animals is of a high order but there is a small number whose welfare is considered poor, should this system be rated more highly than one where *all* the animals have an intermediate standard of welfare? This is considered further in Chapter 10, along with the potential to assess the cost of welfare gains.

Similarly some management actions are applied to all the animals in a flock in order to prevent a number of animals (but not necessarily the whole flock) from experiencing a particular problem – for example tail docking of all lambs in order to prevent a proportion suffering from the serious consequences of fly strike when fly larvae invade the sheep's tissue (myiasis).

Another theme to emerge in this chapter will be questioning the balance of management needs versus tradition, for example the continuance of castrating male lambs in flocks where the lambs are marketed before the breeding season. Methods to monitor and measure how successfully a farmer is achieving welfare goals are not well developed in the sheep sector. Increasingly enterprises are attempting to achieve multiple objectives, of which achieving a high standard of welfare is but one. For any action which is likely to affect the welfare of animals the need to perform the action should always be questioned.

A key influence on the welfare of individual animals is the way in which stockpersons interact with them. In many ways this may be considered to override the specific nature of the individual techniques adopted and will be explored later. The skill of the stockperson and the adoption of proactive rather than reactive approaches are crucial to the delivery of a high welfare management system.

Due to the multiplicity and rapid evolution worldwide of legislation concerning animal welfare, this chapter will not attempt to quote specific standards and requirements in relation to all the techniques discussed, though clearly there are overriding principles involved. Thus it is beyond the scope of this chapter to synthesise the evolving regulation of sheep transport, for example, to provide a unified guide as to what *must* be done to remain within the law. However, it is hoped that the principles described will allow the reader to gain a feel for what *should* be done. Similarly, not every single management activity (housing, for example) will be considered in detail.

One issue which is as yet not possible to resolve from a theoretical point of view is how to relate chronic to acute stress. This has particularly important implications for sheep flocks which are subject to only limited management supervision. It is not currently possible to evaluate with confidence how a sheep (or probably any other animal) would express weighted preferences for experiences in these two categories. Yet some management actions may impose short term, acute stress (e.g. gathering animals, handling them and attending to their feet) in order to resolve a potentially chronic (e.g. footrot) problem. Tackling this question is a major challenge to animal welfare scientists.

8.2 The Evolution of Human Interaction with Sheep

Sheep were the first animals to be domesticated by man for husbandry purposes (Clutton-Brock, 1987), presumably due to their size and temperament as much as their meat and fleece/hide characteristics. Their gregarious, flocking nature facilitated their control and their physical ease of handling meant that management systems could easily evolve to develop the potential for their multiple uses. Over the centuries, selection of individuals with desirable characteristics resulted in a multi-potential animal species with breeds developed to thrive in particular environments. Management systems evolved to reflect a synergistic relationship between the shepherd and the sheep flock (Budiansky, 1994). In exchange for loss of some self-sufficiency traits, sheep were provided with food and protection. Traditionally the shepherd "lived" and moved with the sheep, a practice which can be found today in nomadic regions but there are also remnants of this system in some Mediterranean countries.

Depending on the degree of integration between the shepherd and the flock, the method of shepherding varies. Today sheep farming may still be extremely extensive, for example the transhumance or nomadic pastoralist systems found in some European, Asian or North African countries, where sheep are moved around by a herder. In contrast, over the second half of the twentieth century, many more intensive systems were developed, based on a high degree of pasture management and often a certain degree of housing. These systems and their forerunners necessitated a different kind of human contact, with the stockperson bearing more responsibility for control of resources – feed and shelter – and restriction of natural socialization by the sheep. However, we are now witnessing a move away from such intensive systems in northern Europe, primarily due to economic forces and the cost and availability of labour in rural areas. The public desire for more "natural" production systems has also been influential. It is important not to equate extensive systems with low input systems. Some extensive systems require a considerable management input and are reliant on a high level of knowledge; they are clearly not 'hands off'. However, on an individual sheep basis, human contact will likely be reduced.

Some breeds are now recognised to fulfil specific production functions and some are recognised to be more 'self sufficient' than others, particularly hill and upland breeds. Sheep can also tolerate a range of environmental conditions, although selective breeding has meant that some specialised breeds may not be well adapted to the whole range of conditions experienced by primordial sheep. The docility of sheep (or at least their lack of overt response to management actions in many cases) compared to many other species may have meant that sheep have had to tolerate poorer standards of husbandry.

The interesting concept of gentling has been often used, describing the reduction in general reactivity which follows regular human exposure. It would appear that the effect is relatively non-specific and that it may protect the animal against a range of conditions not previously encountered. While animals that have been 'gentled' may be in a better state to cope with new challenges, it is less clear how the impact of the gentling process itself on the animal is to be assessed. It may also become possible to deliberately habituate sheep to particular activities so making them more tolerant of these activities. The continual presence of the shepherd in some traditional migratory transhumance systems or in the traditional shepherding methods of southern Europe or Asia where the shepherd essentially "lives" with the flock means that the sheep become habituated to him or her and any handling activities are likely to be perceived as less aversive by the sheep, compared to systems in which sheep experience human contact only infrequently.

8.3 Human-Animal Interactions

The quality of human-animal interaction is perhaps key to mitigating against any potential negative experiences sheep may have when they are handled. When sheep have experienced an aversive situation they will be reluctant to experience this again (Rushen, 1986b,c). As a research tool, quantifying this reluctance can be a way of ranking potentially negative experiences. Until recently, there was little information to be found in the scientific literature in relation to the interrelationships of man and farm animals as specific interactive events. Boivin (2003) reviewed this literature in relation to the reaction of sheep to humans and handling but did not find any reports of the specific stockmanship factor. Yet this factor would appear to be fundamental to the way management actions are perceived by the individual animals themselves, with consequent welfare implications. It is now well recognised that the experiences of sheep undergoing management operations will be impacted by the way human-animal interactions occur. From the human side, if the impact on the sheep is aversive, it may mean that repetitions of the same (or related) operation become more difficult. Beausoleil et al. (2002) considered that sheep probably consider humans as predators and thus, from a practical perspective, the stockperson would do well to consider how management techniques are perceived by the sheep. From the perspective of sheep welfare, it is appropriate to adopt management techniques which minimise their impact on individuals.

In many countries, there is a move towards the establishment of larger, more extensively-managed flocks and thus a reduction in human-animal contact. When sheep are handled infrequently there may be a tendency for the stockperson to do more on each occasion. In these situations stockperson interactions with the sheep are more likely to be perceived as aversive and it is harder for the stockperson to maintain individual animal relationships. There is also less chance for positive human-animal relationships to develop. From a practical point of view, this may require different approaches to be adopted. As far as the modification of behavioural responses of animals is concerned, a number of studies have suggested that early life experiences can modulate subsequent responsiveness (e.g. Boivin et al., 2000; Goddard et al., 2000). However, in many of these experimental conditions, the degree of early life manipulation shown to have any effect is significantly more than would be reasonably possible in practice and effects may also be modulated by the presence of the dam (Boivin et al., 2001). A number of studies (with species other than sheep) have combined increased handling in early life with supplementary food provision which would appear to act as a positive reinforcing stimulus. However, there appear to be few similar studies with sheep from which to draw parallel conclusions. Handled sheep generally show less avoidance of humans, though few studies have been conducted in this area (Hargreaves and Hutson, 1990b; Mateo et al., 1991). Hutson (1985) showed that training sheep with a food reward resulted in a reduction in their stress response and improved the speed of handling, an effect which was apparent a year later. The neurological basis for such changes in behaviour is not well understood and even for rodents (where considerable research effort has been expended) it is not clear what modulation of the stress response to handling means in terms of welfare.

The sheep's view of the human and the way the human can impact on the animals they manage in a more general way has been the subject of increasing research over recent years. The general thrust of these studies suggests that early-life experiences can modulate subsequent reactivity to events which may be aversive to the sheep but that there may be sub-sets of animals in any given flock that have a predisposition to being more or less reactive. There is also interest in the use of techniques that may mitigate against the effects of animals remembering aversive experiences at the time they occur – for example providing a feed reward after an aversive event. It is apparent that certain handling activities impose compound stressors on the sheep. Shearing, for example, represents a cluster of activities from gathering to the act of shearing itself. In one study, repeatedly exposing sheep to sham shearing (a procedure resembling shearing save for the removal of wool) had little effect on the measured stress response (Hargreaves and Hutson, 1990e). Clearly knowledge of sheep behaviour will allow the shepherd to be responsive to the sheep but when large mobs are to be dealt with there is little scope for individual care and the processes become essentially mechanical. In these situations, it is important that the system itself is designed to be as sheep-friendly as possible. How to compensate individuals for the loss of individual care is a difficult problem. There may be some scope in selecting breeds or individuals within breeds which have a greater tolerance of the handling processes or react in a less aversive way to handling techniques but this would need to be balanced against the need to retain a degree of responsiveness to facilitate their movement. To some extent this will have happened either directly or indirectly over the years as sheep with a poor tolerance of the process will likely have been eliminated from the flock. Some elements of the aversive experience may never be able to be overcome but in the context of human-animal interactions more advanced training of stockpersons may be beneficial. Whether enhanced human-animal relationships in sheep affect productivity (as might be inferred from studies with other livestock species) remains to be determined.

There have been studies which considered the effect of different stockperson behaviour patterns on animals such as dairy cattle and pigs, though few on sheep. Stockperson behaviour can be modified through training and if it were possible to identify subtle behavioural elements, then human-animal interactions could be improved further. However, some aspects of stockperson behaviour appear to be inherent in their character and a number of studies have attempted to develop psychological profiling for use when selecting stockpeople. This is an important development since it seems that some people may never be ideally suited for this kind of work. So as well as handling techniques, personality traits of stockpersons have been shown to influence a number of livestock species. A contemporary discussion about the influence of attitudes of stockpersons towards animals has been presented by Hemsworth and Coleman (1998). Some behavioural interactions with livestock are likely to be refractory. Thus animals which have become difficult to handle because of a poor earlier experience may engender more negative behaviour by the stockperson during subsequent handling. Similarly, animals that are easier to handle may more likely elicit positive behaviour from the stockperson. Selection of sheep on a genetic basis for traits which fit them for a particular management situation (in addition to production and survival traits) does not appear to be an option that is being considered at present.

It is important to remember (as described earlier in this book) that sheep will respond differently depending on whether they are alone or in a group. While many macro-management activities (e.g. gathering) will be directed at the flock, many of the potentially more aversive activities will be directed at individuals separated from the flock. Ensuring handling facilities are fit for their purpose and effectively augment the natural behaviour of the sheep will benefit both the sheep and their handlers.

8.3.1 Stockperson Training and Selection

What is the best method to improve the skill of stockpersons? What are the key elements to include in a training programme? For example, important factors to affect the welfare of sheep during road transport are the behaviour of the stockpersons who load and unload them and the way the driver drives. While some risks to sheep welfare may occur through deliberate actions or inactions, most are accidental and can be minimised through good training and, if possible, a reward system for the stockpersons which promotes good welfare rather than the quick performance of a task. It may, of course, be a challenge to develop such systems.

There are a number of elements common to a range of livestock care training needs:

(a) Knowledge of sheep behaviour
(b) Knowledge of relevant legislation
(c) Knowledge of the correct care and use of equipment
(d) Knowledge of specific techniques to be performed
(e) Knowledge of specific breed and health problems likely to be encountered
(f) Knowing when to seek help and where to seek help from
(g) Recognition of own limitations and willingness to learn

In addition, if characteristics of the stockperson can have a genuine effect on the human-animal relationship what are these characteristics? Most of the work on selecting livestock workers has been in relation to intensive systems. While many of the principles may be similar, not enough is known about what to look for in relation to more extensive management systems. Stockperson training may be able to overcome some specific flaws in the way human-animal interactions occur. While agricultural training has typically focused on technological aspects (which are often easy to quantify and assess) some of the "soft" skills are clearly important too. Unfortunately, it is not immediately clear what needs to be taught, though a sound understanding of sheep behaviour would be crucial. Stockperson training to improve behavioural interactions could be fundamental to enhancing welfare. This could be achieved through formal training (for example some elements of behavioural theory) or more on-the-job approaches related to animal handling skills and practical demonstrations of how improved human-animal relationships will make work easier and increase job satisfaction. More detailed consideration of this aspect and a review of how change in stockperson behaviour may need to be preceded by a change in beliefs are beyond the scope of this chapter. It is interesting to consider whether, when a number of individuals look after the same group of animals, poor technique practised by one may make the work of the others harder. It has been shown that sheep can distinguish visually between individual sheep and familiar and unfamiliar humans (Kendrick et al., 1995, 2001) and other cues may additionally be shown to be of importance in the way sheep might or might not generalise negative experiences to all humans. The development of a stockperson reward system may be appropriate for specific tasks – for example by careful driving to the abattoir, animal welfare prior to arrival and meat quality thereafter are improved and the driver could receive a bonus based on some quantitative attribute – e.g. fewer bruises. This would enhance welfare in contrast to a bonus payment for rapid delivery. The provision of effective and relevant training in countries where many stockpersons operate independently is a particular challenge.

Additional attributes of an effective stockperson have been suggested. These include good observational skills, empathy with their charges and the ability to plan and think ahead. In many countries certificates of competence in various technological aspects are required: driving long-distance transporters; use of chemicals in sheep dips etc. Some assessment systems require regular up-dating and refresher training. Certificates should be readily revoked if poor practice is identified.

Stockperson safety may be influenced by the effectiveness of the human-animal relationship and the knowledge of the stockperson. Some animals, entire adult rams for example or some ewes with young lambs, can be potentially dangerous. Training must cover the care needed to be exercised by the shepherd and how this defensive approach may affect the human-animal relationship. Clearly a good basic awareness of animal behaviour will allow the stockperson to defuse situations of potential conflict.

8.3.2 The Role of Dogs in Managing Sheep

Dogs can play a large part in the efficient management of a flock of sheep, performing both herding and protection. The types of dog bred for and suited to these tasks differ in their behaviour, though historically one dog may have fulfilled both roles. As natural predators of the sheep, the sheep's primary response to herding dogs is avoidance or distancing. The sheep's fear response and knowledge of flight distances (to dogs and humans) have been used advantageously in gathering mobs and individuals. However, the use of herding dogs in handling areas may be problematic, as sheep will generally turn to face them in confined areas. While social isolation has been shown to cause a substantial heart rate response in sheep, a greater response was seen when sheep were approached by a man with a dog (Baldock and Sibley, 1990). In order to protect the welfare of sheep, dogs must be suitably trained and, in particular, must not be allowed to injure sheep. A useful review of the differences in behaviour between herding (livestock conducting) and protecting dogs is provided by Coppinger et al. (1987).

8.4 Extensive and Low Input Systems

There is currently considerable interest in extensive sheep management systems. The question could be posed: "do we actually 'manage' sheep under these range conditions?" If so, what are the welfare implications of such management and are any resultant welfare compromises acceptable? In very extensive or ranched situations where sheep can range over large areas, all that may happen from a management perspective is a periodic gather and selection of animals for slaughter or release back for breeding. While some may consider this to be far from what is meant by being managed, such systems may require input in terms of breed adoption and the selection of animals for fitness traits to enable them to adequately cope with the system. These aspects are explored in Chapter 6. Care for individuals is very limited if it can be applied at all. Contrast this with the intimate contact between shepherd and sheep which occurs in traditional migratory systems.

There are, in ranching or easy-care systems (for example those developed in New Zealand in the 1970s), possibilities for poor welfare to arise. As individuals will not be inspected regularly, those with injury or disease may go undetected. In some areas there will also be a risk of predation, particularly for young lambs. It is recognised that to be acceptable, such systems crucially depend on the choice of breed or the selective breeding of sheep with desirable characteristics to enable them to survive and essentially look after themselves in these environments (Fisher and Mellor, 2002). The transition to an easy-care system needs to be well managed to avoid problems en route. In the first chapter, the question "Does a low level of inspection carry a welfare cost to the sheep...?" was posed. It could be argued that by using appropriately adapted breeds there is not an unacceptable threat to their welfare. For example, such breeds must show a high resistance to parasitic disease and problems of lameness, and a low incidence of lambing difficulty. Problems with ectoparasites, primarily blowflies, will be related to fleece density, particularly around the tail and anus, so selection for coat characteristics will be important too. It may be possible to move towards an easy-care breed more rapidly through advanced breeding techniques if appropriate traits can be identified and are shown to be highly heritable.

Many countries' welfare legislation and codes of practice contain recommendations about regular inspection. In more extensive situations it is recognised that such an activity is not achievable and here codes recommend that the breeds chosen should be those well-adapted to the prevailing conditions and thus likely to experience fewer problems. However, at times of the year when problems might be anticipated, inspection must occur at a suitable frequency.

Many critics of intensive farming find extensive systems more acceptable, due to the ability of animals to express "natural" behaviour patterns. However, even if this capacity adequately offsets other notional threats to welfare, when animals are handled or transported they may exhibit more extreme reactions due to the novelty of the situation and their lack of exposure to handling activities in general. It is worth considering whether animals from these systems should be subject to the same system of handling, transport and marketing as sheep from more traditional farming systems. However, there is currently insufficient evidence to indicate whether it is appropriate (or possible) to consider any specific actions which could reduce the likely impact of such acute procedures.

8.4.1 Problems in Very Hot Areas

High ambient temperatures and high solar radiation have the potential to challenge the sheep's thermoregulatory systems. If animals lose the ability to thermoregulate, serious welfare (and survival) problems may follow. The ambient temperature above which active heat loss processes occur (e.g. sweating and panting) is defined as the upper critical temperature. An alternative definition of the upper critical temperature is the ambient temperature above which core body temperature begins to rise – this could be used to mark the point at which welfare begins to be compromised (Silanikove, 2000). In domestic animals a rise in body temperature marks the change from a potentially aversive state to a potentially harmful stage. Loss of water through evaporative heat loss may lead to consequences for solute balance. From a management perspective, the adequate provision of water, other nutrients and shade shelter is crucial to avoiding this noxious state. It should also be remembered that animals in different physiological states (e.g. pregnancy) may be less able to withstand heat stress. For sheep, the degree of fleece cover will affect their ability to cope. There is little published information which specifically considers the effect of heat stress on recognised welfare indicators, compared to consideration of the effects of heat on productivity. Management systems which prevent sheep from making behavioural decisions which would reduce their heat stress should be avoided. Thus sheep may choose to rest (or at least reduce their active behaviour) during the hot daytime and graze at night when it is cooler. Indeed high daily temperatures may be partially mitigated if temperatures fall sufficiently at night (Silanikove, 1998). Problems due to heat may be compounded if the area is also very dry and water intake limited.

8.4.2 Problems Due to Cold Conditions

The lower temperature below which the sheep has to increase heat production to maintain thermal balance is defined as the lower critical temperature. Between this and the upper critical temperature lies the thermoneutral zone. Below this lower critical temperature heat production will increase; the point at which heat production is maximal is determined by summit metabolism: thereafter the animal's core temperature will decline if ambient temperature falls still further. Newborn lambs, because of their short, wet fleece and small size, are particularly susceptible to hypothermia (see Chapter 5) and have physiological adaptations to dramatically increase heat production (e.g. through non-shivering thermogenesis fuelled by brown adipose tissue), at least for a limited period immediately after birth. This capability is influenced by the nutritional state of the pregnant ewe and thus the welfare of the neonate is safeguarded through ensuring adequate ewe nutrition. Providing lambing conditions which enhance rapid ewe-lamb bonding (particularly in exposed environments) will ensure that the ewe provides adequate care for the lamb and encourages the lamb to suck. Intake of a sufficient volume of colostrum is crucial to lamb survival, especially under extensive management conditions. Choosing appropriate breeds with strong maternal characteristics will also enhance lamb survival and welfare at this time. Thus hill ewes in particular need to possess good mothering abilities – for example the motivation to keep the lamb close to them. For adult sheep, the degree of fleece cover also influences the lower critical temperature, which is as low as -20°C for a sheep with 20–30 cm of fleece (Slee, 1987). If the fleece is wet and/or conditions are windy with no opportunity for shelter, the lower critical temperature will be higher (Joyce et al., 1966).

8.5 Management Actions

8.5.1 The Balance of Benefit

This section will consider the cost-benefit of a number of common management practices from the individual sheep's point of view. It is not the intention to consider an exhaustive list of actions which may be undertaken, though the general principles may apply equally. While the cost part of the equation (regardless of the size of the cost) will always apply to the individual sheep, the benefit is not always so clearly apportioned. Table 8.1 illustrates an attempt to roughly apportion these benefits. There are clearly many assumptions which may influence the placing of activities

Table 8.1 Suggested beneficiaries of some actions associated with flock management

Benefit primarily to individual sheep	Benefit mostly to the individual sheep	Benefit equally to both individuals and the enterprise or flock management	Benefit mostly to the enterprise or flock management	Benefit primarily to the enterprise or flock management
	Tail docking*	Shearing	Identification	Castration
	Mulesing*	Dipping to prevent scab	Gathering for inspection	Milking
Are there any actions to place here?	Dipping to treat scab	Prophylactic vaccination/preventive medicine	Breeding for fitness traits	Transport
	Prophylactic footcare	Ultrasound scanning		Slaughter
	Treatment of injury and disease			Breeding management

*but see comments in the text

in the table. When the benefit moves towards the economic profit of the enterprise it could be argued that the acceptable cost imposed on the sheep should be less. As with many cost-benefit assessments, when the currencies on either side of the equation are different, the assessment remains subjective. In some cases the benefit cannot be clearly assigned. Consider tail docking for example. If it is assumed that not all individuals with entire tails would suffer from fly strike, the cost-benefit of docking these individuals is weighted in the direction of the cost. If the situation is viewed from the perspective of an animal that otherwise would have been affected, the balance may be in favour of docking. Returning to the notion of flock welfare as an entity, knowledge of local conditions and the likelihood of problems arising may be the only way to allow a more pragmatic balance to be struck. Some problems, such as lameness, present as both economic and welfare issues: resolution can be through a range of activities, each with their own cost to individual sheep (welfare impact) and economic cost to undertake. Solutions will have different long-term efficacy. Interestingly, Table 8.1 does not have any entries in the column indicating benefit only to the sheep. This is because any action that might be undertaken is likely to have at least a marginal benefit for the enterprise: for example provision of optimal nutrition while of value to the sheep should also enhance the profitability of the farming operation.

A further way to consider this is that when the balance of the action is in favour of the enterprise's production goals then only a small welfare cost to the individuals is acceptable. When the balance of the action is in favour of the individuals then a higher welfare cost is tolerable. In many cases a proactive approach is based on the perception of the likelihood of an event occurring. With a proactive approach, the stockperson is in control of the situation and can pay more attention to the likely cost-benefit of any action. When reactive actions are required these may not always allow the best approach to be adopted.

There is also the issue of when to intervene to protect the welfare of individuals or the flock. Some people would consider that any deterioration in the welfare status of a flock requires intervention whilst others would recognise set points at which action should be initiated. This is an ethical issue on which everyone has their own viewpoint. Trying to take the sheep's viewpoint is the challenge for animal welfare science to address and can be approached by attempting to develop a hierarchy of welfare needs. An example of a hierarchy was recently developed through an expert workshop (Waterhouse et al., 2003). When delegates were asked to suggest potential welfare compromises a sheep might consider important, those most frequently identified were related to inadequate nutrition, lameness, ectoparasites, stockperson effects and inadequate facilities and equipment. Another way to look at the whole range of concerns is that some represent day-to-day considerations (e.g. foot problems) whilst others relate to catastrophic events (e.g. severe weather/sheep scab infection).

8.5.2 General Handling and Restraint

There are many general activities which relate to gathering/mustering and handling. The methods adopted are varied and depend on the flock size and the terrain and area over which the animals range. Since disturbance due to gathering will generally have a negative impact on the sheep (and may on occasion result in some injury or other detrimental effects) the methods adopted will ideally exploit the natural behavioural patterns of the sheep. Due regard should also be paid to the prevailing weather conditions and the need to complete the task as efficiently as possible, especially if sheep have limited opportunity to eat, drink and rest for considerable periods. The quality of facilities available to the stockperson will influence the impact of handling on the sheep. Clearly these should be fit for purpose and in good working order. Handling pens and raceways which are designed on good behavioural principles will be of benefit to both the sheep and the shepherd (Grandin, 2000). For example, in enclosed systems sheep should have a clear view of the exit or be able to see the way ahead. Depending on how the sheep are desired to move, slatted sides for pens and some raceways avoid visual isolation of individuals yet solid sides for sections such as drafting races will improve handling by reducing the visual exposure to handlers and reducing balking. If the working environment is good, the stockperson can focus more on the animals' needs (see later).

Sheep are gregarious animals and isolation (especially visual isolation) may be particularly stressful (Kilgour and de Langen, 1970; Price and Thos, 1980; Cockram et al., 1994). Isolated sheep will show significantly increased vocalisation and attempts to break out. Thus management systems should aim to reduce the need to isolate individuals (especially for prolonged periods) for example by designing handling facilities in appropriate ways. Management activities which require sheep to be dealt with as individuals can often minimise the degree of isolation, for example by not using pens with solid sides so allowing the sheep to remain in visual and auditory contact with the remainder of the flock. Commonly, sheep will be restrained either manually or using some form of mechanical device. Restraint is aversive to sheep (Rushen, 1986a) and a range of commonly-performed procedures have been shown to induce increases in plasma cortisol concentrations which are equated with stress (Hargreaves and Hutson, 1990a). The degree of aversion should be minimised through good technique in order to make the technique less difficult to repeat in the future. Catching or pulling sheep by their fleece alone is harmful to them and should not be done. Not only is it likely to be painful to the sheep, if performed shortly before slaughter there is a high probability that damage will lead to the carcase being downgraded or at least require additional trimming (Cockram and Lee, 1991; FAWC, 1994; Knowles et al., 1994). Other activities such as lifting or dragging sheep by their fleece, seen historically as acceptable or not warranting criticism, should not be performed. Horns, if present, can break off if animals are handled roughly or are young. The use of electric goads will be painful. Electro-immobilisation is not considered a humane method of restraint in sheep and it has been shown to be highly aversive (Grandin et al., 1986; Rushen, 1986a).

Well-designed facilities and equipment which are fit for purpose should reduce some negative effects of handling on the sheep and improve the working conditions of the stockperson (see Grandin, 2000). Stockpersons should be aware that different breeds of sheep show different responses to handling (Whateley et al., 1974).

8.5.3 Identification

Ear notching or punching, tattooing and ear tagging are the primary means by which individual sheep have been permanently identified. In some common grazing situations a simple flock mark is sufficient to allow ownership to be proven. Individual identification assumes an additional value when epidemic disease and human health are concerned. As indicated in Table 8.1, identification is of minimal welfare value to the individual sheep. There may be some value to other sheep in subsequent generations if selection for disease resistance or other fitness traits such as maternal behaviour is practised. Thus since the individual sheep bears the cost of the technique (e.g. the pain associated with insertion of an ear tag, with a small chance of subsequent infection or fly worry) with little if any benefit, it is especially important to ensure that the optimum technique is chosen and performed proficiently. For example tags should be inserted at the correct location, avoiding significant blood vessels, and not introduced on days when fly activity is high (Hosie, 1995; Edwards and Johnston, 1999). The requirement in some countries to apply tags to facilitate disease control and movement regulation may require additional consideration on welfare grounds. Modern developments such as the use of Passive Integrated Transponder (PIT) tags, either placed subcutaneously (usually at the base of the ear) or introduced via intra-ruminal boluses, may offer advantages as they are less invasive but suffer from the drawback of requiring some form of reader device to interrogate them. Electronic systems do, however, require less physical handling to subsequently identify individual sheep. Electronic identification using PIT tags can also be incorporated into conventional ear tags and thus a dual visual and electronic system can be used. With no battery to fail, PIT tags will function effectively for the life of a sheep. When using conventional ear tags one should be aware that some have the potential to cause more damage to the ears than others (Edwards and Johnston, 1999) and may have different retention rates. This potential for tag loss, which is more likely for extensively-kept sheep, may result in a requirement for double tagging. There may also be a small loss of intra-ruminal identification boluses.

8.5.4 Lambing Intervention

Lambing time is often one of the key points at which human-animal interactions have an especially important impact. In more intensive systems, the stockperson has regular contact which allows timely intervention if difficulties arise. Recent observations suggest that, in some extensive systems, human intervention at lambing time may have mixed effects: for sheep unused to human involvement, such intervention may prolong parturition, potentially disadvantage the ewe and reduce lamb survival through a variety of mechanisms (Fisher and Mellor, 2002). Factors such as preparedness on the part of the shepherd and allowing the ewe-lamb bond to develop are critical. Much will depend on the breed and the system but there is clearly the potential to either enhance or reduce welfare at this time; neonatal survival is a widely-recognised welfare issue – with death being a severe consequence. This issue has been covered extensively earlier, in Chapters 3 and 6, together with a consideration of the welfare impact of more intensive systems which do not, for example, allow the needs of a ewe to seek isolation at the time of parturition to be fulfilled.

8.5.5 Weaning

While the natural weaning generally occurs around the time of re-breeding, many lambs will be weaned earlier. This is particularly the case for milk sheep. The age at which lambs are weaned may affect their welfare. A range of weaning strategies has been investigated in terms of behavioural and physiological responses of lambs (Napolitano et al., 1995, 2003; Orgeur et al., 1998; Boivin et al., 2001; Sevi et al., 2003). There does not seem to be a consensus view of the effects, though in the most recent study (Sevi et al., 2003) it was found that a gradual separation of ewe and lamb was not a suitable strategy when lambs were to be artificially reared. Since weaning may be a stressful experience for lambs, it is not surprising that infectious disease may be a problem at this time, through a reduction in the efficacy of the immune system (immunosuppression). Weaning practices may also result in the mixing of unfamiliar groups of lambs, thereby exposing individuals to new disease challenges.

8.5.6 Castration

This is a technique which provides benefit primarily to the management of the enterprise, through avoiding unwanted mating and the birth of early lambs or lambs of poor genetic merit (Table 8.1). As adults, castrated males are easier to handle than entire sheep (this may deliver a minor benefit to the individuals concerned of being less stressed during handling) and are less prone to fighting. More generally, castrated animals may also have more desirable carcase characteristics and in some situations such animals may be sold when older and heavier without a decline in carcase quality. Lambs born early through unplanned mating may have poorer survival chances, and mating of relatively immature females may prejudice their welfare. Castration has been regularly practised in a number of other farmed species in the past, notably pigs and cattle. In the modern pig industry with rapidly-growing strains of pig, castration is no longer performed in most countries and fears that boar taint from entire animals would deter purchasers have not materialised. In beef cattle, castration continues to be practised, mainly on human safety grounds (as animals are retained until a later stage of maturity) but bull beef systems do exist. As the growth rate benefit of entire males can be financially advantageous in any sheep breed, it is hard to understand why castration of male lambs destined for market prior to sexual maturity is still practised even if in other situations a case can be advanced. Genetic selection of faster growing lambs also reduces the imperative for castration. It has been estimated that in New Zealand around 40% of male lambs are left entire (Tarbotton et al., 2002) and there is a move in this direction in some European production systems. In extensive systems with little human-animal contact, castration will generally not be feasible.

If castration is undertaken, since the cost to the individual (handling stress and acute and chronic pain) may not always be offset by sufficient benefit (see tail docking later), the onus is on reducing the individual impact as much as possible, ideally to zero. This may be done through the use of good technique and effective analgesia, effective both during the procedure and for the duration of the likely subsequent chronic pain. A range of techniques can be practised. These are based on tradition and available resources. Castration at a later age allows farmers to select rams for breeding on the basis of their conformation. The ease of gathering the lambs themselves can be a serious constraint in extensive systems. In many cases legislation dictates what is acceptable and a considerable amount of contemporary work is aimed at determining best practice. In many cases specific methods are time-limited by legislation, though there is little information available on the way pain perception develops in the young animal. There are thus insufficient data to determine whether it is acceptable to perform techniques without anaesthetic at an early age even if anaesthesia is to be used shortly thereafter.

A range of techniques can be applied. Common ones include the use of rubber (elastrator) rings to restrict the blood supply to the scrotum and its contents or the surgical approach using a knife to open the scrotum to allow the testicles to be removed by traction. The commonly-practised elastrator method has been shown to be acutely painful to lambs (Mellor and Murray, 1989; Molony et al., 1993) and there is evidence that chronic pain occurs too (Kent et al., 2000). Recognition of the aversive effect of this technique has led to the recommendation that the technique should not be performed in lambs less than 24 hours old, whilst the ewe-lamb bond is being established. The crushing of the spermatic cord within the scrotal neck using a bloodless castration device (Burdizzo) is less frequently undertaken in sheep, compared to cattle (Hosie et al., 1996). Since castration causes pain (Mellor and Murray, 1989) and both acute and chronic responses (Molony et al., 1993, 2002), research effort has been directed at identifying methods which cause least damage and pain and routes by which effective analgesia across the required time spectrum can be delivered.

One factor minimising the impact of castration is the skill and competence of the operator. These can interact with the technique selected: it may require a highly technically skilled stockperson to perform certain techniques. Issues of training and assessment of competence are important here. In an industry where, in the UK at least, the tendency is towards fewer "professional" shepherds with an increasing proportion of flocks being cared for by the owner/farmer, continuing education and skill updating may require imaginative approaches to achieve a good uptake.

8.5.7 Mulesing

This is a technique whereby some of the folded skin around the breech and tail regions of Merino and Merino-type animals (which have highly-folded skin in this area) is surgically removed. The contraction of the resultant wool-free scar tissue reduces the likelihood of subsequent faecal soiling (dags) and fly strike. While the likelihood of fly strike is permanently reduced following healing, the cost to the individual of a procedure such as this without the use of anaesthetic is probably quite high and fly worry could occur during the healing period. There is considerable opposition to the operation. If effective anaesthesia/analgesia could be provided the balance of the cost to the sheep would be improved. Post-mulesing aversion to humans over an extended period is evidence that the procedure has a long-lasting negative effect (Chapman et al., 1994). Selective breeding for reduced fleece/skin folding in the breech area as an animal welfare goal continues to be exploited and may (eventually) give the same benefits (Scobie et al., 1999). Developing systems and adopting strains of sheep which reduce the level of internal parasites and thereby faecal soiling would also reduce the incidence of fly strike.

8.5.8 Tail Docking

Docking of lambs' tails is considered to deliver a welfare benefit, since lack of the lower part of the tail reduces the risk of blowfly strike as the breech area is less likely to become soiled by faeces and urine (French et al., 1994). (In some flocks, counting the removed tail ends is the method used to determine the number of lambs born.) There are many situations where tail docking is not undertaken – in extensive hill flocks for example where fly strike is less of a problem and the tail may provide added thermal insulation in cold weather. Otherwise it remains a common technique, even though French et al. (1994) indicated that it should not be assumed that failure to dock would inevitably lead to an increase in fly strike. Fly strike (and subsequent myiasis) is considered to be a very painful and debilitating condition which, if untreated, may result in death. Thus it represents a serious welfare problem which must be addressed and is considered unacceptable by many farmers and members of the public.

In Table 8.1 tail docking is categorised as a procedure capable of delivering benefit to the sheep, the proviso being that there is a cost to the sheep of performing the technique. However, many would consider it an unacceptable mutilation. Acute pain during the procedure (Kent et al., 1993; Molony et al., 1993) and the potential for chronic pain and discomfort while the tail stump heals mean that it should only be considered in situations where blowfly strike is likely to occur. A long-term welfare cost is the occasional development of a neuroma at the tail tip – though whether this causes pain is unclear. In attempting to construct a cost-benefit analysis of the technique there is also the complication that not all individuals are likely to suffer from strike and thus for these individuals the cost clearly outweighs any advantage. For the remainder, there remains the difficulty of weighing the effect of the procedure on the lamb against the potential benefit; as noted previously, while these may be difficult to equate for production benefits, when a degree of pain is associated with each it may be easier to come to a conclusion (Fig. 8.2). The method used to dock the lambs will be crucial when making this assessment. If a pain-free technique could be used then clearly the technique could be unequivocally advocated in areas where fly strike is a problem. Studies have shown that there is a difference in the pain elicited by different techniques employed, including the use of elastrator rings, cauterization using a hot docking iron or the combined use of a Burdizzo and elastrator ring, together with the use of a range of analgesic treatments (Graham et al., 1997).

If docking is to be practised then it must be performed using the least painful method in order to shift the balance in favour of docking versus myiasis (Fig. 8.2b). Lester et al. (1991) and Graham et al. (1997) found the use of a hot iron to be the least painful method. Currently analgesia is not usually provided. The development of improved, simple-to-use methods of analgesia which address issues of both acute and chronic pain would shift the balance towards docking being performed. (At a different level, productivity of the flock should improve if strike does not occur – though there remains a small chance that docked animals may be affected. There will be also be less labour required to treat affected individuals, offset against the work required to dock the lambs). The balance also swings back in the direction of not

a) Tail docking is considered to have a higher cost than the chance and potential impact of fly strike

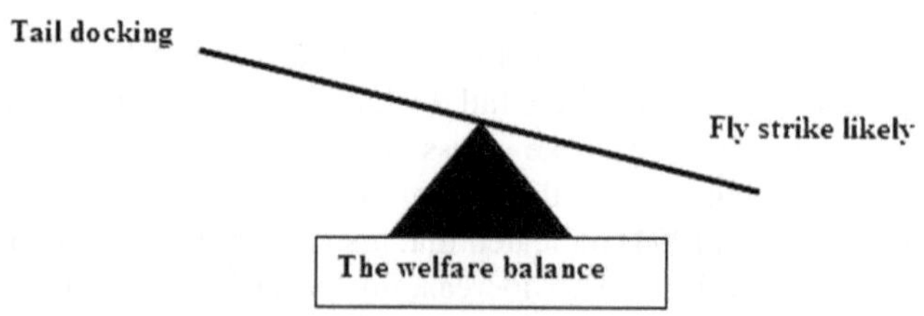

Tail docking		Fly strike
	Costs	
Pain associated with the technique: acute and chronic		Pain (possibly death) associated with strike
	Benefits	
Significantly reduced chance of strike		– but not every animal will be affected

b) Pain associated with tail docking is considered to have a lower cost than the chance and potential impact of fly strike

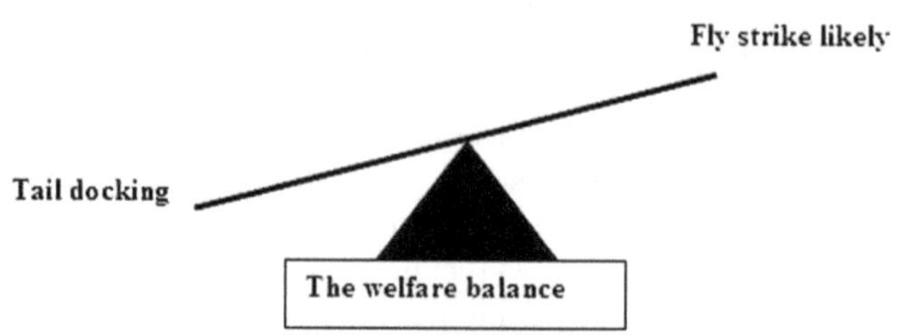

Fig. 8.2 A theoretical assessment of the welfare utility of tail docking

docking if effective measures to control intestinal parasitism can be implemented since wool soiling by loose faeces will be reduced. Such measures, particularly in extensively-managed flocks, will increasingly rely on the use of sheep with genetic resistance to parasites. Conversely, while docking may to some extent be seen as a palliative measure in response to management problems, it should not be viewed as the best response. A better, long-term solution would be the selection and breeding of animals with a reduced fleece in the breech, so reducing the opportunity for faecal soiling to occur, or breeding sheep with shorter tails (Scobie and O'Connell, 2002).

8.5.9 Transport and Marketing

Within the European Union, animal transport has been one of the areas where considerable attention has been focused by legislators, animal welfare activists and researchers. Why should this be the case, bearing in mind that transport occurs for a very short period of any sheep's life? One reason may be that to the public at large and welfare advocates in particular, transport is an area where any conditions that are less than optimal and may jeopardise animal welfare can often be witnessed. While mortality and injury are dramatic and unequivocal indicators of poor welfare at this time, inadequate journey conditions may leave larger numbers of animals hungry, thirsty and fatigued (Cockram and Mitchell, 1999). In addition, for animals destined for slaughter, transport effects may result in problems with carcass downgrading (through bruising for example) or poor meat storage characteristics. If animals have inadequate muscle glycogen stores prior to slaughter, there will be inadequate conversion to lactate, muscle pH will not decline sufficiently and the keeping quality of the resultant meat will be poor. From the researchers' perspective transport is an area amenable to study and for which politicians have made research funding available. Consequently various conditions for (primarily) road transport have been described, where contemporary scientific research is having an impact through prescribed conditions (See Cockram and Mitchell (1999) and the European Commission (2002) report for reviews). These conditions relate to loading densities, journey duration, provision of food and water, resting times, inspection etc. Yet these basic factors are not always easy to quantify in practice – it is not easy to look at a pen of sheep particularly within a large commercial transporter and report the stocking density. What is permitted sometimes relates to the standard of the vehicle used. The age of animals is also important and in a number of countries the transport of heavily pregnant ewes and the very young, particularly unweaned lambs, is prohibited. It is also important that animals travel in familiar groups if possible. Cast (aged/old) ewes are also less suited to long road journeys.

A wide range of measurements of physiological, behavioural and immunological parameters has been made to generate an accurate description of the sheep's response to transport. These measurements have been related to aspects of food deprivation and dehydration, physical exertion and fear. In many areas quantitative data can be generated. Studies have been conducted during commercial journeys (e.g. Knowles et al., 1994) or in more controlled settings where different putative stressors have been singled out for *a priori* investigation (e.g. Cockram et al., 1996). It is perhaps the psychological aspect which is more difficult to assess yet the animal's perception and awareness of a novel situation and its response are key to any successful management event, especially if it is likely to be repeated. Thus, returning to the human factor, one aspect which has been recognized as of being of importance is the "quality" of driving (Lambooij et al., 1999). Particular risk factors are rapid acceleration and deceleration, sharp cornering and unattended vehicles in which sheep become overheated during hot weather. Yet if driving conditions permit and space provision is adequate, sheep will readily lie down and ruminate during a long journey. Drivers should always keep in mind that the driving standard needs to be higher than for human passengers in a road vehicle, supported as they are in seats and who can usually see the road ahead to anticipate vehicle movement to maintain adequate balance. While possibly difficult to quantify, a recent research project in the UK has considered the effects of driver behaviour and road type on sheep behaviour (Cockram et al., 2004). A video programme aimed at improving driver awareness of the impact of the driver's action on the sheep is now available (University of Edinburgh Animal Welfare Research Group and the Roslin Institute, 2003). Such a resource is complementary to existing training requirements in some countries and should significantly improve sheep welfare in transit. Rewarding drivers based on improved sheep welfare (for example via an improvement in meat quality – Grandin, 2000) is to be encouraged, though currently rare.

The human factor will also be of significance during loading and unloading, times when the welfare of the sheep is particularly at risk. Loading and unloading can be improved by having well-designed facilities, avoiding over-steep ramp angles, allowing appropriate lighting differentials from inside to the outside of the vehicle and providing well-trained stockpersons. It has been suggested that it is the novelty of the conditions inside the vehicle or in the unloading area which may cause difficulties at this time rather than the procedure itself – physiological measurements support this interpretation (Broom et al., 1996; Parrott et al., 1998). The availability of adequate lighting within the transporter is also important to allow inspection to be made at appropriate intervals during the journey in order to identify sick or injured animals; in multi-decked vehicles such inspections can pose problems, which should be addressed. The alternative of unloading animals solely for the purpose of inspection brings with it its own set of problems. It is important that the responsible person, usually the driver, knows what to do should a sick or injured animal be identified during a mid-journey inspection. This must be considered before the journey begins. For example, the driver may need to be trained in methods of humane destruction where animals are deemed unable to complete the journey.

Conditions for sea transport are covered by legislation, whether sheep remain on a road transporter (e.g. on a roll-on/roll-off (RORO) vessel) or when loaded 'loose' in pens in the ship's hold or on open decks in stacked pens. There are occasional news reports of the occurrence of problems on long sea voyages. In most cases, a short sea crossing forms an integral part of a longer road journey. For long sea journeys, inappetence would appear the greatest threat to their welfare (Higgs et al., 1991; Phillips, personal communication, see also Chapter 6).

Under European conditions, many sheep will arrive at a market following a road journey. Here they will encounter unfamiliar conditions, stockpersons and animals. It is thus important that the road journey delivers the sheep to the market in as good a condition as possible in order for them to be able to cope with these new experiences. Movement from the unloading area to the various market areas will be facilitated by good design and experienced handlers who understand and exploit the natural behaviour of the sheep. Behavioural evidence for sheep having problems coping with what is required in this situation include remaining motionless or failing to move forwards in the raceway, lying down, backing off or running away or excessive vocalisation. Since sheep can be transported considerable distances both to and from markets it may be appropriate to indicate the maximum duration of the entire process or place more weight on the consideration of journey structure (Murray et al., 2000).

In relation to transport it is only recently becoming recognized that the background or experience of sheep may be important in fitting them to cope with this experience. Pre-transport inspections should identify animals clearly unfit to travel. Sheep from extensive management systems may rarely have encountered humans and thus may be expected to react differently to the novelty of these experiences compared to animals from more intensive systems. Arguably the welfare of the former animals may be more at risk during the entire process. For marketing purposes, the use of electronic 'virtual' auctions, as sometimes practised for cattle, does not seem to be a current option for slaughter or fattening stock. If this were a possibility it would also deliver disease control benefits but would not facilitate the common practice of producing suitably sized batches of animals (e.g. for subsequent fattening). Markets also provide stockpeople with a valuable social event, so important in retaining the fabric of dispersed rural communities.

Clearly disease control requirements will necessitate appropriate cleaning and disinfection of vehicles and communal market areas between consignments. Such measures will be dictated by knowledge of important endemic or exotic diseases – both in relation to the control of disease within a flock and the prevention of disease entering the flock. The need for such measures and the ability of transported and marketed sheep to spread exotic disease were starkly illustrated by the foot and mouth disease outbreak in the United Kingdom in 2001. Since the stress of transport may result in immunosuppression, infected animals may release more organisms and uninfected animals may be more susceptible at this time.

8.5.10 Slaughter

The events surrounding slaughter impose a wide range of challenges on the sheep, commencing with gathering and transport. The whole process from unloading, lairage, movement to the stunning pen and the act of slaughter itself offers many possibilities for the sheep's welfare to be compromised. Many people hold the view that if animals are to be slaughtered for human consumption that this should be done as close to the point of production as possible, with carcases rather than live animals being moved (e.g. FVE, 2001). Some of the regulations recently implemented across the European Union to safeguard human health may be at the expense of sheep welfare; for example the decline in the number of small, local slaughterhouses in the UK means that the average journey duration for slaughter animals has increased. However, improvements in slaughterhouse design and operation can better take into account animal needs.

On arrival at the slaughterhouse, sheep will generally be unloaded into a lairage area where *ante-mortem* inspection can be carried out and individual sheep showing particular problems following the journey singled out for immediate slaughter. The holding period is also provided to allow animals to recover from the rigours of the journey although it is not clear how recovery is in their interest, compared to immediate slaughter. It has been suggested that physiological events due to transportation may have detrimental effects on meat quality if slaughter immediately follows the journey. However, retention in the lairage has a welfare cost to the sheep which needs to be weighed against any perceived gain. The holding of animals in a lairage has benefits for the efficient running of large slaughterhouses where the smooth flow of animals has a major economic effect.

Provision of food, water and shelter from the weather (possibly the availability of bedding too) are important considerations and will have a significant impact on the well-being of the sheep, particularly if the period between arrival and slaughter is prolonged. It is important that food and water are offered in familiar form; for example sheep coming from an extensive hill system will probably be unfamiliar with automatic watering systems and may be reluctant to drink. Thus holding conditions should be such as to allow the sheep the maximum opportunity to rest, and to some extent recover from the preceding journey.

Movement around the abattoir can present problems. Designs which facilitate natural movement patterns are clearly advantageous, yet a number of facilities encounter movement problems which could be easily resolved. Design of raceway systems should be based on the behaviour of sheep and, for example, have few blind corners and encourage forward movement. Leader sheep ('Judas' sheep) are used in a number of situations (Bremner et al., 1980). These are individuals, usually resident within the slaughterhouse, which lead groups of sheep on a route to the pre-slaughter area. The natural following behaviour of sheep makes this a relatively easy system to institute, though in some countries it is not permitted for the leader animals to reside in the abattoir (for disease control purposes) and so this system cannot be adopted.

A number of measurements made *post mortem* can provide evidence of pre-slaughter conditions: carcase bruising or other injuries can give an indication of transport or lairage conditions through the ageing and location of lesions. Attainment of a low muscle ultimate pH (pHu) attests to adequate pre-slaughter muscle glycogen reserves, which may be poor in hungry animals or those which needed to expend energy (e.g. standing for long periods) during the journey. Animals which experience stress prior to slaughter also have reduced glycogen reserves. Passing any financial loss associated with down-graded carcases on to the producer and transporter is a powerful mechanism through which to improve some aspects of pre-slaughter management. Technological improvements can be made to reduce the number of aversive experiences in the pre-slaughter period: the advent of electronic identification has the potential to reduce the need for physical handling to inspect ear tags for example.

Since it has been suggested for some time that some methods of slaughter could be more humane than others (e.g. Gregory and Wotton, 1984), research continues to identify the most humane methods of slaughter. Killing methods should be chosen so as to cause the least possible suffering to the individual, be this when sheep are killed for human consumption or to relieve pain and suffering, for example when dealing with a casualty animal. The UK Farm Animal Welfare Council (FAWC) identified five basic principles to be observed at this time (FAWC, 2003):

1. pre-slaughter handling facilities which minimise stress;
2. use of competent well trained, caring personnel;
3. appropriate equipment which is fit for the purpose;
4. an effective process which induces immediate unconsciousness and insensibility or an induction to a period of unconsciousness without distress; and
5. guarantee of non-recovery from that process until death ensues.

In the slaughterhouse situation, stunning generally precedes bleeding out. In some countries the sheep is subject to a religious slaughter method (e.g. Shechita or Halal); stunning can also precede religious slaughter but this is not universal and in some countries no exemption is given for religious slaughter. The lack of pre-slaughter stunning was considered unacceptable by FAWC (2003). The period between stunning and bleeding should be minimised to ensure that loss of blood results in death before consciousness is regained (head only electrical stunning is reversible for example (Velarde et al., 2002)). Similarly, rapid bleeding is important and can be achieved through good technique; the competence of operators and efficient functioning of equipment are important. Contrary to previously held views, recent work has failed to produce evidence to suggest that a sheep is distressed by witnessing stunning and slaughter (Anil et al., 1996).

8.5.10.1 Emergency Slaughter

In an emergency it may be necessary and desirable to kill a sheep immediately to prevent further suffering or to control disease outbreaks. A variety of on-farm practices occur including the use of a free bullet or shotgun. On-farm practices are generally more difficult to regulate than killing in a facility where inspection and supervision occur. Whatever method is used, it must be done competently and be suitable for the size and age of the animal concerned. Difficulties often arise in determining whether a casualty animal is fit to be transported to a slaughterhouse (or indeed a place of treatment) without this leading to additional suffering. Veterinary advice may be sought in an attempt to resolve difficulties. When sheep are transported it is necessary to plan for the eventuality of having to kill humanely an animal severely injured during the journey and unfit to travel further.

8.5.11 Management Practices to Maintain Health and Well-Being

Most people would advocate the use of preventive medicine and other pre-emptive actions as best practice to reduce the likelihood of problems occurring. It is always possible to develop a flock health plan that will identify hazards to the health and well-being of the sheep. This should be based on the history of the flock, local conditions and threats to biosecurity. Although many health and welfare considerations are intertwined (in so far as any animal experiencing disease is in a poorer welfare state) with increasing knowledge it should be possible to identify specific threats to welfare in the plan and consider measures to reduce the likelihood of these occurring. As with the more detailed analysis of the potential benefit of tail docking previously presented, the overall welfare gain of any preventive strategy needs to be assessed. There will be economic implications if the financial cost of a particular preventive strategy is high, possibly precluding an alternative action being taken: for example the cost of vaccinating the flock against a particular disease may preclude expenditure on supplementary feeding. Such issues of great practical importance are addressed in more detail in the next chapter. Foot care will be needed to address lameness, which is one of the most easily and commonly recognised problems affecting sheep. Similarly, the control of gastrointestinal worms, liver fluke and ectoparasites is crucial to maintaining good welfare. Details have been given already in Chapter 5. Training shepherds in best practice and good technique is crucial.

For sheep which are handled infrequently, for example those kept under extensive or range conditions, there is generally the necessity to "do more" on each handling occasion. Gathering animals has a number of potentially negative effects: there may be physical injuries to individuals, social groups may be disrupted (lambs may become detached from their mothers) and by concentrating animals together infectious disease may be more readily spread between animals. The immune status of animals is also thought to be poorer at this time, so exacerbating the situation by lowering the threshold of the amount of infectious agent needed to initiate infection. It has been suggested that due to changes in stress hormone profiles at this time, immune responses (e.g. to vaccination) may vary in relation to the novelty of the situation (Goddard et al., 2000). Additionally, diseased animals experiencing physiological stress may increase the rate of shedding of infectious organisms, thereby exposing others to a greater challenge. Transporting sheep may induce similar effects, putting animals at greater risk of disease during and after the journey. Mixing batches of young animals also has the effect of exposing animals to novel disease agents, an effect often compounded by coincident weaning stress. Thus actions aimed at improving sheep health may be counteracted by effects on natural disease defence mechanisms.

The widely-used technique of condition scoring allows an objective assessment to be made of the fatness of a sheep. It is a valuable aid to management, providing information beyond that of simply weighing individuals. It was suggested by FAWC (1994) that any flock with a significant number of individuals at a condition score of less than 1.5 must be regarded as demonstrating inadequate care and welfare. Some may consider that exceptions to this would need to be made, for example in cases of severe drought or other unpredictable events.

8.5.12 Dipping

The control of ectoparasites is one of the major challenges to any sheep enterprise as has been discussed in Chapter 5. While a number of spray race systems and the development of pour-on or injectable compounds appear to offer suitable therapeutic regimens in some situations, the use of a plunge dip in which the sheep is immersed for a minimum period, during which time the head must be briefly submerged, is sometimes the only effective way to ensure control of the parasite is achieved. There appear to be no research results available that comment on the welfare implications for the individual sheep. However, it is recognised that it is inadvisable to dip sheep when they are hot, tired and panting and thirsty or fully-fed. Depending on whether the process is to act as a prophylactic treatment or to control a disease outbreak, there will be compensatory benefit to the sheep themselves to set against the stress associated with dipping.

8.5.13 Advanced Breeding Techniques

The reproductive behaviour of sheep has been described earlier (see Chapter 13); it is the seasonal physiology of sheep (at least at temperate latitudes) which drives the annual management cycle. Some techniques result in increased fecundity through hormonal manipulation or the selection for breeding of superior individuals. Other management actions result in lambs being born at times of the year when the availability of natural pasture may be poor. Consequently in such systems, it is not so much the breeding technique itself but the need for adequate provision of feed and other resources to ewe and lamb that will need special attention. Yet lambing at times of the year when grass growth is suboptimal has not been thought to necessarily result in poorer welfare, though no definitive studies have been undertaken (For a review of this topic see Fisher, 2004. Low lamb birth weight may be the most important issue). It is beyond the scope of this chapter to consider all the methods available by which to manipulate the breeding season – for example, the use of melatonin implants or intra-vaginal sponges/pessaries (e.g. Controlled Intravaginal Drug Release (CIDR) devices) which release progestagens, resulting in oestrus synchronization following their removal. In general, the advanced breeding techniques described in this section are restricted to a limited range of pedigree flocks; these should have the ability to assess the impact of the techniques on recipient animals.

A range of new techniques has been developed in the last 10–20 years, mostly aimed at providing an accelerated rate of genetic improvement. The ability to identify superior sires and to distribute their genetic gain as widely as possible has led to the effective development of a range of artificial insemination, superovulation and embryo transfer techniques. Since these generally have delivered financial benefits to the enterprise the welfare impact may not always have been adequately addressed – there are few reports which consider the actual welfare implications of these procedures (e.g. Murray and Ward, 1993), though the issues were addressed in a report of a committee established to consider the ethical implications of emerging technologies in farm animal breeding (HMSO, 1995). For example, though an effective technique in skilled hands, laparoscopic embryo transfer (McKelvey et al., 1985) requires the sheep to be restrained, inverted in a cradle and subject to a minor surgical procedure. This follows artificial oestrus synchronisation.

In order to safeguard welfare, in many countries laparoscopic intrauterine insemination is required to be undertaken by appropriately trained veterinarians. The degree of intervention and duration of this procedure have prompted work on non-surgical, trans-cervical (intra-vaginal) approaches which, though technically challenging and requiring greater volumes of semen than those used for intrauterine insemination, may have a lesser impact on the recipient ewe; again only a suitably trained veterinarian may perform trans-cervical AI in sheep in some countries. To date this approach is also less effective when using frozen semen. Whatever the technique, the risk of infection may always be present and, in itself, may constitute a threat to welfare. The impact on the male is through the collection of semen. This is usually achieved using an artificial vagina (it may be necessary to treat a teaser ewe with hormones to bring her into oestrus at the required time). In some cases electro-ejaculation may be performed. As well as having a recognised welfare impact on the male, the quality and quantity of the resulting semen are sometimes inferior and the technique is generally restricted to use in veterinary examinations. Despite the fact that many people find electro-ejaculation to be an unacceptable technique, Stafford et al. (1996) found it to be no more aversive to sheep than part-shearing.

A further technique, ultrasound pregnancy diagnosis and the prediction of foetal numbers, has become widely practised in many areas (Russel and Goddard, 1995). This has the potential to result in enhanced welfare. Knowledge of the foetal load allows adequate nutrition to be provided in the last third of pregnancy, so preventing undernourished ewes and lambs. Thus the incidence of pregnancy toxaemia is significantly reduced. Similarly, the avoidance of over nutrition of ewes carrying single lambs reduces the incidence of dystocia. Shepherds also report that a major benefit of the procedure is the ability to group together ewes carrying different numbers of lambs, so facilitating management at lambing time. It allows housing of ewes carrying triplets, or the co-location of ewes with singletons and triplets so easing fostering of lambs from triplet-bearing ewes. The use of sector scanner technology allows the act of scanning to be performed on standing ewes and the only welfare costs relate to gathering and handling (see earlier).

8.5.14 Shearing

The fleece of the sheep is potentially a valuable commodity but in some recent years the value of the fleece from the majority of breeds of sheep has barely covered the cost of shearing (excluding specialized fine-wool flocks). Removal of the fleece at the start of the summer is beneficial to sheep welfare and thus if the price of wool is low, the action of shearing becomes more economically slanted towards an activity performed for the welfare of the individual sheep (and consequently may not be performed on economic grounds if prices are poor). However, it has been shown that as well as wool removal itself (which is the most stressful component), the combination of actions involved with shearing, including isolation and up-ending, are aversive to sheep (Hargreaves and Hutson, 1990c,d). Shearing must be conducted skillfully to ensure that injuries to the sheep (primarily cuts) are kept to an absolute minimum. In systems where the production of wool is not viewed as an economic proposition, breeding strategies to produce sheep with reduced wool or the use of existing breeds with reduced fleece can be considered, though due regard must be paid to the potential loss of thermal insulation properties. Winter shearing of sheep allows housing at greater animal density but should not be undertaken in the absence of housing if adverse weather conditions are expected. Shearing in the summer should ideally not occur if inclement weather is forecast.

8.6 Conclusion

The symbiotic ontogeny of the relationship between humans and sheep has developed based on the productivity of individual flocks through appropriate management practices and resource inputs. Many of these practices are now being considered from a welfare perspective and each activity would benefit from being reassessed in terms of firstly its necessity and thereafter its potential welfare cost and benefit. While the cost should always be considered from the individual sheep's perspective, the benefit can be apportioned between the individual sheep, other sheep in the flock and the producer.

Acknowledgment I would like to thank Joyce Kent for allowing me access to her report to the Winston Churchill Memorial Trust (December, 2002) entitled "Elective Surgeries (Mutilations): A comparison of the legislative control of their use in farm animals in the United Kingdom, America and New Zealand".

References

Anil, M.H., Preston, J., McKinstry, J.L., Rodway, R.G. & Brown, S.N. (1996) An assessment of stress caused in sheep by watching slaughter of other sheep. *Animal Welfare* 5: 435–441.

Baldock, N.M. & Sibley, R.M. (1990) Effects of handling and transportation on the heart rate and behaviour of sheep. *Applied Animal Behaviour Science* 28: 15–39.

Beausoleil, N., Stafford, K. & Mellor, D.J. (2002) Do sheep regard humans as predators? In: *Animal welfare and behaviour: From science to solution. Programme of an international conference*. Hamilton, New Zealand, p. 29.

Boivin, X. (2003) The human-animal relationship, stockmanship and extensive sheep production. In: Ed. Goddard P. *Improving sheep welfare on extensively managed flocks: Economics, husbandry and welfare*. Proceedings of a workshop. Aberdeen. pp. 11–18.

Boivin, X., Nowak, R. & Terrazas Garcia, A. (2001) The presence of the dam affects the efficiency of gentling and feeding on the early establishment of the stockperson-lamb relationship. *Applied Animal Behaviour Science* 72: 89–103.

Boivin, X., Tournadre, H. & Le Neindre, P. (2000) Hand feeding and gentling influence early-weaned lambs' attachment responses to their stockperson. *Journal of Animal Science* 78: 879–884.

Bremner, K.J., Braggins, J.B. & Kilgour, R. (1980) Training sheep as "leaders" in abattoirs and farm sheep yards. *Proceedings of the New Zealand Society of Animal Production* 40: 111–116.

Broom, D.M., Goode, J.A., Hall, S.J.G., Lloyd, D.M. & Parrott, R.F. (1996) Hormonal and physiological effects of a 15 hour road journey in sheep: Comparison with the responses to loading, handling and penning in the absence of transport. *British Veterinary Journal* 152: 593–604.

Broom, D.M. & Johnson, K.G. (1993) *Stress and Animal Welfare*. Chapman Hall, London.

Budiansky, S. (1994) A special relationship: The co-evolution of human beings and domestic animals. *Journal of the American Veterinary Medical Association* 204: 365–368.

Chapman, R.E., Fell, L.R. & Shutt, D.A. (1994) A comparison of stress in surgically and non-surgically mulesed sheep. *Australian Veterinary Journal* 71: 243–247.

Clutton-Brock, J. (1987) *A natural history of domesticated animals*. Cambridge University Press, 208pp.

Cockram, M.S., Baxter, E.M., Smith, L.A., Bell, S., Howard, C.M., Prescott, R.J. & Mitchell, M.A. (2004) Effect of driver behaviour, driving events and road type on the stability and resting behaviour of sheep in transit. *Animal Science* 79: 165–176.

Cockram, M.S., Kent, J.E., Goddard, P.J., Waran, N.K., McGilp, I.M., Jackson, R.E., Muwanga, G.M. & Prytherch, S. (1996) Effect of space allowance during transport on the behavioural and physiological responses of lambs during and after transport. *Animal Science* 62: 461–477.

Cockram, M.S. & Lee, R.A. (1991) Some preslaughter factors affecting the occurrence of bruising in sheep. *British Veterinary Journal* 147: 120–125.

Cockram, M.S. & Mitchell, M.A. (1999) Role of research in the formulation of 'rules' to protect the welfare of farm animals during road transportation. In: Eds. A.J.F. Russel et al. *Farm animal welfare – who writes the rules?* Occasional publication of the BSAS No 23 – British Society of Animal Science, Edinburgh, UK, pp 43–64.

Cockram, M.S., Ransom, M., Imlah, P., Goddard, P.J., Burrells, C. & Harkiss, G.D. (1994) The behavioural, endocrine and immune responses of sheep to isolation. *Animal Production* 58: 389–399.

Coppinger, R., Glendinning, J., Torop, E., Matthay, C., Sutherland, M. & Smith, C. (1987) Degree of behavioral neoteny differentiates canid polymorphs. *Ethology* 75: 89–108.

Edwards, D.S. & Johnston, A.M. (1999) Welfare implications of sheep ear tags. *The Veterinary Record* 144: 603–606.

European Commission (2002) The welfare of animals during transport (details for horses, pigs, sheep and cattle). *Report of the Scientific Committee on Animal Health and Animal Welfare*. 130pp.

FAWC (1994) *Report on the Welfare of Sheep*. MAFF Publications, London. 40pp.

FAWC (2003) *Report on the Welfare of Farmed Animals at Slaughter or Killing*. Part 1: Red Meat Animals. Defra Publications, London, 72pp.

Federation of Veterinarians of Europe (2001) Transport of Live Animals. *FVE Position Paper*.

Fisher, M.W. (2004) A review of the welfare implications of out-of-season extensive lamb production systems in New Zealand. *Livestock Production Science* 85: 165–172.

Fisher, M.W. & Mellor, D.J. (2002) The welfare implications of shepherding during lambing in extensive New Zealand farming systems. *Animal Welfare* 11: 157–170.

French, N.P., Wall, R. & Morgan, K.L. (1994) Lamb tail docking: A controlled field study of the effects of tail amputation on health and productivity. *The Veterinary Record* 134: 463–467.

Goddard, P.J., Fawcett, A.R., Macdonald, A.J. & Reid, H.W. (2000) The behavioural, physiological and immunological responses of lambs from two rearing systems and two genotypes exposed to humans. *Applied Animal Behaviour Science* 66: 305–321.

Graham, M.J., Kent, K.E. & Molony, V. (1997) Effects of four analgesic treatments on the behavioural and cortisol responses of 3-week-old lambs to tail docking. *The Veterinary Journal* 153: 87–97.

Grandin, T. (2000) *Livestock Handling and Transport*. 2nd. Edition. Ed. T. Grandin. Wallingford, CABI.

Grandin, T., Curtis, S.E., Widowski, T.M. & Thurmon, J.C. (1986) Electro-immobilization versus mechanical restraint in avoid-avoid choice tests for ewes. *Journal of Animal Science* 62: 1469–1480.

Gregory, N.G. & Wotton, S.B. (1984) Sheep slaughtering procedures. II. Time to loss of brain responsiveness after exsanguinations or cardiac arrest. *British Veterinary Journal* 140: 354–360.

H.M.S.O. (1995) *Report of the Committee to Consider the Ethical Implications of Emerging Technologies in the Breeding of Farm Animals*. HMSO, London. 77pp.

Hargreaves, A.L. & Hutson, G.D. (1990a) The stress response in sheep during routine handling procedures. *Applied Animal Behaviour Science* 26: 83–90.

Hargreaves, A.L. & Hutson, G.D. (1990b) The effect of gentling on heart rate, flight distance and aversion of sheep to a handling procedure. *Applied Animal Behaviour Science* 26: 243–252.

Hargreaves, A.L. & Hutson, G.D. (1990c) An evaluation of the contribution of isolation, up-ending and wool removal to the stress response to shearing. *Applied Animal Behaviour Science* 26: 103–113.

Hargreaves, A.L. & Hutson, G.D. (1990d) Changes in heart rate, plasma cortisol and haematocrit of sheep during a shearing procedure. *Applied Animal Behaviour Science* 26: 91–101.

Hargreaves, A.L. & Hutson, G.D. (1990e) Some effects of repeated handling on stress responses in sheep. *Applied Animal Behaviour Science* 26: 253–265.

Hemsworth, P.H. & Coleman G.J. (1998) Attitudes of stockpeople. In: Eds. P.H.Hemsworth & G.J. Coleman. *Human Livestock Interactions: The Stockperson and the Productivity and Welfare of Intensively Farmed Animals*. CABI. pp. 62–75.

Higgs, A.R.B., Norriss, R.T. & Richards, R.B. (1991) Season, age and adiposity influence death rates in sheep exported by sea. *Australian Journal of Agricultural Research* 42: 205–214.

Hosie, B. (1995) Problems with the use of ear tags in sheep. *The Veterinary Record* 137: 32.

Hosie, B., Carruthers, J. & Sheppard, B.W. (1996) Bloodless castration of lambs: Results of a questionnaire. *British Veterinary Journal* 152: 47–55.

Hutson, G.D. (1985) The influence of barley food rewards on sheep movement through a handling system. *Applied Animal Behaviour Science* 14: 263–273.

Joyce, J.P., Blaxter, K.L. & Park, C. (1966) The effect of natural outdoor environments on the energy requirements of sheep. *Research in Veterinary Science* 7: 341–359.

Kendrick, K.M., Atkins, K., Hinton, M.R., Broad, K.D., Fabrenys, C. & Keverne, B. (1995) Facial and vocal discrimination in sheep. *Animal Behaviour* 49: 1665–1676.

Kendrick, K.M., da Costa, A.P., Leigh, A.E., Hinton, M.R., & Pierce, J.W. (2001) Sheep don't forget a face. *Nature* 414: 165–166.

Kent, J.E., Jackson, R.E., Molony, V. & Hosie, B.D. (2000) Effects of acute pain reduction methods on the chronic inflammatory lesions and behaviour of lambs castrated and tail docked with rubber rings at less than two days of age. *The Veterinary Journal* 160: 33–41.

Kent, J.E., Molony, V. & Robertson, I.S. (1993) Changes in plasma cortisol concentration in lambs of three ages after three methods of castration and tail docking. *Research in Veterinary Science* 55: 246–251.

Kilgour, R. & de Langen, H. (1970) Stress in sheep resulting from management practices. *Proceedings of the New Zealand Society of Animal Production* 30: 65–76.

Knowles, T.G., Maunder, D.H.L. & Warriss, P.D. (1994) Factors affecting the incidence of bruising in lambs arriving at one slaughterhouse. *The Veterinary Record* 134: 44–45.

Lambooij, E., Broom, D.M., von Mickwitz, G. & Schutte, A. (1999) The welfare of farm animals during transport. In: Ed. F.J.M. Smulders. *Veterinary Aspects of Meat Production*. pp. 113–128, E.C.C.A.M.S.T.

Lester, S.J., Mellor, D.J., Ward, R.N. & Holmes, R.J. (1991) Cortisol responses of young lambs to castration and tailing using different methods. *New Zealand Veterinary Journal* 39: 134–138.

Mateo, J.M., Estep, D.Q. & McCann, J.S. (1991) Effects of differential handling on the behaviour of domestic ewes (*Ovis aries*). *Applied Animal Behaviour Science* 32: 45–54.

McKelvey, W.A.C., Robinson, J.J. & Aitken, R.P. (1985) A simplified technique for the transfer of ovine embryos by laparoscopy. *The Veterinary Record* 117: 492–494.

Mellor, D.J. & Murray, L. (1989) Changes in cortisol responses of lambs to tail docking, castration and ACTH injection during the first seven days after birth. *Research in Veterinary Science* 46: 392–395.

Molony, V., Kent, J.E. & McKendrick, I.J. (2002) Validation of a method for the assessment of an acute pain in lambs. *Applied Animal Behaviour Science* 76: 215–238.

Molony, V., Kent, J.E. & Robertson, I.S. (1993) Behavioural responses of lambs of three ages in the first three hours after three methods of castration and tail docking. *Research in Veterinary Science* 55: 236–245.

Murray, K.C., Davies, D.H., Cullinane, S.L., Eddison, J.C. & Kirk, J.A. (2000) Taking lambs to slaughter: Marketing channels, journey structures and possible consequences for welfare. *Animal Welfare* 9: 111–122.

Murray, R.D. & Ward, W.R. (1993) Welfare implications of modern artificial breeding techniques for dairy cattle and sheep. *The Veterinary Record* 133: 283–286.

Napolitano, F., Annicchiarico, G., Caroprese, M., DeRosa, G., Taibi, L. & Sevi, A. (2003) Lambs prevented from suckling their mothers display behavioural, immune and endocrine disorders. *Physiology and Behavior* 78: 81–89.

Napolitano, F., Marino, V., DeRosa, G., Capparelli, R. & Bordi, A. (1995) Influence of artificial rearing on behavioural and immune responses of lambs. *Applied Animal Behaviour Science* 45: 245–253.

Orgeur, P., Mavric, N., Yvore, P., Bernard, S., Nowak, R., Schaal, B. & Levy, F. (1998) Artificial weaning in sheep: Consequences on behavioural, hormonal and immuno-pathological indicators of welfare. *Applied Animal Behaviour Science* 58: 87–103.

Parrott, R.F., Hall, S.J.G. & Lloyd, D.M. (1998) Heart rate and stress hormone responses of sheep to road transport following two different loading responses. *Animal Welfare* 7: 257–267.

Price, E.O. & Thos, J. (1980) Behavioural responses to short-term social isolation in sheep and goats. *Applied Animal Ethology* 6: 331–339.

Rushen, J. (1986a) Aversion of sheep to electro-immobilisation and physical restraint. *Applied Animal Behaviour Science* 15: 315–324.

Rushen, J. (1986b) The validity of behavioural measures of aversion: A review. *Applied Animal Behaviour Science* 16: 309–323.

Rushen, J. (1986c) Aversion of sheep for handling treatments: Paired choice experiments. *Applied Animal Behaviour Science* 16: 363–370.

Russel, A.J.F. & Goddard, P.J. (1995) Small ruminant reproductive ultrasonography. In: Ed. P.J. Goddard. *Veterinary Ultrasonography.* CABI. pp. 257–274.

Scobie, D.R., Bray, A.R. & O'Connell, D. (1999) A breeding goal to improve the welfare of sheep. *Animal Welfare* 8: 391–406.

Scobie, D.R. & O'Connell, D. (2002) Genetic reduction in tail length in New Zealand sheep. *Proceedings of the New Zealand Society of Animal Production* 62: 195–198.

Sevi, A., Caroprese, M., Annicchiarico, G., Albenzio, M., Taibi, L. & Miscio, A. (2003) The effect of a gradual separation form the mother on later behavioural, immune and endocrine alterations in artificially reared lambs. *Applied Animal Behaviour Science* 83: 41–53.

Silanikove, N. (1998) Effects of heat stress on the welfare of extensively managed domestic ruminants. In: Goddard P. (Ed.) *Improving sheep welfare on extensively managed flocks: Economics, husbandry and welfare*. Proceedings of a workshop. Aberdeen. pp. 29–34.

Silanikove, N. (2000) Effects of heat stress on the welfare of extensively managed domestic ruminants. *Livestock Production Science* 67: 1–18.

Slee, J. (1987) Sheep. In: Ed. Johnson, H.D. *Bioclimatology and the adaptation of livestock.* Elsevier science, Amsterdam. pp. 229–244.

Stafford, K.J., Spoorenberg, J., West, D.M., Vermunt, J.J., Petrie, N. & Lawoko, C.R.O. (1996) The effect of electro-ejaculation on aversive behaviour and plasma cortisol concentration in rams. *New Zealand Veterinary Journal* 44: 95–98.

Tarbotton, I., Bray, A. & Wilson, J. (2002) Incidence and perceptions of cryptorchid lambs in 2000. *Proceedings of the New Zealand Society of Animal Production* 62: 195–198.

University of Edinburgh Animal Welfare Research Group and the Roslin Institute (2003) *Careful Driving and Animal Welfare. Behaviour of Sheep during Transit.* Video Production by the University of Edinburgh Media and Learning Technology Service.

Velarde, A., Ruiz-de-la-Torre, J.L., Roselló, C., Fàbrega, E., Diestre, A. & Manteca, X. (2002) Assessment of return to consciousness after electrical stunning in lambs. *Animal Welfare* 11: 333–341.

Waterhouse, A., Colgrove, P., Vipond, J. & Goddard, P. (2003) Identification of key issues – a summary of the final group session. In: Ed. Goddard P. *Improving sheep welfare on extensively managed flocks: Economics, husbandry and welfare*. Proceedings of a workshop. Aberdeen. pp. 73–77.

Whateley, J., Kilgour, R. & Dalton, D.C. (1974) Behaviour of hill country sheep breeds during farming procedures. *Proceedings of the New Zealand Society for Animal Production* 34: 28–36.

Chapter 9
The Economics of Sheep Welfare

C.E. Milne, A.W. Stott, and J.M. Santarossa

Abstract This chapter aims to provide an introduction to some of the economic issues surrounding the welfare of farmed sheep. Typically, ethical issues, or welfare indicators measured by natural scientists (physiological, physical and behavioural), gain precedence in discussion, and economic aspects are considered to be side issues. Moreover, economics is assumed, by some, to be a means of justifying profit generation at the expense of welfare – which it is not. Economics studies human behaviour as a relationship between ends and scarce means that have alternative uses. It is central to the achievement of acceptable levels of animal welfare, which are identified by the natural scientist and agreed through ethical debate. The actions of a wide range of individuals and groups influence the level of welfare that a sheep can experience during its life. These include farmers, hauliers, slaughtermen, consumers, non-consumers and government. Their roles and interactions are introduced in this chapter by addressing some questions relating to the supply of, and demand for, goods and services such as sheep meat and wool. Practical examples are provided to illustrate points and reference is made to high welfare and organic systems. The potential impact that imposition of high welfare standards may have on the sustainability of businesses within the sheep industry, and consequences for sheep welfare are raised

Keywords Sheep welfare · Economics · Supply · Demand

9.1 Introduction

When it comes to the welfare of animals – including sheep, economics is not the first thing that most people think about. Usually it is the more familiar ethical issues, or welfare indicators measured by natural scientists (physiological, physical and behavioural). Economics is perceived by some as a method of justifying profit generation at the expense of animal welfare – which it is not, or an unfathomable subject.

C.E. Milne
Land Economy Group, SAC, Aberdeen, UK

During this chapter the role that economics has to play in achieving acceptable animal welfare standards will be introduced. You will discover that you probably already knew some economic theory without recognising it. Throughout the chapter we will be looking at the welfare of farmed sheep – that is sheep which are being kept to produce goods/services for the benefit of people.

Economics is 'the science which studies human behaviour as a relationship between ends and scarce means which have alternative uses' (Robbins, 1945). As our 'means' (resources) are limited, or 'scarce', it is preferable to use them to best effect. The two criteria used in economics to evaluate 'best effect' are the efficiency of resource use (maximising output from the resources used), and how equitably they are distributed amongst society. Economists investigate two types of question, those that relate to the *supply* of goods and services and those relating to society's *demand* for goods and services.

Looking at supply, three specific questions that can be asked are:

1. What goods and services should be produced from the resources available?
2. How much of them should be produced?
3. How should they be produced?

So choices need to be made as to whether resources will be best used for sheep welfare for example, or something else such as environmental protection or food safety. As individuals we will all have our own answer to this question – this affects our demand for goods and services. Even though there is a demand for sheep welfare (or any other good), that does not mean that it will be supplied – we all want things that we cannot or do not get.

Economists therefore also look at demand questions such as:

(a) How do individuals in a society communicate their demand (needs/wants) to producers of goods and services?
(b) What incentives do they provide to encourage producers to meet their demands?

Poor communication of demands, inadequate incentives and other factors can all be barriers that hinder suppliers (sheep farmers, hauliers, slaughtermen, etc.) from meeting the demand for sheep welfare. So by finding answers to these questions economists seek to identify these barriers and ways that they can be overcome, so that supply meets demand as closely as possible.

In summary then, ethical debates can determine accepted sheep welfare standards, natural scientists can establish methods of measuring that standard and monitoring practice, and economics helps us to achieve those standards through matching supply with demand.

In this introductory section there are two further economic concepts that need to be covered. Firstly, the concept of a *marketplace*. The marketplace is central to economics, this is not an actual 'place', what economists mean by the marketplace is the trading of goods and services where prices are adjusted in accordance with supply and demand. The second concept is that of *margins*. The term 'marginal' is used by economists in association with 'costs', 'benefits', and a range of other economic expressions. It refers to the *additional* (or extra) costs, benefits, etc. that arise from a change in production or consumption, e.g. marginal costs or marginal benefits.

Economics is a large subject area, only a portion will be touched upon in this chapter and the explanations of complex issues are kept simple in this overview. The main topics to be covered are the demand for sheep welfare, how that demand is communicated to suppliers and the incentives given to producers to meet that demand. Our focus on demand ensues from the current trend towards demand-led (in place of supply-led) supply chains for food, and other, goods.

9.2 The Demand for Sheep Welfare

Why people want what they want is a subject studied by psychologists as well as economists and some of the knowledge developed separately in these two subject areas is now coming together (see Willock et al., 1999) and for a wider perspective on the interaction between economics and psychology see Wilkinson (2008). In economics, the concept of preferences is central to demand theory and classical economics assumes that people act to maximise their preferences. In economics, preferences are represented by a utility function that ranks individual preferences for goods or services. To give an example consider preferences for feeding a sheep whose welfare was being compromised by poor nutrition (starvation), giving the sheep additional food would be ranked higher than doing nothing.

Measuring the preferences of an individual are difficult as they are not clearly visible. In addition, since factors such as attitudes and perceptions vary between individuals, preferences are not constant across a population. For example let us consider a thin sheep; someone with a very protective attitude towards sheep welfare is likely to rank the provision of additional food higher than a person who does not like or care about sheep. Secondly, the ranking of preferences can vary depending upon the quantities of a good or service being considered. Continuing with the thin sheep example, providing it with a large quantity of food would rank more highly than providing it with a small quantity of food. An illustrative utility function of this is given in Fig. 9.1 where it can be seen that the level of preference, represented by utility, diminishes as the quantity of food provided increases. Thus as the quantity of food provided increase from 1½ to 2 units the marginal utility (U^2–U^1) is greater than that gained when raising the quantity fed from 2 to 2½ units (U^3–U^2). Keeping these features of individual preference in mind and that farmed sheep are kept for the goods/services they provide for man (meat, wool, land management etc.) this leads us to our purchasing behaviour as consumers.

As consumers we have a wide array of goods and services to select from (and since money is a limited resource, we cannot have everything we might wish). For example, we could choose to have beef or lamb or beans as the protein source in our diet. In making such choices we do not try to measure the utility we would gain from each, rather we seek to identify our preference for combinations of goods (in this

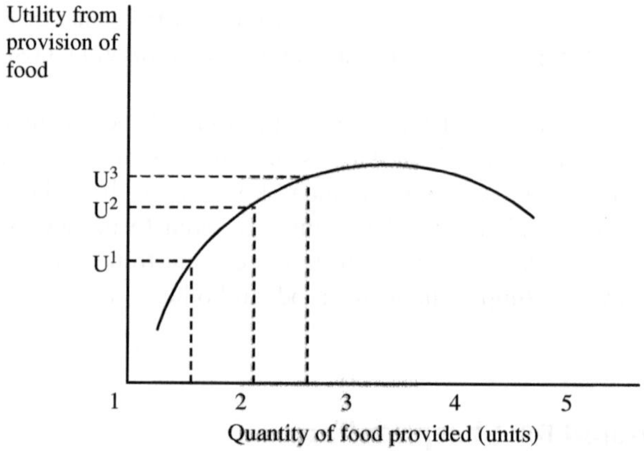

Fig. 9.1 An illustrative utility function: the total and marginal utility of additional food provision for a sheep

example lamb, beef, beans). Thus the fact that utility cannot be quantified does not prevent us from making choices – we make them by comparison. Economics adopts the same approach and seeks to explain the actions of individuals according to their *indifference* between alternative actions. For example, the amount of beef and/or lamb that would be equally preferable can be quantified and shown graphically as an indifference curve like that shown in Fig. 9.2.

At any point along the indifference curve, the combination of beef and lamb will yield the same level of utility e.g. the utility gained at point F (combination c + d), will be the same as at point G (combination a + b).

In reality the choices we make are not as simple as choosing between two or three actions that can substitute for each other, such as consuming beef or lamb or beans. Choices are also made between how much of our income to spend on food (for nutritional purposes) versus something else, such as leisure pursuits or ethical

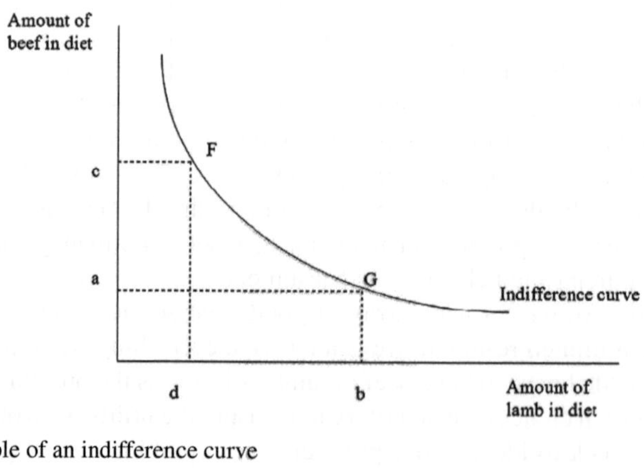

Fig. 9.2 Example of an indifference curve

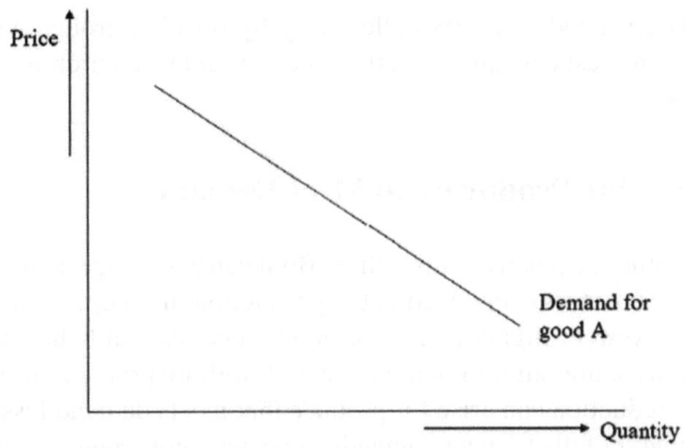

Fig. 9.3 A demand curve

actions (e.g. giving to charity or purchasing food produced in high welfare systems). An indifference curve could therefore be drawn between purchasing lamb from a high welfare system and providing food for people caught up in a famine in Africa (through the donation of funds to charity). In conclusion then, the demand for high welfare lamb is influenced not only by our animal welfare concerns but also by the alternative choices that modify our preferences.

This leads us to *demand curves*. These show the relationship between the quantity of a good demanded and its price and this can also be illustrated graphically as in Fig. 9.3. As the price of a good increases, then the quantity demanded will fall[1]. Thus demand curves have negative slopes as shown in Fig. 9.3, where quantity demanded responds positively to price decreases. This can be seen in the marketplace where, when the price of sheep products (meat, wool, etc.) or animal welfare attributes is low, the quantity demanded will be greater than when the price is high.

9.3 Communicating Demand

Having explored the concepts of preferences and indifference curves that lie behind the demand for sheep welfare (or any other good), how is that demand communicated to suppliers? The first way is the one that we have just considered – through the purchasing actions of consumers in the market place. What is bought, and not bought, by consumers in turn determines what retailers and processors buy from farmers and other producers. (This also forms the basis of the main way in which incentives are provided to producers to provide goods that are demanded which we will consider in the next section.) Members of society can also communicate their

[1] The indifference of the consumer to the good and alternative goods changes, and greater utility can be gained from purchasing other goods.

demand for certain goods/services collectively by forming groups with a specific point of view or interest e.g. animal welfare, human rights, environmental groups or a political party.

9.4 Incentives for Producers to Meet Demand

Incentives can either be positive or negative. Both can encourage action in a particular direction. For businesses involved in the production of sheep, positive incentives can come in the form of higher prices for goods/services that better meet demand. For example, a price premium for lamb from high welfare production systems. Conversely, a price reduction can arise for produce that meets demand less well.

Where the market fails to meet demands of society, governments may intervene. There are a number of reasons why markets can fail to be effective in achieving desirable production levels of a good or service. This includes characteristics of the good/service itself such as whether it is a 'public' or 'merit' good[2], as might be argued for animal welfare (see McInerney (2004) for a fuller description as applied to animal welfare or economics textbook for a general description). Government interventions can take the form of subsidies, taxes, regulation and education. Regulation is the primary mechanism currently adopted by UK government to encourage producers to adopt acceptable animal welfare standards. These operate by setting minimum standards, which legislation enforces by enabling prosecution if they are breached. For example regulations set minimum welfare standards for sheep at slaughter, during haulage and handling and during production on farm. Fines imposed as a consequence of breaches in agreed standards act as a financial incentive (albeit negative) to producers.

Finally, there are also non-financial incentives to consider. While businesses must generate a profit (a topic we will cover later) business managers gain satisfaction (utility) from other factors, such as self-esteem. The actions they take to maximise their utility may therefore not be solely driven by the expected financial consequences. The farmer, for example, may take actions that improve the welfare of his sheep, such as treating a case of foot-rot, because he gains satisfaction from having well-cared-for animals.

9.5 The Supply of Goods and Services

Producers supply goods and services in response to the demand communicated to them and incentives provided. Here issues of money and profit arise, and the reasons why businesses produce, transport, etc. sheep. Businesses exist only while they are

[2] A public good is one in which exhibits non-rivalry and non-excludability attributes. That is where consumption by one individual does not diminish the amount that is available for others to use and it is not possible to effectively exclude any individuals from its use. A classic example is public street lighting. A merit good is one which it is deemed to be intrinsically valuable or socially important and which will be under-provided by the market, for example education or health care.

profitable, if a loss is generated then the business is spending more than it has coming in, and that is no more sustainable by a business than it is by an individual. The profit for many farmers is equivalent to a salary or wage[3], for larger businesses it can be the 'dividend' which shareholders receive in return for investing their capital. If farmers (and other small traders such as livestock hauliers) are unable to generate sufficient profit to meet the needs of their family then they will have to find alternative ways of earning a living. With bigger businesses, if the shareholders are dissatisfied with the level of dividend they are getting they may withdraw their capital from the business and invest it elsewhere – and without capital the business cannot continue. So there is a minimum level of profit required by the owner(s) of a business, which economists refer to as the 'normal rate of profit'.

Profit, in simplistic terms, is the difference between all the business costs and the value of the goods/services produced during a fixed time period (usually one year). (Farmers produce goods such as sheep, vets, hauliers and slaughtermen provide a service, hence economists talk about goods and services.) A loss is generated where the costs of production exceed the value of what is produced. This is undesirable in terms of making good use of limited resources – it would not be wise, for example, to convert scarce resources into a good, e.g. lamb, if its final value was less than that of the resources used to produce it. So, from a resource-use point of view, the fact that businesses will not be able to operate at a loss is desirable to society as a whole.

There is a relationship between the price of a good and the quantity that will be supplied, this is known as the *supply curve.* The higher the price the more producers will supply (the incentive increases because the owners of a business gain additional monetary reward with which they can purchase goods and services that give them utility). *Supply curves* therefore have positive slopes, as illustrated in Fig. 9.4. In this example, price y represents the point at which the average costs of production

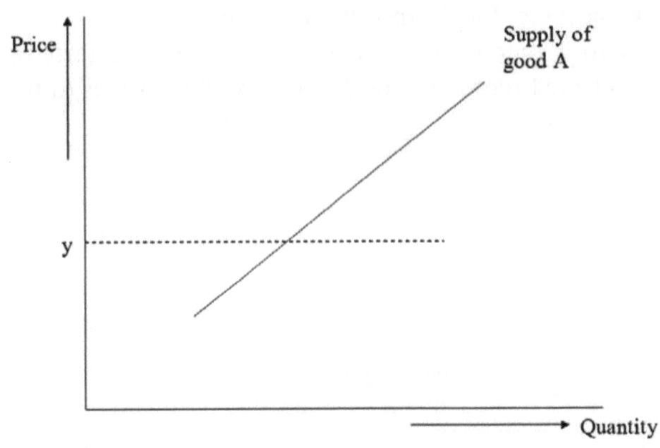

Fig. 9.4 A supply curve

[3] The return they receive for the money (capital) they have invested in the business and for their time and effort, both manual and managerial.

equal the value of the product sold and below this businesses will stop supplying the good/service.

Trading in the marketplace establishes an equilibrium point between supply and demand through adjustments in the price – a price is established at which both the needs of suppliers and demands of consumers are met. This is illustrated in Fig. 9.5.

These are some of the basic economic concepts that apply to the welfare of sheep as to any other good or service. It is worth pausing for a moment then to consider a few of these in practice.

Think for a minute about how *you* communicate the sheep welfare standard you wish farmers, hauliers, and slaughtermen for example, to apply. Do you communicate with him/her at all? If you purchase lamb, woollen garments etc. without differentiating in any way between whether they have been produced in high welfare systems then the message you communicate, as a consumer, is that sheep welfare does not matter. Furthermore, if you are unwilling to pay a higher price for goods produced in high welfare systems then you are not providing any incentive to producers to act in such ways. Hence, where high welfare adds costs, suppliers are unlikely to produce sheep in such ways since it would reduce their profit (or salary in the case of many farmers).

Perhaps the first question that really needs to be asked is, as a consumer, can you selectively purchase in this way – is there adequate information at point of sale for you to be able to do so? For much of the time consumers are unable to differentiate between products from high welfare systems – have you ever seen it on the sales description of a woollen garment? Where the consumer cannot selectively purchase goods (or services) that meet their demand then it is clear that their ability to communicate with suppliers through the market is constrained. This is one example of why a market may fail and why government intervention may be justified to ensure that society's demands, e.g. for sheep welfare, can be met.

The second question then is do you communicate your preferred sheep welfare standards to your elected representative? For some the answer to this will clearly

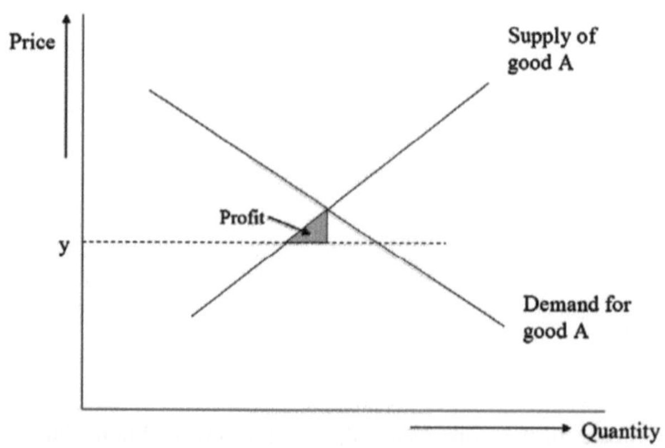

Fig. 9.5 Market equilibrium

be yes, either through membership of a group that lobbies government, by direct communication with a local Member of Parliament (MP), or perhaps by using your electoral vote to support a candidate MP that supports animal welfare issues. For others the answer is likely to be no, in which case either you are content with the current standards of sheep welfare or you are sufficiently content not to take any action. The outcome of not communicating your preferences, however, may not be what you expect since the views of those who are communicating their demands are likely to prevail.

Assuming that people are communicating their sheep welfare demands, are producers meeting that demand? Again some mixture of answers is likely to be given. Where the answer is no, the reason in some cases will be that the incentives (positive or negative) provided to producers are not sufficient to adjust their actions. More detailed examination is thus needed of the costs and benefits of sheep welfare to those involved in their care – farmers, vets, hauliers, slaughtermen, etc.

9.6 Costs and Benefits of Sheep Welfare

Recalling the economic concept of utility where satisfaction can be lost (a cost) or gained (a benefit), for an individual (including business managers) both financial and non-financial costs and benefits arise from actions that affect the welfare of sheep. Also, as explained earlier, to ensure continuation (i.e. economic sustainability) businesses must generate 'sufficient' profit to meet the requirements of owners. For any business there will be a course of action that has the potential to maximise profitability. This course of action can be modified to achieve a higher level of non-financial benefits (e.g. sheep welfare standards) down to the point at which 'sufficient' profit is generated. Actions to maintain or enhance sheep welfare that simultaneously maintain or improve business profitability are therefore likely to be adopted for financial reasons and/or non-financial reasons. That is, there is sufficient financial incentive for them to be adopted by individuals whether they have sheep welfare concerns or not. However, actions that would maintain or enhance sheep welfare, but which reduce the level of profit generated (i.e. where the financial costs exceed the financial benefits), will only be adopted to the point of 'sufficient' profit as shown in Fig. 9.6.

Though this is a simple and logical notion, determining whether the financial benefits exceed the financial costs or visa versa can be difficult in practice. Two examples will be used to demonstrate this using a *partial budget* as a framework. These examples are the implications of blowfly strike in sheep and haulage of store lambs to an auction market.

A partial budget can be used to assess the financial impact of a small change in the way something is done. It compares the expected financial outcomes of two alternatives, e.g. taking action or taking no action to prevent blowfly strike. The financial outcomes are divided into four categories and presented as shown in Table 9.1 below.

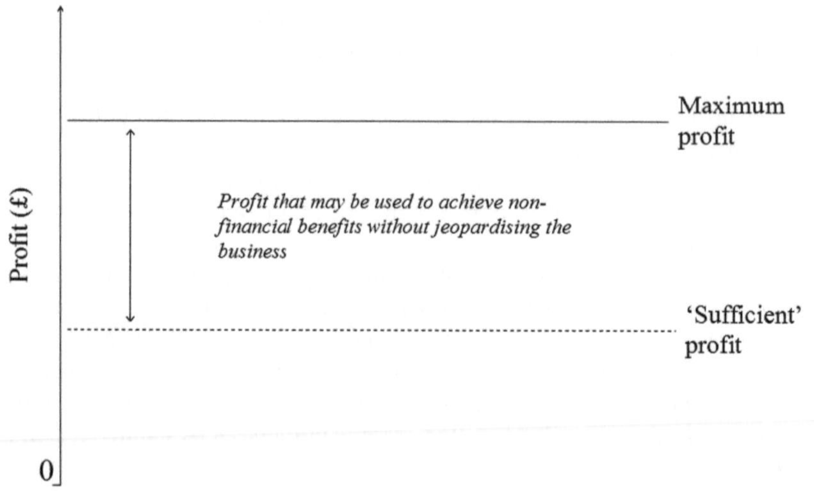

Fig. 9.6 Maximum and 'sufficient' profit

The *totals* of each side of the partial budget (losses and benefits) will be equal once the net gain or loss are entered – there will be a net gain if F>E, or net loss if E>F. All losses and gains are estimated on an average annual basis. As the budget is prepared ahead of making the change, they are also *expected* values and not actual. Only revenue or costs that **change** are included. If there is a net gain then adopting the proposed change is expected to be financially beneficial.

Within the 'cost' categories, a further subdivision is made between *variable* and *fixed* costs. Variable costs tend to vary directly with the size (or scale) of the enterprise and can be accurately attributed to individual enterprises. For example, the cost of the veterinary medicine used to prevent blowfly strike would be a variable cost – the amount used will vary directly with the number of sheep in a flock and its cost can be accurately attributed to the sheep enterprise. Fixed costs by comparison are difficult to allocate to an enterprise and do not vary directly in relation to the scale of an enterprise. They include depreciation of buildings and machinery, fuel, interest and regular labour. Take labour for gathering a flock as an example. If the sole purpose of gathering the sheep is to prevent blowfly strike then all the labour involved can be allocated to this task. However, if the sheep are also wormed at the same gathering, only part of the labour cost is attributable to the prevention

Table 9.1 A partial budget framework

Losses	£	Benefits	£
Revenue lost	A	Extra Revenue	C
Extra costs	B	Costs saved	D
Total losses (A+B)	**E**	**Total benefits (C+D)**	**F**
Net gain		*Net loss*	
Total		*Total*	

of blowfly strike, and it can be difficult to decide how much of the total labour to allocate to each task. (For a more detailed description refer to texts on farm management such as Warren, 1998 or Turner & Taylor, 1998.) Variable costs are generally easier to identify and quantify than fixed costs but both are equally important in the analysis. It is particularly important not to ignore the cost of regular labour devoted to animal health/welfare on the grounds that it is already present and paid for and/or the activities concerned are normal farming practice (McInerney, 1996). As labour is a scarce resource (see arguments above), there is usually a gainful activity that must be neglected in order to carry out the activity of interest. The net loss associated with this diversion of labour must be properly accounted for and is known as the *opportunity cost* of labour. It may not equate with the wage rate paid.

For items in all four categories the change in the quantity of output or input must be estimated and a value put on it (price per kg, tonne, hour etc.). Errors in estimating the expected quantity or price can be made. It is therefore important that the expected net gain or loss is interpreted in light of likely risks – the probability that the outcome will be different from the average on which the budget is constructed. For example, quantifying the expected reduction in a lamb's liveweight gain following blowfly strike can be problematic. (Note also that this is the realm of the natural scientist, and is an example of the linkage between natural science and economics.) Equally will the price per kg for that lamb when it comes to sale in several months time be affected? Thus, if there is a large net gain and a low risk of the expected outcome not matching that used in preparing the budget then it is likely to be worth adopting the change proposed. Alternatively if there is a small net gain and/or a high risk that the expected outcome will be different (worse) than that used in preparing the budget then adopting the proposed change may not be worthwhile. The two worked examples (blowfly strike and store lamb haulage to auction market) illustrate some of these points.

9.6.1 Partial Budget for the Control of Blowfly Strike

Blowfly strike can cause serious reductions in animal welfare; the natural scientists may describe this in behavioural, physical and physiological terms. The physical measures that might be identified by the natural scientist are also the measure of the quantity of output that might be lost in the construction of the partial budget. For this example let us assume that we have a flock of 100 ewes with a 160% lambing (i.e. ewes produce an average 1.6 lambs each) and for the sake of simplification ignore the rams. Our start point is taking no action to prevent blowfly strike, and we want to know whether it would be beneficial to change to a new situation where we use veterinary medicines to prevent strike – and this is 100% effective. An estimate of whether the proposed change would be financially beneficial can be made by preparing a partial budget. The information required is:

1. The number of ewes and lambs that are expected to be struck if no action is taken (assume that 5% of ewes and lambs will be affected).

2. The reduction in liveweight gain that affected lambs are likely to incur and the expected price per kg (assume that for every lamb struck 2 kg of liveweight gain is lost and each kg has a value of £0.80.)
3. For ewes, instead of lost liveweight the value of the additional food required to regain any liveweight loss may be more appropriate than valuing the liveweight reduction directly, as this is what the farmer will actually incur (assume that for every ewe that is struck, an additional £2 of feed is required to regain the lost weight.)
4. The labour required to gather and treat affected animals or to prevent strike (assume that to eradicate strike from affected sheep (8 lambs and 5 ewes) requires an additional 2 h of labour for gathering and treatment and as preventative treatment can be applied when the flock is in for worming only 1½ h of extra labour is required. The opportunity cost of the labour is assumed to be £5/h.).
5. The cost and volume of veterinary medicine required to treat or prevent strike (assume it is £0.80/dose in both cases for ewes and £0.40/dose for lambs).
6. Any cost items will also incur an interest charge – if you spend more, then either you will incur an interest charge if it is borrowed money or lose interest you might otherwise have obtained if it is drawn from savings. This is also an *opportunity* cost (equivalent to the argument above concerning the opportunity cost of labour). In this example, interest will be assumed to be 7% on all costs over a period of 4 months – a third of a year (the time until the lambs may be sold or ewes recover).

Using these assumptions Table 9.2 gives the partial budget for taking action to prevent blowfly strike. A net loss of £110.62 is projected for taking action to prevent blowfly strike in place of treating stock once struck. This expected financial outcome must then be interpreted in light of the risks. It represents the approximate value of

Table 9.2 A partial budget for taking action to prevent blowfly strike compared to taking no preventative action

Losses	£	Benefits	£
Revenue lost		**Extra revenue**	**12.80**
None		8 lambs × 2 kg/lamb @0.80/kg	
A: Total revenue lost		**C: Total extra revenue**	**12.80**
Extra costs		**Costs saved**	10.00
Veterinary medicine to prevent		Feed to 5 ewes @ £2	10.00
blowfly strike in all sheep		Veterinary medicine to treat 8	10.40
100 ewes @ 0.80	80.00	lambs and 5 ewes @ 0.80 per dose	
160 lambs @ 0.40	64.00	Labour 2 h @ £5/h	10.00
Labour 1½ h @ £5/h	7.50	Interest of feed and veterinary	0.48
Interest on vet medicine for 4 months @ 7%	2.80	medicines (£20.40 × 7%/3)	
B: Total extra costs	**154.30**	**D: Total costs saved**	**30.88**
Total losses (A+B)	**154.30**	**Total benefits (C+D)**	**43.68**
Net gain		*Net loss*	*110.62*
Total	**154.30**	***Total***	**154.30**

three sheep so if three sheep (ewes or lambs) were to die or lose their whole value as a result of strike the financial decision would change and a net gain would arise from preventative treatment.

As mentioned there are also non-financial costs and benefits. Having completed a partial budget, one way of looking at this is for the decision-maker to consider is £111 worth paying for the benefits to sheep welfare and to remove the risk of the potential need to treat struck sheep later?

9.6.2 Haulage of Store Lambs to Auction Market

The second example is that of haulage of lambs to an auction market. The benefit of good haulage conditions (space/animal, water availability, rest periods etc.) is maintaining the maximum quantity of marketable sheep at the end of the process. Equally, the quality of stock at the end of a transfer can affect their market price – if lambs arrive with faeces-covered fleeces, dehydrated and exhausted they are unlikely to achieve as high a price as they might otherwise. Thus, while good welfare during transportation cannot add value to sheep, it can prevent loss of value – hence there is a financial benefit that can be obtained from good welfare practices.

Keeping this very simple, let us assume that if the lambs in this case are transported in poor welfare conditions, arrive with dirty fleeces and tired, then the expected price that buyers will bid for them at the auction market is reduced by 10%. For a batch of 100 lambs we wish to know how much we can pay for extra welfare in haulage. In other words when we build the partial budget by identifying the expected gain from good welfare we will know how much extra cost we can afford before breaking even. The partial budget might look like that in Table 9.3. In this example, the extra cost, which could be paid for better welfare conditions during haulage, is £250 or £2.50 per lamb.

9.6.3 Cost Benefit Assessments

In the prevention of blowfly strike case, the expected cost of preventing any loss of sheep welfare was greater than the expected benefits. As the non-financial benefits

Table 9.3 Partial budget to establish break-even point for welfare of store lambs in transit to auction market

Losses	£	Benefits	£
Revenue lost		**Extra revenue**	
		10% of sale value @ £25/lamb × 100 lambs	250
Extra costs	?	**Costs saved**	
Total losses		**Total benefits**	**250**
Net gain		*Net loss*	
Total	**250**	***Total***	**250**

for any individual farmer are indeterminate (since utility is a measure of individual preferences), there is no certainty as to whether preventative actions will be taken – since this would be financially disadvantageous. Returning to our question about incentives, it would be possible for this situation to be reversed or equalised if consumers paid a higher price for lamb produced in systems where preventative veterinary medicines were used. For our example, if the average lamb were sold at 42 kg and 80 p/kg over the 160 lambs produced by the flock, an additional 1.7 p/kg would cover the extra cost of the preventative treatment. Alternatively, the cost of the veterinary medicine could be subsidised to reduce the extra cost incurred – a cost of 14 p/ewe and 7 p/lamb for the preventative veterinary medicine would bring the partial budget close to a break-even position. Either of these approaches could achieve the prevention of the loss of sheep welfare from blowfly strike and theoretically they would cost society the same – though only consumers of lamb would pay if the market price premium (additional x p/kg) approach operated.

Earlier, the question of whether consumers can and do purchase on the basis of the welfare standards under which sheep are kept was raised. There is a potentially wide array of desirable sheep welfare actions that those caring for live sheep could adopt. In the example discussed here, the estimated market price premium required (1.7 p/kg) is a small amount, but if higher welfare standards were to be adopted overall, this could become a far higher figure. Quality assurance schemes began on this basis, certain 'quality' standards had to be met by producers and a higher price to cover those costs was sought from consumers in exchange. The alternative is to provide a subsidy (e.g. on the price of veterinary medicines) that would cover the costs of actions that improve welfare standards. Many people are against the provision of subsidy, in this case the disadvantages include the risk that farmers could overuse the veterinary medicines since they are low cost, and this would increase the cost to the taxpayers in society who would ultimately be providing the finance. Where there is insufficient benefit arising from actions that improve or protect sheep welfare there are two potential outcomes – either the producer must accept a reduced level of profit (and risk being replaced by farmers who won't) or the level of sheep welfare is likely to be lower. This choice will be returned to in the last section in this chapter on sustainability.

9.7 High Welfare and Organic Production Systems

The economic concepts remain the same for high welfare and organic as for traditional production systems, but two additional points are worth noting:

(a) the guarantee provided to the consumer,
(b) the problem of conflicting objectives.

9.7.1 The Guarantee to the Consumer

An assurance label is one of the attributes that food products may have. Typically these assure the type of production system from which the food originated such as organic or high welfare (such as RSPCA 'Freedom Food'). An assurance scheme ensures that certain minimum standards and rules of production have been adhered to, giving the consumer confidence in the product with regards to applicable issues. For some issues, the minimum standards and production rules within an assurance scheme may be identical to the minimum regulatory standards that apply to all producers. In such respects, 'assured' produce will then be the same as produce from non-scheme members, but consumers may still have greater confidence in the 'assured' produce. Thus effectively a new product can be defined with the additional attribute of coming from an 'assured' farm. Strictly speaking, this product will have its own demand curve and, if examined, this would be expected to lie to the right of the demand curve for the original product. The distance between the two demand curves will reflect the consumer's willingness to pay a higher price for the additional attribute ('assurance'), as illustrated in Fig. 9.7.

For other issues, such as use of veterinary medicines in organic systems, assurance scheme rules are different from the legal requirements. The consumer expects the assured product to be differently produced from competing goods in this case, and, where consumers demand the additional attributes, a higher price would be expected for those products. It is not logical to assume however that other farmers cannot and do not operate to high welfare standards, for example. Some non-scheme members may also undertake the same actions, but the consumer has no assurance of this.

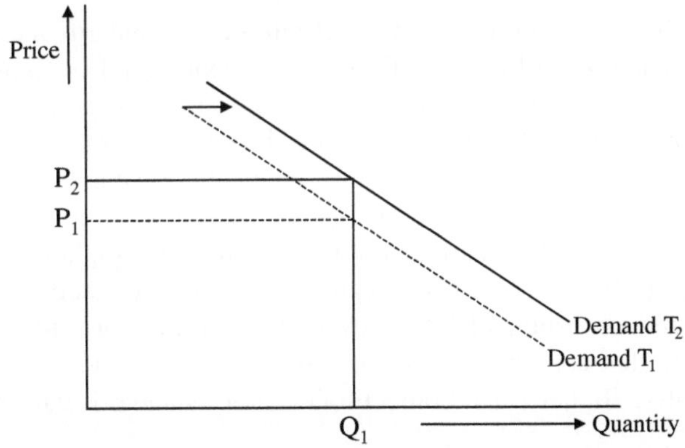

Fig. 9.7 The demand curve to the right of the original demand shows what consumers are willing to pay for the new product

9.7.2 The Problem of Conflicting Objectives

In many assurance schemes, including organic, there is more than one objective, and with multiple objectives there is the potential for conflict. Taking organic systems as an example, there are animal care and environmental objectives (Defra, 2003). On occasions veterinary medicines can be necessary to control disease and protect animal welfare and this is allowed for within the rules that govern organic farming practices (Defra, 2003). However, in some cases the medicines can be harmful in the environment e.g. avermectins for internal or external parasite control in sheep (Strong, 1993), which are excreted in faeces. In such cases a compromise must be made between animal health/welfare and environmental objectives, and abiding by all the production principles all of the time becomes impossible. The dilemma is just another example of the resource allocation problem at the heart of economics. However, a common misconception is that organic farming is a 'gold standard' that unlike alternatives is somehow immune from such dilemmas. The reality is that organic farming provides a specific set of rules that may lead to alternative resource allocations and therefore a different balance of outcomes than other farming systems. Given that these rules are designed to work with nature and natural cycles rather than dominate them (Lampkin, 1990) it is particularly important for organic farmers to understand the interacting systems that govern its resource allocation process.

9.8 Sustainability

Sustainability is a much used term nowadays, but it can mean a number of things. Here it is used to describe 'the ability to continue' and will specifically refer to the ability of those involved in sheep production to continue to produce sheep to a particular welfare standard.

As noted earlier, regulation is used to set minimum standards including that of animal welfare, and this will increase the costs of production from a previous level. If the same minimum standards do not apply to competing producers, then they could produce and hence sell their goods more cheaply. Under these conditions and where the consumer has a greater preference for cheap food e.g. lamb, than for food (lamb) produced in higher animal welfare systems then over time, for the economic reasons covered earlier in this chapter, those producing in the higher welfare systems will gradually stop. In other words the regulations will have made those businesses uncompetitive and unsustainable. Moreover, the welfare standards for sheep (or other animals) in that country will not be improved, as production will just move to another country. Higher (than competitor) welfare standards can therefore only be sustained if:

(a) consumers are willing to and do pay the additional costs of production that arise; and/or

(b) Governments subsidise high welfare producers/production systems to the value of the additional production costs. (In this case it is all taxpayers that bear the cost.).

9.9 Conclusions

The welfare of a sheep is determined by the actions of all the people it might encounter – the shepherd, the farmer, the vet, the haulier, the slaughterman, etc. It is also determined by the actions and inactions of consumers – whether they demand and are willing to pay for products that have come from high welfare systems – and by government interventions in the market. The relationships between these different parties are complex. Economics, as demonstrated by the topics which are touched upon in this chapter, is at the core of making sure that the sheep welfare standards, agreed through ethical debate and quantified as a result of research by the natural scientists, are attained.

References

Defra (2003). Compendium of UK organic standards. Defra, London, UK.

Lampkin, N. (1990). *Organic Farming*. Farming Press, Ipswich, UK.

McInerney, J. (1996). Old economics for new problems – livestock disease: Presidential address. *Journal of Agricultural Economics*, **47**, 295–314.

McInerney, J. (2004). Animal welfare, economics and policy: Report on a study undertaken for the Farm & Animal Health Economics Division of Defra. DEFRA, UK. http://statistics.defra.gov.uk/esg/reports/animalwelfare.pdf

Robbins, L. (1945). An essay on the nature & significance of economic science. 2nd edn. MacMillan & Co. Ltd., London. http://www.mises.org/books/robbinsessay2.pdf

Strong, L. (1993). Overview: the impact of avermectins on pastureland ecology. *Veterinary Parasitology*, **48**(1–4), 3–17.

Turner, J., Taylor, M. (1998). *Applied Farm Management*, 2nd edn. Blackwell Science Ltd., Oxford, UK.

Warren, M. (1998). *Financial Management for Farmers and Rural Managers*, 4th edn. Blackwell Science Ltd., Oxford, UK.

Wilkinson, N. (2008). *An Introduction to Behavioural Economics*. Palgrave MacMillan, Basingstoke, UK.

Willock, J., Deary, I.J., Edwards-Jones, G., Gibson, G.J., McGregor, M.J., Sutherland, A., Dent, J.B., Morgan, O., Grieve, R. (1999). The role of attitudes and objectives in farmer decision making: Business and environmentally orientated behaviour in Scotland. *Journal of Agricultural Economics*, **50**(2), 286–303.

Chapter 10
Sheep Welfare: A Future Perspective

A.B. Lawrence and J. Conington

Abstract There is every indication that animal welfare will continue to be a major issue affecting livestock farming in the future. At the same time sheep farming is under pressure worldwide to become a lower-input farming enterprise. The drive to lower inputs and, in particular, restrictions on labour could have important implications for animal welfare by reducing levels of care and perhaps altering the traditional human-animal 'contract'. In addition, climate change will impose further challenges, such as altering the range of parasites, increasing environmental stress and through exposure to more extreme weather. Science and technology have important roles to play in maintaining and improving welfare under these scenarios. Animal breeding, for example, can be used to breed for health and welfare traits such as increased resistance to the disease, improved lamb survival and appropriate adjustments in behavioural temperament. New developments in breeding also open up possibilities of selecting for 'robustness' or increased adaptation to a range of environmental conditions. Objective approaches to assess animal welfare can be used both to provide farm assurance about animal welfare, but also as management tools by farmers to identify welfare issues and to train stockpeople in the recognition of poor welfare. Lastly, use of sensor technologies provide the potential for remote monitoring of welfare in extensively managed animals, such as sheep, to replace more traditional shepherding.

Keywords Human-animal contract · Climate change · Animal breeding · Robustness · Disease resistance · Welfare assessment · Technology

10.1 Introduction

Concern for animal welfare is not a new phenomenon. Religious and philosophical writings demonstrate that humans have been considering our relationships with other animals for 1000's of years (e.g. Lawrence, 2007a; Chapter 1). However, the

---

A.B. Lawrence
Sustainable Livestock Systems Group, SAC, Edinburgh, UK. EH26 0PH

period since the 2nd World War has seen a dramatic heightening in public concern over farm animal welfare with much of this directed at the welfare implications of intensive forms of animal production (Harrison, 1964; Brambell, 1965). Extensively managed species, such as sheep, have largely escaped public attention possibly because they are seen to be living 'natural' lives and as such to be experiencing good levels of welfare. However, there is evidence that attitudes to sheep welfare are changing (e.g. Advocates for Animals, 2004). This is, perhaps in part, because sheep have been visibly involved in situations that have raised wider public concern, such as transportation of live animals (e.g. the Cormo Express debacle; Wright & Muzzatti, 2007) and the 2001 Foot and Mouth outbreak in the UK. At the same time sheep farming is having to adjust to a range of influences (the reform of the Common Agricultural Policy (CAP) in the EU is just one example) that are likely to bring about substantial changes in sheep farming systems (e.g. Goddard et al., 2006). The aim of this chapter is to take a future perspective on sheep welfare by looking at factors that will shape sheep farming and the welfare of sheep within farming systems of the future.

10.2 Wider Society

Currently there are high levels of interest and concern over animal welfare continuing the trend established with the publishing of the Brambell Report (Brambell, 1965). There are a number of indications to suggest that this concern for animal welfare has gradually changed from being a largely European issue to becoming a global concern. For example, a recent MORI poll found that over 90% of respondents from the UK, China, Vietnam and Korea agreed with the statement 'we have a moral duty to minimise animals' suffering as much as possible' (IPSOS MORI, 2005). A 2003 Gallup poll found that 71% of US respondents agreed 'that animals need some protection', and 62% agreed with the 'passing of strict laws to protect farm animals' (Swanson, 2004). A recent Eurobarometer survey (Eurobarometer, 2005) found a broad awareness of animal welfare issues but also reported substantial variation in response to questions on farm animal welfare in relation to variables such as country, gender, previous visits to farms and the species of farm animal. Interestingly three-quarters of the sample believe that they could improve animal welfare through their purchasing behaviour, but just over 50% claimed not to consider animal welfare when purchasing meat. This paradox may be explainable through the deficiencies in using labels to identify 'welfare friendly' systems especially in new Member states (Eurobarometer, 2005).

There is no single reason that can explain the continued growth in public concern for animal welfare. One possibility is that public interest in animal welfare is underpinned, in part, by a growing acceptance that animals are sentient (i.e. that they have the capacity 'feel' in a way that is analogous to human experiences). Historically there has been a strong relationship between belief in animal sentience and human sensitivity to animal suffering (e.g. Lawrence, 2007a). Belief in animal sentience is not a new idea as studies of religious ideas demonstrate (e.g. Preece & Fraser, 2000). Since the 1700s, however, animal sentience has been given greater rational support through research which has demonstrated the many biological similarities between humans and animals (e.g. Chandroo et al., 2004). Scientific evidence relating to animal sentience may help advance the moral status of animals by heightening the 'moral intensity' we associate with animal welfare issues (e.g. Bennett et al., 2002). Recent work has shown that belief in animal sentience is important in determining attitudes to animals (Knight et al., 2004; Phillips & McCulloch, 2005). This suggests that further scientific support for animal sentience should increase the moral importance of animal welfare through widening societal acceptance and belief in animal mind. In other words the more we believe that animals may suffer the more seriously we will take animal welfare (New Scientist, 2005).

This increasing public concern for animal welfare, has been a driver for important changes. For example, animal welfare legislation in the EU and the UK has accepted in law that animals (or at least vertebrates) are sentient. The EU Treaty of Amsterdam states 'The ... Contracting Parties ... desiring to ensure improved protection and respect for the welfare of animals as sentient beings' (EU, 1997). Similarly, the Explanatory Notes to the new UK Animal Welfare Bill read that 'The Act will apply only to vertebrate animals, as these are currently the only demonstrably sentient animals' (House of Commons, 2005). Animal sentience underpins UK government policy on animal welfare, to quote: 'welcome to the Animal Welfare section of the Defra web-site. Like many of you looking at this site, we recognise that animals are sentient beings – not merely commodities' (Defra, 2006). Australia has published an Animal Welfare Strategy (Australian Government, 2005), the UK an Animal Health and Welfare Strategy (Defra, 2004) and very recently the EU has launched an Animal Welfare Action Plan (EU, 2006). These governmental strategies signify a change of direction at least for European animal welfare, as they tend to move away from legislation as the main mechanism for animal welfare improvements and emphasise more 'collective action' on behalf of all parties (including consumers) with interests in animal welfare.

Other recent activities have included a global conference on animal welfare organised by the OiE (World Organisation for Animal Health; OiE, 2004). The OiE has also published a scientific and technical review of animal welfare (Bayvel et al., 2005) and has drawn up international standards for specific issues such as land and sea transport (OiE, 2006). Farm assurance schemes which incorporate animal welfare standards are virtually a requirement in order for producers to find markets for their products (FAWC, 2005), and many large corporations which deal with animals now have developed their own welfare standards often with input from applied behaviour scientists (e.g. Grandin, 2006).

In summary, public concern over animal welfare shows every indication of becoming an established global socio-political issue. In part, this may reflect a greater appreciation of the similarities between animals and humans resulting from scientific research into animal psychology and behaviour. As a result animal sentience (the capacity for animals to 'feel') is now recognised in law in the UK and within the EU. Other effects of this public concern are specific pieces of legislation to phase out controversial systems for farming animals, government-level action plans to increase participation in health and welfare improvements and an increase in farm assurance schemes that aim to ensure at least minimum legal welfare standards. It is likely that the future will see more of the same and that sheep farming should not feel any more protected from this public scrutiny than other livestock industries.

10.3 Sheep Farming Systems

In recent years livestock farming has experienced a range of pressures including increased global competition, greater public scrutiny of farming and, in Europe, the reform of the Common Agricultural Policy, all of which will continue to effect change in farming systems (e.g. Sorenson et al., 2006). In some sectors these pressures have led to greater intensification, but in general the pattern with sheep production has been for a move towards lower input systems. Perhaps the best example of this is sheep production in New Zealand (see also Chapter 6). Whilst the period following removal of farming subsidies in the 1970s led to generally more intensive farming, the sheep industry tended to contract and move to lower input systems (Macleod & Moller, 2006). One aspect of this trend has been the development of so-called 'easy-care' sheep breeds, which are intended to be able to survive, grow and reproduce with minimal human assistance even in relatively harsh environments (e.g. Fisher, 2003; see also later).

In Europe a combination of factors, including reform of the CAP, is creating a rather similar environment for sheep farming to that experienced in New Zealand when it abandoned farming subsidies. The purpose of CAP reform is, in part, to reduce the direct support for production of agricultural products, and to encourage farming to deliver 'public goods' in the form of care for the environment and acceptable levels of animal welfare (e.g. Gohin, 2006). Countries have some discretion as to how they implement these changes. For example, Scotland has chosen both to subsidise farmers for maintaining acceptable standards of animal welfare (so-called Tier 1 support), and also to provide additional funding ('modulated' from existing production-based subsidy) to promote health planning and other preventative actions to improve on-farm health and welfare (Tier 2 support; see SEERAD, 2006). So far, around 4,000 farms have taken up the Tier 2 'Health and Welfare Programme' option, and data are being collected to assess the effectiveness of this approach in terms of improved flock health (SEERAD, 2007).

A number of studies have looked at the potential impact of these changes in European farming policy on farming systems and the environment and there has also been some consideration of their impacts on animal welfare (see Stott et al., 2005). Generally, the reduction in direct subsidies will not be sufficiently offset by the subsidies for public 'goods' (so called agri-environment measures), to prevent there being a negative impact on the financial performance of livestock farming (Matthews et al., 2006). A possible outcome of this increased economic pressure will be to create a more dichotomous livestock industry with some producers seeking to make the most of market opportunities ('market-led managers') whilst others will seek different opportunities in the form of maximising income from the agri-environment schemes ('environmental managers'; Oglethorpe, 2005). Combined with the effects of CAP reform the general economic climate is likely to drive increases in flock sizes (Sorenson et al., 2006) and also put pressure on the economic efficiency of the business (e.g. through reductions in labour input) especially where farmers are seeking to maximise returns from the market. The future is also likely to see an increase in the numbers of part time farmers (e.g. Lohr & Park, 2003). Even though the precise conditions or 'drivers' may vary it is suggested that similar trends to dichotomous production systems will also be observed in other regions such as Asia (e.g. Devendra, 2007).

In terms of the impact of these changes in farming systems on sheep welfare we might expect to see the following:

(a) The potential divergence in farming systems could have an influence on the relationship between the farmer and his/her animals. Where farmers continue to be financially rewarded for production (market-led management), we can expect the traditional 'human-animal contract' to be maintained where humans balance their exploitation of animals by husbanding and caring for them (e.g. Rollin, 2004). However where farmers begin to be rewarded in other ways such as for using animals to maintain the environment (environment-led management) then we might see important changes to the status of domestic sheep. In part this could come about because farmers are no longer rewarded for production and hence issues, such as disease, which affect production are no longer be prioritised. Such changes could also arise because sheep cease to have the status of farm animals and are seen more as being 'feral' or even 'wild'. Generally we have less of a feeling of obligation to feral or wild animals compared to domestic species. It is interesting that in New Zealand such a change in the human-animal contract appears to have already happened with the development of easy-care breeds. The rationale for these breeds is based upon natural selection producing better adapted animals, that the terrain and the cost of labour necessitate a reduction in labour input and because it is argued that human intervention can have detrimental effects at lambing (Fisher, 2006; Fisher & Mellor, 2002). There is other evidence to support the view that human intervention can reduce some of the adaptive capacities of sheep (Dwyer & Lawrence, 2005). However, there is still a need to consider the overall impact on welfare of such changes to the status of sheep and the consequential lack of supervision and care of animals (e.g. Goddard et al., 2006).

(b) Where there is pressure to increase flock sizes and economic efficiency there could also be adverse impacts on sheep welfare. Reduction in labour input is again likely to be a central issue. In a recent study of farmers' opinions on health and welfare, farmers themselves recognise the need to have more labour available for dealing with health issues, such as ectoparasite infections (Morgan-Davies et al., 2006). Indeed analysis of the effect of management strategies on economics and welfare indicates that whilst labour is a key input to sheep welfare it has at the same time negative financial implications on the business (Stott et al., 2005). The increase in part-time farming is also of potential concern either because farmers' will be even more constrained in the time they can give to sheep husbandry or because the people coming into part-time farming lack the necessary skills and training to ensure animal health and welfare. Other potential impacts could be to expose deficiencies in infrastructure such as buildings or handling facilities due to poor economic returns and a lack of investment. A Scottish Agricultural College review of handling facilities for cattle in Scotland found that these were often poorly designed and in need of upgrading to ensure handler safety and cattle welfare (Turner et al., 2003).

In summary the future should see a greater variety of sheep farming systems where some farms become focused on market returns whilst others seek to maximise return from environment based subsidy schemes. Various pressures are likely to drive forward lower-input sheep farming systems principally by reducing labour inputs whilst at the same time flock sizes are likely to increase. These trends are likely to affect sheep welfare by reducing the care provided to farmed sheep, perhaps combined with an adjustment to the human-animal contract where we feel less obliged to look after sheep on the basis that they are closer to being feral or wild animals.

10.4 Climate Change

Despite the seriousness of the threat of climate change (e.g. Kurukulasuriya et al., 2006), it is only relatively recently that research has begun to explore the potential effects of global warming on farming systems and as yet there has been little consideration of the impacts on animal welfare. Given that the effects of climate change will be region specific it is clearly too vast a subject for us to make little more than general comments on possible major effects on welfare. The most obvious changes that could directly impact on sheep welfare would be: (a) changing patterns of disease, including the ranges of sheep parasites (e.g. Colebrook & Wall 2004); (b) increased heat stress with consequential effects on biological functioning (e.g. King et al., 2006), including during transport of sheep to markets and abattoirs; (c) changes to patterns of forage production (e.g. Hopkins & Wilkins, 2006) interacting adversely with the seasonal reproductive cycle of sheep; (d) increased variability of weather, including the likelihood of extreme precipitation events (e.g. Boroneant et al., 2006) or prolonged drought; (e) slow response times from farmers in adapting to these changes (e.g. Menzel et al., 2006), further exacerbating impacts on welfare. These are only preliminary 'guesses' about how climate change might impact in welfare. There is a clear need now for research on climate change to embrace the potential effects on animal welfare as part of the development of integrative approaches to modelling climate change impacts and potential adaptations (e.g. Rivington et al., 2007). An important issue for the future will be to explore conflicts between actions necessary to ameliorate or adapt to climate change and animal welfare.

10.5 The Role of Science and Technology

The indications are that sheep farming will tend to move generally to lower-input systems yet with a trend to larger flocks. These trends if anything place even greater importance on the role that science can play in ensuring sheep health and welfare in the future. We will focus on the role of science and technology in finding solutions to welfare problems, on assessment of sheep welfare and how technology might shape the farm of the future.

10.5.1 Finding Solutions to Health and Welfare Issues Through Selective Breeding

Arguably one of the most significant shifts in the application of science to animal welfare in recent years has been the acceptance that animal breeding can be used to better adapt animals to current systems of production (e.g. Lawrence et al., 2004). There appear to be particular opportunities to improve sheep welfare through breeding in the future, especially given the trends in sheep systems we have already identified.

Selective breeding refers to the practice of choosing animals with superior qualities to become parents of the next generation. There is no 'best' animal for all situations, as the animal that is best adapted to one environment may not be well adapted to different circumstances. The key to natural, as well as 'artificial', selective breeding is the amount of variation in a population for any measurable characteristic (or 'trait'). Selecting animals for breeding that lie at the relative extremes of a population mean for a trait (usually greater than 2–3 standard deviations) will generally 'improve' that trait in the desired direction.

In practice, farmers want animals with good performance in more than one characteristic, such as survival, growth and litter size. For this reason, it is important to know the relationships between the different traits to understand how these traits vary with one another. For instance, an animal breeder may wish to know whether survival will increase or decrease if animals are selected to be heavier at a given age. In animal breeding, the correlation between traits is often a cause concern for animal welfare. In other animal species, such as dairy cows, there is considerable evidence that selection for milk production has a negative correlation with traits that are important for animal welfare such as lameness and mastitis (e.g. Koenig et al., 2005). Taking such genetic correlations more into account can mean it is possible to select for animals that are good producers as well as having a propensity to be healthy (see also similar effects seen in dairy sheep as discussed in Chapter 6).

In animal breeding, predicted values are used to estimate an animal's genetic propensity for a given trait, or traits. Animals are usually chosen to be parents of the next generation on the basis of estimated breeding values (EBVs). Breeding values combine information from relatives' performance to help predict an animal's own breeding value for a given trait. This is particularly important for so-called 'sex-linked' genes, such as litter size, that are only recorded on female sheep. If a farmer wanted to select a ram that would have prolific daughters, then having a breeding value for litter size is particularly useful to this breeder.

Breeding values can also be thought of as a 'common currency' with which to compare one animal against another. They are most valuable when animals are compared with others from different farms that participate in a group-breeding scheme, such as Sire Reference Schemes. This is because the known environmental influences that affect animal performance, such as grass supply and animal management, are taken into account when flocks are evaluated together in such a scheme, using the Best Linear Unbiased Prediction methodology (BLUP; Henderson, 1975).

In the past, selective breeding has mainly focussed on production related traits. However in recent years there has been greater emphasis placed on traits relevant to welfare. For example, the use of selection indices in hill breeds that includes lamb survival as a trait of the ewe has led to improvements in the number of lambs reared compared to control-line ewes (Conington et al., 2001). New methods to apportion the source of genetic variation to lamb survival including the importance of maternal genes have also been described (Sawalha et al., 2007). The b3-adrenergic receptor (involved in the regulation of energy balance) has been found to have 6 alleles in the sheep (Forrest et al., 2003). Significant associations have been found between the frequency of different alleles, ability to withstand cold and lamb survival in Merino sheep (Forrest et al., 2006), suggesting that assessments of allele frequencies may help reduce neonatal deaths from hypothermia, at least in this breed.

For the future, decreases in labour inputs could be offset by breeding for sheep that are better adapted to their environment with a minimum degree of human support. An 'easy care' composite breed has been developed in Wales which has several characteristics that set it apart from other, more traditional, breeds including that it sheds its fleece annually, a trait of economic importance where the fleece itself lacks value (www.easycaresheep.com). Even though a full breed comparison trial has yet to be established, it is claimed that this breed needs no attention at lambing and is generally well-suited to low-labour units. However, as pointed out earlier there is a need for a systematic risk analysis of the challenges to sheep welfare in easy care systems with minimum human interventions. The particular attributes required for easy care sheep need to be identified, quantified and bred for (e.g. reduced wrinkling in the breech of Merinos; Scobie et al., 2005).

It is also possible to improve the welfare-related characteristics of existing pure breeds. For example, breeding goals for maternal breed types, such as the Scottish Blackface, can be applied to improve survival in lambs and performance up to weaning without incurring greater labour inputs (see Conington et al., 2001, 2006). In addition, selecting sheep with improved resistance to disease could reduce the requirement for human interventions to support sheep health. In New Zealand and Australia, established breeding programmes that include faecal egg count as a breeding objective are currently being implemented. In the UK, breeding values for faecal egg count were available for the first time in 2001 although they are not yet integrated into the overall breeding objective. Although selection against one species of pathogen or parasite can result in enhanced immunity against others little is known about the genetic relationships with other common diseases, such as footrot and mastitis, although this is starting to be addressed (Conington et al., 2007). In addition, there is a dearth of information about the relationship between resistance to disease and other animal performance characteristics. The primary requirement for including disease resistance in breeding programmes is knowledge of the extent of genetic variation in resistance and the genetic correlations between resistance and all other important traits. Therefore, considerable research effort is being expended to estimate these quantities (Bishop et al., 1996; Bishop & Stear, 1999; Raadsma, 2000).

Breeding for resistance to other major sheep diseases has shown that there is genetic variation for flystrike, facial eczema and footrot (Raadsma, 2000), and for mastitis in dairy sheep (Mavrogenis et al., 1995). Diseases that are difficult to measure can be selected against using molecular information. In the UK, the National Scrapie Plan is based on selective breeding only from non-susceptible genotypes in an attempt to eliminate the disease (Defra, 2002). In New Zealand, efforts are underway using microsatellite markers to identify regions of the genome that influence particular phenotypic characteristics related to parasite resistance (McEwan & Kerr, 1998). Such quantitative trait loci (QTLs) are expected to be soon available for breeding programmes for several other diseases. A recent example is breeding for resistance to footrot. Footrot is the most common cause of lameness in sheep, causing pain, reducing mobility, and inhibiting feed intake, with negative consequences on production traits and animal welfare (see also Chapter 5). If left untreated, sheep lose body condition and are predisposed to other diseases. Colostrum and milk production in ewes are also affected and rams with bad feet are often unwilling to mount ewes at mating time. Footrot is highly contagious and is easily transmitted from sheep to sheep, even by sheep that show no disease symptoms. In particular, the new, severe form of lameness, *Contagious Ovine Digital Dermatitis* (CODD) poses a potential threat to the sheep industry. In a survey of farmers' practices and attitudes towards footrot (Wassink & Green, 2001, Hosie and Vipond, personal communication), more than 90% of sheep farmers had seen footrot in their sheep in the past year and 31% considered that 6% or more of their flock were affected with footrot. However, recent results from the foot examinations of over 9,000 sheep on farms across the UK in 2005 and 2006 suggest that footrot prevalence (including scald – the precursor to footrot) is 21.5%, 29% and 50% for Scottish Blackface, Texel and Mule sheep respectively (Conington et al., 2007). The cost of footrot from lost productivity and treatment is reported to be over £24M in the UK (Nieuwhof & Bishop, 2005) and therefore the impetus to breed sheep that are resistant to footrot will not only address the issue of animal welfare but also its economic impact to the farmer. A new footrot resistance gene test has been developed in New Zealand (Hickford, 2000) and is currently available to breeders to identify genetically resistant strains of sheep for breeding via Ovita, the New Zealand company that undertakes commercial genome screening in practice. Efforts are currently underway at SAC in collaboration with the New Zealand team to develop a programme for similar genetic tests that will be applicable to UK sheep breeds.

Breeding for desirable behavioural qualities can also have potential benefits for improved animal welfare. Humans have arguably had less impact on sheep behaviours than some other species (e.g. see Chapter 2), largely due to the more extensive farming methods under which the majority of sheep are kept. However, breeds that have been traditionally kept in small flocks under intensive management, such as the Suffolk breed, provide evidence that human intervention can adversely affect behavioural adaptations, particularly those related to early neonate and maternal responses (Dwyer & Lawrence, 2005). There is also good evidence that both early lamb and maternal behaviours are under genetic control (e.g. Cloete et al., 2002; Lambe et al., 2001), opening up the possibility of using these as proxy traits to improve lamb survival. Depending on the breed there may be a requirement to adjust the selection pressure on lamb or maternal behaviours. Other sheep behaviours that may benefit sheep welfare in extensive pastures could include shelter-seeking, foraging and temperamental traits, for example reduced flightiness. Different genotypes of sheep and cattle have different preferences for grazing particular grass species or land areas (see review by Hohenboken, 1986). Using genotypes that have preferences for some indigenous plant species and not others is important for pasture management geared towards environmental benefit, and also for flora diversity. Breeding sheep that have particular preferences for different forage types could benefit sheep that are managed in organic sheep systems using forages with anthelmintic properties, such as those containing condensed tannins (Athanasiadou et al., 2002). In relation to temperament traits, research in Australia has demonstrated that selection for 'calmer' temperament can have positive impacts on lamb survival (Blache & Ferguson, 2005). Behavioural characteristics are difficult and expensive to measure but, potentially, could be ideal candidates for selection using QTLs in the future. However, to our knowledge, no known QTLs for behavioural characteristics are currently being used in breeding programmes for sheep, although considerable research is underway in other animal species to identify QTLs for behavioural traits, such as feather pecking in hens (Keeling et al., 2004).

We have already discussed that narrow breeding goals in the past have led to a deterioration in some traits, such as mastitis and lameness in dairy cattle. It is critical that breeding indexes for sheep in the future also include a wider set of breeding goals than are currently available, to prevent any future deterioration in health and welfare. This is starting to happen with the introduction of breeding for resistance to *teladorsagia circumcincta* and scrapie resistance, and with the recent introduction of longevity and lamb survival to maternal sheep indexes. However, there is still a lack of knowledge about the feasibility and desirability of selecting for other welfare-related traits including behaviour. Future research should focus on this issue to develop selection programmes that improve sheep health and welfare. A concept that is currently receiving attention in this sense is that of breeding for robust animals (e.g. Kanis et al., 2004). The idea of breeding for robust animals can be essentially the same process as broadening breeding goals, for example to improve the health and welfare of animals (i.e. there is no specific 'robustness' trait); in this sense breeding for 'easy-care' could be said to be breeding for robustness.

However, in one important sense breeding for robustness is different to existing breeding approaches and that is where it incorporates the sensitivity of genotypes to different environmental conditions (e.g. Knap, 2005). Such 'environmental sensitivity' measures are in effect estimates of genotype by environment interactions and in principle could be developed to provide breeding values for the sensitivity of individual sires to different farming systems. Currently research is exploring the possibilities for inclusion of environmental sensitivity measures in breeding programmes (e.g. Haskell et al., 2007). The likely trend for divergent systems in sheep production and the challenge of climate change suggest that such measures of environmental sensitivity may become increasingly important in the future.

10.5.2 Assessing the Welfare of Sheep

The Brambell Report (Brambell, 1965) placed particular emphasis on protecting both the physical and mental wellbeing of animals. Following this, scientists began to develop a range of approaches for objectively assessing animal welfare, including the development of approaches specifically intended to provide an understanding on the animals' perspective and mental state (e.g. the use of preference testing; Hughes & Black, 1973). Up until the 1990s much of this research was conducted in experimental settings, often using invasive or time-consuming methodologies. More recently there has been a significant move to assessing on-farm welfare. There are a variety of reasons for assessing welfare on-farm, including for farm assurance, as a management tool for farmers, or for training of stock people, and different measures may be used depending on the purpose of the assessment. There has been progress in applying measurement tools to assess welfare in the field (e.g. Blokhuis, 2004), however there remain important issues still to be resolved before we have a sound approach to on-farm welfare assessment.

(a) A critical issue is that many measures that are proposed for on-farm welfare assessment have not been subjected to a rigorous validation (see also Lawrence, 2007b). This is particularly the case where measures are intended to give insight into animals' mental state. A good example of this is the use of condition scoring which is a widely used management tool in livestock farming (e.g. Arthington et al., 2007). It has also been proposed that condition score be used as a means of ensuring that farm animals are not subject to excessive hunger (Keeling & Vessier, 2005). However there has been no systematic attempt to validate condition score as a measure of hunger. It is common sense that very low condition scores are indicative of poor physical welfare and also very likely mental suffering as well. However, there is very little information available to judge at what point low condition scores begin to impact significantly on welfare, including the range of condition scores that are commonly used in management systems, although research is beginning to address these issues (e.g. Schutz et al., 2006). Similar arguments apply to the use of behavioural approaches. For example, approach and avoidance is often used to assess human-animal interactions but again we have only a limited understanding of how to interpret different responses to a human approach. It is often assumed that avoidance of humans indicates fear and poorer welfare, an assumption largely based on experimental studies using aversive stimulation (electronic shock) to create avoidance behaviour (e.g. Hemsworth, 2003). However, again we lack the understanding to interpret approach-avoidance in less extreme conditions and also the more subtle responses to human approach. Approach-avoidance measures may be particularly difficult to interpret in animals, such as extensively managed sheep, that rarely have contact with humans. Similar points have been made by Rushen (2003) who pointed out that we lack the understanding to interpret the welfare implications for even simple measures, such as lying behaviour.

(b) There are many untested assumptions surrounding the assessment of on-farm welfare. For example, it is assumed but never tested, that a sufficiently accurate assessment of welfare can be made on a very short visit lasting no more than a few hours, and that an assessment of welfare in the short-term tells us something meaningful about welfare over longer-periods of time. It is also often assumed that production measures are not good indicators of welfare, based on experience of intensive farming of pigs, poultry and dairy cattle where individual welfare can be poor despite the flock or herd performance being acceptable (e.g. Dawkins, 1980). However the limits of this assumption have also never been properly explored.

These assumptions are highly relevant to the question of how to assess on-farm sheep welfare. This assessment will be constrained by only being able to sample a small number of sheep, or requiring that the sheep are gathered and penned and by only being able to carry out assessments in one season. There is no evidence to suggest that such sampling protocols will provide a representative assessment of welfare in other seasons or in the grazing environment in which sheep spend the majority of the year. Further it seems possible that production measures could be validated as welfare measures for sheep, given the lack of human interventions, and consequently a closer link between physical functions and welfare than happens for intensively managed species.

If sheep welfare assessment protocols are to provide objective and accurate results then we believe that these issues need to be addressed with the understanding that the development of properly validated welfare measures from concept to practice can take many years of investment (see Lawrence, 2007b).

10.5.3 Other Technological Solutions for the Future

Technology is likely to play an important part in the future development of sheep farming and also in protecting animal welfare. However, as labour input is likely to be a major future constraint in sheep farming then technologies that can substitute for traditional husbandry (stockperson time) are likely to find their way into use in a variety of forms. These are likely to include: (a) electronic ID and rapid throughput handling systems that provide automatic registration of sheep identity and other measured traits (e.g. body weight; flight speed as a measure of temperament (e.g. Muller & von Keyserlingk, 2006); (b) use of electronic tagging combined with remote monitoring equipment to provide information on the movements and activity of sheep flocks including the possibility of remote monitoring for health and welfare (e.g. recording of visits to key resources such as drinkers); (c) use of virtual fencing and intelligent stock handling systems to allow more flexible movement and restraint of animals, and to prevent wildlife accessing resources (such as food and water) but permits stock to gain access via a transponder; (d) the use of DNA-based technologies combined with genetic evaluations to estimate parentage, genetic parameters and presence of deleterious or advantageous genes on a far wider scale than at present. All of these possibilities have the capacity to have both positive and adverse effects on welfare. For example, remote monitoring of activity, if validated carefully and used responsibly, could provide improved surveillance of 'behavioural' changes indicative of health or welfare problems. Conversely, lack of validation and or improper use could result in sheep being exposed to potentially serious neglect. These technologies provide potential solutions, but only if developed and used appropriately.

10.6 Conclusions

Sheep farming, like many livestock industries, is undergoing significant change at present due principally to competitive pressures and reform of farm policy in Europe. These effects are driving sheep farming towards lower input systems and particularly a reduction in levels of labour input. New Zealand sheep farming has already gone down this route and forms an interesting case study for other countries. In particular, the development of so-called 'easy care' breeds and systems presents a potential solution to counter the pressure on labour inputs. It seems that, as practised in New Zealand, the development of such systems requires a change in the 'human-animal' contract with humans no longer providing traditional levels of care. The advantage of this approach is that appropriate breeds of animal are developed that are well suited to the local conditions. The disadvantage may be that there is little or no safety net to protect sheep welfare especially when weather or foraging conditions are poor, situations which may become more common with climate change. It seems that Europe has yet to commit fully to this form of sheep farming. Sheep farmers in Europe are moving to lower-input systems and easy-care breeds are being farmed but there appears to be uncertainty about how far we can go in reducing human interventions and still maintain animal welfare (e.g. Goddard et al., 2006). There is a risk in today's environment of greater public scrutiny of raising adverse public reaction if reduced levels of care significantly impact on animal welfare (e.g. Advocates for Animals, 2004). Hence there are great opportunities in recruiting appropriate developments in science and technology to counter these pressures and, in particular, the likely reductions in labour inputs. We believe that animal breeding has a great deal to offer in this sense by identifying measurable traits to improve health and general adaptation to the environment, including estimation of the sensitivity of genotypes to specific farming systems or conditions ('robustness'). This selection can be targeted at specific traits and does not necessarily require a change to composite easy-care breeds. We also believe it is critical to invest in the development of measures to provide valid assessments of short and long-term welfare states of extensively managed sheep. Lastly there will be potential to use electronic technologies, particularly those focused around electronic identification, to provide more remote sensing of sheep that could balance against the downward pressure on labour provided that appropriate validation and operating procedures are carried out.

References

Advocates for Animals (2004) Silent Lambs: A Report on Sheep Welfare in Scotland. Advocates for Animals, Edinburgh, UK

Arthington, J.D., Roka, E.M., Mullahey, J.J., Coleman, S.W., Muchovej, R.M., Lollis, L.O. & Hitchcock, D. (2007) Integrating ranch forage production, cattle performance, and economics in ranch management systems for southern Florida. *Rangeland Ecology & Management* **60**: 12–18.

Athanasiadou, S., Arsenos, G. & Kyriazakis, I. (2002) Animal health and welfare issues arising in organic ruminant production systems. In: *Organic Meat and Milk from Ruminants* (Eds. I. Kyriazakis & G. Zervas), Proceedings of the Joint Hellenic Society for Animal Production and British Society for Animal Science Conference Athens, Greece 4–6 October. EAAP publication No. 106 pp. 39–56.

Australian Government (2005) The Australian Animal Welfare Strategy. The Department of Agriculture, Fisheries and Forestry, Commonwealth of Australia, Canberra.

Bayvel, A.C.D., Rahman, S.A. & Gavinelli, A. (Eds), (2005) Animal welfare: global issues, trends and challenges. OIE, Paris.

Bennett, R.M., Anderson, J. & Blaney, R.J.P. (2002) Moral intensity and willingness to pay concerning farm animal welfare issues and the implications for agricultural policy. *Journal of Agricultural and Environmental Ethics* **15**: 187–202.

Bishop, S.C., Bairden, K., McKellar, Q.M., Park, M. & Stear, M.J. (1996) Genetic parameters for faecal egg count following mixed, natural, predominantly Ostertagia circumcincta infection and relationships with live weight in young lambs. *Animal Science* **63**: 423–428

Bishop, S.C. & Stear, M.J. (1999). Genetic and epidemiological relationships between productivity and disease resisitance: gastro-intestinal parasite infection in growing lambs. *Animal Science* **69**: 515–524.

Blache, D. & Ferguson, D. (2005). Boost lamb survival – select for calm ewes. Sheep Updates, Department of Agriculture, Western Australia.

Blokhuis, H.J. (2004) Recent developments in European and international welfare regulations. *Worlds Poultry Science Journal* **60**: 469–477.

Boroneant, C., Plaut, G., Giorgi, F., Bi, X. (2006) Extreme precipitation over the Maritime Alps and associated weather regimes simulated by a regional climate model: Present-day and future climate scenarios. *Theoretical and Applied Climatology* **86**: 81–99.

Brambell, F.W.R. (1965) Report of the Technical Committee to Enquire into the Welfare of Animals Kept under Intensive Livestock Husbandry Systems. HMSO, London.

Chandroo, K.P., Duncan, I.J.H. & Moccia, R.D. (2004) Can fish suffer?: perspectives on sentience, pain, fear and stress. *Applied Animal Behaviour Science* **86**: 225–250.

Cloete, S.W.P., Scholtz, A.J., Gilmour, A.R. & Olivier, J.J. (2002) Genetic and environmental effects on lambing and neonatal behaviour of Dormer and SA Mutton Merino lambs. *Livestock Production Science* **78**: 183–193.

Colebrook, E. & Wall, R. (2004) Ectoparasites of livestock in Europe and the Mediterranean region. *Veterinary Parasitology* **120**: 251–274.

Conington, J., Nieuwhof, G.J., Hosie, B.D., Bishop S.C. & Bünger L. (2007) Breeding for resistance to footrot in sheep. *Proceedings of the 58th Congress of the European Association for Animal Production* August 26–29, Dublin Eire.

Conington, J., Bishop, S.C., Lambe, N., Bunger L. & Simm, G. (2006) Testing new selection indices for sustainable hill sheep production – lamb growth and carcass traits. *Animal Science* **82**: 4: 445–453

Conington, J., Bishop, S. C., Grundy, B., Waterhouse, A. & Simm, G. (2001). Sustainable, multi-trait selection indexes for UK hill sheep. *Animal Science* **73**: 413–424

Dawkins, M.S. (1980) *Animal Suffering: the Science of Animal Welfare*. Oxford University Press, Oxford, UK.

Defra (2002) National Scrapie plan for Great Britain. Department for Environment, Food and Rural Affairs (UK). http://www.defra.gov.uk/corporate/regulat/forms/Ahealth/nsp/nsp1.pdf (accessed August, 2007)

Defra (2004) Animal Health and Welfare Strategy for Great Britain. Defra, London.

Defra (2006) http://www.defra.gov.uk/animalh/welfare/default.htm. (accessed August 2007).

Devendra, C. (2007) Perspectives on animal production systems in Asia. *Livestock Science* **106**: 1–18.

Dwyer, C.M. & Lawrence, A.B. (2005) A review of the behavioural and physiological adaptations of hill and lowland breeds of sheep that favour lamb survival. *Applied Animal Behaviour Science* **92**: 235–260.

Eurobarometer (2005) Attitudes of consumers towards the welfare of farmed animals. Special Eurobarometer Report 229.

EU (2006) *Improving Animal Welfare: EU Action Plan Adopted*. EU Document IP/06/04. EU, Brussels.

EU (1997) Treaty of Amsterdam Amending the Treaty on European Union. EU Official Journal C340, Brussels, November 1997.

FAWC (2005) *Report on the Welfare Implications of Farm Assurance Schemes*. Farm Animal Welfare Council, London.

Fisher, M. (2003) New Zealand farmer narratives of the benefits of reduced human intervention during lambing in extensive farming systems. *Journal of Agricultural and Environmental Ethics* **16**: 77–90.

Fisher, M.W. (2006) Life science ethics. *Journal of Agricultural and Environmental Ethics* **19**: 199–201.

Fisher, M.W. & Mellor, D.J. (2002) The welfare implications of shepherding during lambing in extensive New Zealand farming systems. *Animal Welfare* **11**: 157–170.

Forrest, R.H., Hickford, J.G.H., Hogan, A. & Frampton, C. (2003). Polymorphism at the ovine beta(3)-adrenergic receptor locus: associations with birth weight, growth rate, carcass composition and cold survival. *Animal Genetics* **34**: 19–25.

Forrest, R.H., Hickford, J.G.H., Wynyard, J., Merrick, N., Hogan, A. & Frampton, C. (2006). Polymorphism at the b3-adrenergic receptor (ADRB3) locus of Merino sheep and its association with lamb mortality. *Animal Genetics* **37**: 465–468.

Goddard, P., Waterhouse, T., Dwyer, C. & Stott, A. (2006) The perception of the welfare of sheep in extensive systems. *Small Ruminant Research* **62**: 215–225.

Gohin, A. (2006) Assessing CAP reform: Sensitivity of modelling decoupled policies. *Journal of Agricultural Economics* **57**: 415–440.

Grandin, T. (2006) Progress and challenges in animal handling and slaughter in the US. *Applied Animal Behaviour Science* **100**: 129–139.

Haskell, M.J., Brotherstone, S. & White, I.M.S. (2007) Characerisation of dairy farm environments and their effect on cow lifespan. *Journal of Dairy Science* 90: 5316–5323.

Harrison, R. (1964.) *Animal Machines*. Stuart, London.

Hemsworth, P.H. (2003) Human-animal interactions in livestock production. *Applied Animal Behaviour Science* **81**: 185–198.

Henderson, C.R. (1975) Best Linear Unbiased estimation and prediction under a selection model. *Biometrics* **31**: 423–447.

Hickford, J. (2000) Development of a sustainable method of natural footrot control. Report to the NZ Meat Research and Development Council. Lincoln University.

Hohenboken, W.D (1986) Inheritance of behavioural characteristics in livestock. A review. *Animal Breeding Abstracts* **54**: 623–639.

Hopkins, A. & Wilkins, R.J. (2006) Temperate grassland: key developments in the last century and future perspectives. *Journal of Agricultural Science* **144**: 503–523.

House of Commons (2005) *The Animal Welfare Bill: Explanatory Notes*. The United Kingdom Parliament, UK.

Hughes, B.O. & Black, A.J. (1973) Preference of domestic hens for different types of battery cage floor. *British Poultry Science* **14**: 615.

IPSOS MORI (2005) Asian nations share British concern for animals. IPSOS MORI, http://www.ipsos-mori.com/polls/2005/ciwf.shtml.

Kanis, E., van den Belt, H., Groen, A.F., Schakel, J., & De Greef, K.H. (2004) Breeding for improved welfare in pigs: a conceptual framework and its use in practice. *Animal Science* **78**: 315–329.

Keeling, L, & Vessier, I. (2005) Developing a monitoring system to assess welfare quality in cattle, pigs and chickens. In: *Science and Society Improving Animal Welfare. Welfare Quality Conference Proceedings*, (Ed. A. Butterworth), Brussels, pp. 46–50.

Keeling, L., Anderson, L., Schutz, K.E., Kerje, S., Fredriksson, R., Carlborg, O., Cornwallis, C.K., Pizzari, T. & Jensen, P. (2004) Chicken genomics: Feather-pecking and victim pigmentation. *Nature* **431**: 645–646.

King, J.M., Parsons, D.J., Turnpenny, J.R., Nyangaga, J., Bakari, P. & Wathes, C.M. (2006). Modelling energy metabolism of Friesians in Kenya smallholdings shows how heat stress and energy deficit constrain milk yield and cow replacement rate. *Animal Science* **82**: 705–716.

Koenig, S., Sharifi, A.R., Wentrot, H., Landmann, D., Eise, M. & Simianer, H. (2005) Genetic parameters of claw and foot disorders estimated with logistic models. *Journal of Dairy Science* **88**: 3316–3325.

Knap, P.W. (2005) Breeding robust pigs. *Australian Journal of Experimental Agriculture* **45**: 763–773.

Knight, S., Vrij, A., Cherryman, V. & Nunkoosing, K. (2004) Attitudes towards animal use and belief in animal mind. *Anthrozoos* **17**: 43–62.

Kurukulasuriya, P., Mendelsohn, R., Hassan, R., Benhin, J., Deressa, T., Diop, M., Eid, H.M., Fosu, K.Y., Gbetibouo, G., Jain, S., Mahamadou, A., Mano, R., Kabubo-Mariara, J., El-Marsafawy, S., Molua, E., Ouda, S., Ouedraogo, M., Sene, I., Maddison, D., Seo, S.N. & Dinar, A. (2006) Will African agriculture survive climate change? *World Bank Economic Review* **20**: 367–388.

Lambe, N.R., Conington, J., Bishop, S.C., Waterhouse, A. & Simm, G. (2001) A genetic analysis of maternal behaviour score in Scottish Blackface sheep. *Animal Science* **72**: 415–425.

Lawrence, A.B. (2007a) What is animal welfare? In: *Fish Welfare* (Ed. E. Branson), Blackwell Publishing, UK.

Lawrence, A.B. (2007b) Applied Animal Behaviour Science: Past, present and future prospects. *Applied Animal Behaviour Science* (in press).

Lawrence, A.B., Conington, J. & Simm, G. (2004) Breeding and animal welfare: practical and theoretical advantages of multi-trait selection. *Animal Welfare* **13**: S191–S196.

Lohr, L. & Park, T.A. (2003) Improving extension effectiveness for organic clients: Current status and future directions. *Journal of Agricultural and Resource Economics* **28**: 634–650.

MacLeod, C.J. & Moller, H. (2006) Intensification and diversification of New Zealand agriculture since 1960: An evaluation of current indicators of land use change. *Agriculture Ecosystems & Environment* **115**: 201–218.
Matthews, K.B., Wright, I.A., Buchan, K., Davies, D.A. & Schwarz, G. (2006) Assessing the options for upland livestock systems under CAP reform: Developing and applying a livestock systems model within whole-farm systems analysis. *Agricultural Systems* **90**: 32–61.
Mavrogenis, A.P., Koumas, A., Kakoyiannis, C.K. & Taliotis, C.H. (1995) Use of somatic cell counts for the detection of sub-clinical mastitis in sheep. *Small Ruminant Research* **17**: 79–84.
McEwan, J.C., Mason, P., Baker, R.L., Clarke, J.N., Hickey, S.M. & Turner, K. (1992) Effect of selection for production traits on internal parasite resistance in sheep. *Proceedings of the New Zealand Society of Animal Production* **52**: 53–56.
McEwan, J.C. & Kerr, R.J. (1998) Further evidence that major genes affect host resistance to nematode parasites in coopworth sheep. *6th World Congress on Genetics Applied to Livestock Production, Armidale, Australia* **27**, 335–338.
Menzel, A., Von Vopelius, J., Estrella, N., Schleip, C., Dose, V., 2006. Farmers' annual activities are not tracking the speed of climate change. *Climate Research* **32**: 201–207.
Morgan-Davies, C., Waterhouse, A., Milne, C.E., Stott, A.W., 2006. Farmers' opinions on welfare, health and production practices in extensive hill sheep flocks in Great Britain. *Livestock Science* **104**: 268–277.
Muller, R. & von Keyserlingk, M.A.G. (2006) Consistency of flight speed and its correlation to productivity and to personality in Bos taurus beef cattle. *Applied Animal Behaviour Science* **99**: 193–204.
New Scientist (2005) Animals and us: Forward to the revolution. *New Scientist* **2502**: 42.
Nieuwhof, G.J. & Bishop, S.C. (2005). Costs of the major endemic diseases of sheep in Great Britain and the potential benefits of reduction in disease impact. *Animal Science* **81**: 23–29
OIE (2004) Global Conference on Animal Welfare: an OIE Initiative. OIE, Paris.
OIE (2006) Terrestrial Animal Health Code. OIE, Paris. http://www.oie.int/eng/normes/mcode/en_titre_3.7.htm.
Oglethorpe, D.R. (2005) Livestock production post CAP reform: implications for the environment. Animal Science **81**: 189–192.
Phillips, C.J.C. & McCulloch, S. (2005) Attitudes of students of different nationalities towards animal sentience and the use of animals in society, with implications for animal use in education. *Journal of Biological Education* **40**: 17–24.
Preece, R. & Fraser, D. (2000) The status of animals in Biblical and Christian thought: A study in colliding values. *Society and Animals* **8**: 245–263.
Raadsma H.W. (2000) Genetic aspects of resistance to ovine footrot. In: *Breeding for disease resistance in Farm Animals* (Eds. R. Axford, S. Bishop, F. Nicholas & J. Owen), CABI. pp. 219–241, Oxford, UK.
Rivington, M., Matthews, K.B., Bellocchi, G., Buchan, K., Stockle, C.O. & Donatelli, M. (2007) An integrated assessment approach to conduct analyses of climate change impacts on whole-farm systems. *Environmental Modelling & Software* **22**: 202–210.
Rollin, B.E. (2004) Annual meeting keynote address: Animal agriculture and emerging social ethics for animals. *Journal of Animal Science* **82**: 955–964.
Rushen, J. (2003) Changing concepts of farm animal welfare: bridging the gap between applied and basic research. *Applied Animal Behaviour Science* **81**: 199–214.
Sawalha, R.M., Conington, J., Brotherstone, S. & Villanueva, B. (2007) Analysis of lamb survival of Scottish Blackface sheep. *Animal* **1**: 151–157
SEERAD (2006). *Land Management Contract Scheme. Notes for Guidance.* Scottish Executive, St Andrews House, Edinburgh, UK.
SEERAD (2007). Land Management Contract Scheme. http://www.scotland.gov.uk/News/Releases/2005/08/25151946.
Scobie, D.R., Young, S.R., O'Connell, D. & Eythorsdottir, E. (2005). Skin wrinkles of the sire adversely affect Merino and halfbred pelt characteristics and other production traits. *Australian Journal of Experimental Agriculture* **45**: 1551–1557.

Schutz, K., Davison, D. & Matthews, L. (2006) Do different levels of moderate feed deprivation in dairy cows affect feeding motivation? *Applied Animal Behaviour Science* **101**: 253–263.

Sorenson, J.T., Edwards, S., Noordhuizen, J. & Gunnarsson, S. (2006) Animal production systems in the industrialised world. *Revue Scientifique Et Technique-Office International Des Epizooties* **25**: 493–503.

Swanson, J. (2004) Husbandry and welfare of livestock in pasture based systems. Animals in the Food System Conference, C.S. Mott Group for Sustainable Food Systems, Michigan State University.

Stott, A.W., Milne, C.E., Goddard, P.J. & Waterhouse, A. (2005) Projected effect of alternative management strategies on profit and animal welfare in extensive sheep production systems in Great Britain. *Livestock Production Science* **97**: 161–171.

Turner, S.P., Lawrence, A.B. & Lowman, B. (2003). Handling beef cattle: Identifying research needs and knowledge transfer opportunities to improve human safety and animal welfare. SAC, Edinburgh, UK.

Wassink, G.J. & Green, L.E. (2001). Farmers' practices and attitudes towards footrot in sheep. *Veterinary Record* **149**: 489–490.

Wright, W. & Muzzatti, S.L. (2007) Not in my port: The "death ship" of sheep and crimes of agri-food globalization. *Agriculture and Human Values* **24**: 133–145.

Index

U

V

W

MIX
Papier aus verantwortungsvollen Quellen
Paper from responsible sources
FSC® C105338

If you have any concerns about our products,
you can contact us on
ProductSafety@springernature.com

In case Publisher is established outside the EU,
the EU authorized representative is:
Springer Nature Customer Service Center GmbH
Europaplatz 3, 69115 Heidelberg, Germany

Printed by Libri Plureos GmbH
in Hamburg, Germany